IEE Power Engineering Series 4
Series Editors: Dr A.T. Johns
G. Ratcliff
Prof. A. Wright

CW00554005

INDUSTRIAL MICROWAVE
HEATING

INDUSTRIAL MICROWAVE HEATING

A.C.Metaxas & R.J.Meredith

PETER PEREGRINUS Ltd on behalf of the Institution of Electrical Engineers

Previous volumes in this series

Volume 1   Power circuit breaker theory and design
                Edited by C.H. Flurscheim
Volume 2   Electric Fuses
                A. Wright and P.G. Newbery
Volume 3   Z transform electromagnetic transient analysis in high
                voltage networks
                W. Derek Humpage

Published by Peter Peregrinus Ltd., London, UK.

ISBN 0 906048 89 3

Printed in England by Short Run Press Ltd., Exeter

Contents

Principal list of symbols

a, a', a_0	dimension (of waveguide broad, dielectric specimen broad or semi-major axis of ellipse or aperture broad, off-centre position of dielectrics in TM_{11n} cavity)
$a_1 - a_4$	input voltages to directional coupler
α	attenuation constant
$\alpha_T, \alpha_p, \alpha_m$	diffusivities (thermal, pressure, mass)
α_{mv}	diffusion coefficient of vapour flow
$\alpha_t, \alpha_a, \alpha_e, \alpha_d, \alpha_{MW}$	polarisabilities (total, atomic, electronic, dipolar, Maxwell–Wagner)
α_r	number of revolutions/sec
A_c, A_s	area (cross-sectional, surface)
A	general constant, energy fraction
b, b'	dimension (of waveguide narrow, semi-minor axis, aperture narrow or dielectric specimen)
$b_1 - b_4$	output voltages from directional coupler
b_i	normalised iris susceptance, i.e., (B/Y_0)
β	phase constant
β'	coupling factor
$\boldsymbol{B}$	flux density
B, B_a	susceptance (general, aperture)
$\gamma, \gamma_2, \gamma', \gamma_c$	propagation constant (general, specimen filled line, in periodic waveguides, in conductor)
c	velocity of light
c_a	specific moisture capacity of vapour phase
c_v, c_p	specific heat (constant volume, constant pressure)
C_0	constant
C_{14}	voltage coupling coefficient in loop resonator

C, C_p, C_c	capacitance (general, between parallel plates, cavity)
Γ	coupling factor in dielectric loaded applicator
$d, d_i, d_s, d_0, d_0', d_c$	distance (general, iris width, meander interpass length, meander pass length, load/waveguide distance in horn applicators, TE_{10n} cavity length
D_p	penetration depth
D	charge density vector
D_e	electron diffusion coefficient
e	eccentricity
ϵ_0	permittivity of free space
ϵ^*	relative complex permittivity, $\epsilon' - j\epsilon''$
ϵ_a	absolute permittivity ($= \epsilon_0 \epsilon^*$)
$\epsilon', \epsilon_s, \epsilon_\infty$	(relative dielectric constants)
$\epsilon_{eff}'', \epsilon_d'', \epsilon_c''$	relative loss factors (effective, dipolar, conductive)
ϵ_i	emissivity
ϵ_v	ratio of vapour flow to total moisture flow
$E, E_{rms}, E_{av}, \hat{E}, E_i, E_{max},$ E_{res}, E^*, E_z, E_b and $E_m,$ $E_s, E_+, E_-, E_e, E_{zm}, E_{ext}$	electric field strength (general, r.m.s., average, peak, in dielectric, maximum, resultant, conjugate, axial in cylindrical cavity, breakdown and maintenance of a gas, in metal, forward, reflected, effective, maximum longitudinal in elliptical cavity, external)
δ	loss angle
δ_T, δ_p	gradient coefficient (thermal, pressure)
δn	number of modes at a given frequency
δ_s	skin depth
δv	incremental volume
δP	localised power density
Δf	bandwidth ($\Delta\omega/2\pi$)
f, f_0, f_a	frequency (general, resonant, operating in waveguide)
f_c, f_{c2} and $f_{c1}, f_{cc}, f_{cc\epsilon}$	cutoff frequencies (general, upper and lower of pass band, in cutoff waveguide, in cutoff waveguide carrying dielectric)
F	frequency independent constant
$g(t)$	fraction of orientation polarisations
$g(M), g'(M), g(U_a)$	functions of moisture and energy barrier
g_1, g_0	capital costs of microwave equipment in £/kW, unit cost of electricity in p/kWh
g_{con}, g_{mw}	energy running costs per hour (conventional equipment, of microwave equipment)

G	conductance
h	height (general, cylindrical or elliptical cavity, slab in waveguide or oven)
$h_T, \bar{h}_T$	heat transfer coefficient (general, mean)
H_h	enthalpy
$H, H_t, H_{rms}, H_+, H_-,$ $H_{max}, H_n, H_\phi, H_{res}$	magnetic field strengths (general, tangential, r.m.s., forward, reflected, maximum, nornal, circumferencial, resultant)
θ_0	angle between E and I vectors (power factor $\cos\theta_0$)
θ	phase angle
η	refractive index
η_v	viscosity
i, i_e, i_c	currents (general, resistive, capacitive)
J, J_c	current density (general, conductive)
J_n	Bessel function of the first kind and of order of n
k, k_0	general constant, frequency exponent in ϵ''_{dc} equation
k_b	Boltzmann's constant
K, K_0	constant in Bessel function and $K_0 = Kr$
l, l_e	distance (general, effective cylindrical cavity length), integer
L_j	electron loss process
L, L_c	inductance (general, cavity)
L_h	latent heat of vaporisation
Λ	characteristic diffusion length
$\lambda, \lambda'_0, \lambda_g, \lambda_c, \lambda_0$	wavelengths (general, free, waveguide, cutoff, resonant)
$M, M', M'', M''', M_0, M_c,$ M_l and $M_v, M_i,$ and $M_f,$ M_∞	moisture content (dry basis, wet basis, pick-up basis, solids basis, equilibrium, critical, liquid and vapour, initial and final, constant)
M_a, M_{ad}	mass (general, dry)
M'_p	magnetic polarisability
$\mu_0, \mu^*, \mu_a, \mu_c, \mu', \mu''_{eff}$	permeability (free space, complex, absolute $(= \mu_0\mu^*)$, wall, real, effective magnetic loss factor)
μ	dipole moment
$n_0, n_g, n_t, n_c, n_a, n_u$	efficiencies (overall, generation, transfer, coupling, applicator, utilisation)
n	integer
n_e, n_{e0}	electron concentration (general, initial)
N	turns ratio

N_0	number of stirrer blades
ν, ν_c	frequency (dipole oscillation $(1/\tau)$, collision)
p	pressure, integer
p_r^2	ratio of dielectric constants in cylindrical cavity heater
P_i	electron production process
$P, P_w, P_s, P_{av}, P_m, P_{in}, P_0,$ P_{ex}, P_{sc}, P_{th}	power dissipated (general, dielectric, surface, average, mains, input into cavity, incident on cavity's aperture, in external circuit, microwave source power, theoretical)
$\boldsymbol{P}, \boldsymbol{P}_{max}$	polarisation field vector (instantaneous, maximum)
$\boldsymbol{\rho}$	Poynting's vector
q	charge
$Q, Q_0', Q_0, Q_L, Q_{ext},$ Q_{Leff}, Q_d	quality factor (general, isolated empty, isolated with dielectric, loaded, external, effective, dielectric workload)
Q_h	quantity of heat
Q_i	charge production due to an external source
r, r_q	radius (general, dipole)
R_w, R_d, R_0, R_c	radii (dielectric, dielectric filled cavity, empty cavity, partial filled cavity)
$R', R_c', R_s', R_a', R_G', R_p',$ R_L'	resistance (general, cavity, skin, aperture coupling network, generator, parallel plate capacitor, load)
ρ	density
$\rho, \rho^*, \rho_e, \rho_0$	reflection coefficient, its conjugate, effective, resonant
σ	conductivity
σ_T	Stefan Boltzmann's constant
$S, S_0, S_{1/2}$	VSWR (general, minimum, 3 dB)
$\boldsymbol{S}'$	surface integral
t	thickness of slab in waveguide or time variable
τ'	fraction of energy
$\tau, \tau_e, \tau_0, \tau_{mean}$	relaxation time constant (general and Debye, effective, in double well theory, mean)
T_v	voltage transmission coefficient in loop resonator
$T, T_g, T_w, T_s, T_c, T_{bp}$	temperature (general, dry bulb, wet bulb, surface, critical, boiling point)
U, U_s, U_a	energy (general, stored, activation)
v	volume (ratio (V_L/V_c), fraction of conducting material)

v_w	web speed
V, V_L, V_c, V'_c	volume (general, dielectric, cavity, clearance volume ($V_c - V_L$) in multimode applicator)
V, V_+, V_-, V_b and V_{str}	voltage vectors (general, forward, reflected, onset for gas breakdown and streamer)
w	web width
$W, W_w, W_d, W_{wp}, W_{we},$ W_{te}	weight (general, water, dry matter, water pick-up, water at quilibrium, total weight at equilibrium)
ϕ	phase angle
ϕ_0	$= \tanh(\alpha d_c)$
χ	electric susceptibility
x, x_0, x_i, x_d, x'_1 or x'_2, x_1	distance (general, first min from s/c sample, charge separation in dipole, between discs in disc loaded cavity, slot separations for broadside waveguide centre-line, distance of slab from narrow waveguide face)
$x_{l,m}$	mth root of $J_l(x) = 0$
$\omega, \omega'_0, \omega_0, \omega_{lmn}, \omega_{clmn},$ $\hat{\omega}$	angular frequency (general, resonant unperturbed, resonant perturbed, pertaining to a particular mode, cutoff of a particular mode, complex)
y	distance
y_n	number of years
Y, Y_0, Y_{in}	admittance (general, characteristic, cavity input)
Y_n	Bessel function of the second kind and of order n
z, z_c, z_0	distance (general, of cutoff choke, cavity length above or below that which gives resonance in TE_{10n} cavity)
z_i, z_b	rate of production per electron (general, at breakdown)
$\mathscr{X}$	spreading factor in Debye response
Z, Z_{in}	impedance (general, input)
Z_0, Z_{0s}, Z_{02}	characteristic impedance (general, stub, specimen filled line)

Preface

This book forms an attempt to bring together the theory and practice of industrial microwave heating. Since the publications on the subject of microwave heating by D. A. Copson (1962, 1975), H. Püschner (1966) and E. C. Okress (1968) during the nineteen-sixties there has been a noteable absence of a book dealing with this topic. During this period there has been much research and development into the subject, reported in numerous publications spanning countless scientific journals. Moreover, the theory and practice of microwave heating itself involves many disciplines from electrical and process engineering to physics and material sciences. We felt, therefore, that such fragmented information on dielectric theory and properties of materials, design of equipment and the state of the art in applications relevant to the manufacturing industry should be collated, updated, extended and presented as a single reference volume.

Microwave heating was given special emphasis at the Electricity Council Research Centre (ECRC) about ten years ago where a unit was established to carry out research and development on the future of industrial applications. One of us (A.C.M.) would like to acknowledge the positive role played by J. Lawton, DSc, of the Central Electricity Research Laboratories, Leatherhead, in introducing him to the subject of industrial microwaves and wishes to thank all his colleagues in the electrophysics group at ECRC, particularly Drs. J. L. Driscoll, P. L. Jones and T. Farrell, also R. Morrow of CSIRO, Lindfield, for many valuable discussions and from whom he has learned a great deal. It was through contacts and collaboration with Magnetronics Ltd. in those early days that brought together the authors of this book.

It would be impossible to acknowledge all the people who have in one way or another helped us with the essence of the book. However, we would particularly like to thank Mr. G. Ratcliff of the ECRC for encouraging us to write this book, for reading the manuscript and for making helpful suggestions and to Dr. R. Perkin and Mr. W. Baker of ECRC for reading and commenting on part of the manuscript. We owe a great debt to our colleagues in industry at large who have openly discussed their processes with us and to the many engineers in the Electricity Area Boards for initiating many industrial contacts and thus keeping us aware of the problems and requirements for new industrial plant. We found stimulating the many discussions we have had over the years with countless fellow researchers in the field of industrial microwaves during the annual IMPI (International Microwave Power Institute) Symposium and whose publications in the literature have made a valuable contribution to this book. To this we would particularly like to thank Dr. R. Schiffman of R. Schiffman Associates, New York, Dr. S. Stuchly of the University of Ottawa, Mr. G. Freedman of Raytheon Co, Massachusetts, Dr. A. L. VanKoughnett of the Communications Research Centre, Ottawa, Drs. W. Wyslouzil and S. Kashyap of the NRC, Ottawa, Dr. S. Nelson of the U.S. Dept. of Agriculture, Georgia, Dr. M. Stuchly of the Radiation Protection Bureau, Ottawa, Mr. J. Gerling of Gerling Laboratories, California, Mr. B. Krieger of Cober Electronics, Inc., Connecticut, Mr. K. Ogura of Toshiba, Japan, our Swedish colleagues Dr. T. Ohlsson of the SIK Institute, Göteborg, Dr. P. Risman of Microtrans, AB, Huskvarna, Mr. B. Edin of Scanpro AB, Bromma and many others.

One of us (R.J.M.) would particularly like to thank Mr. M. P. Tahany, Chairman of Magnetronics Ltd., for his support and his many colleagues and staff, in particular Mr. J. Mitton, Mr. G. W. Geffery and Mr. G. A. Swann for frequent and helpful discussions. In addition he would like to thank his many customers who have contributed so much of the industrial environment. To his wife, Judy, R.J.M. owes a special debt for her support and encouragement in this project.

It would be impossible and invidious to mention all those who have contributed to industrial microwave heating technology over many years but nonetheless the very valuable contribution of Dr. R. Dunsmuir, Dr. J. E. Curran and J. R. G. Twistleton of the former B. T. H. Research Laboratory, Rugby, deserve special mention for the development of the 25 KW, 900 MHz magnetron and associated equipment, also Mr. H. B. Taylor for his valuable contribution to microwave applicator design.

We would also like to extend our thanks to Mr. H. Barber of Loughborough University, Mr. R. Shute of Microwave Heating Ltd., Luton, Mr. P. Giles and Mr. K. Ike of Microwave Ovens Ltd., Shirley, Dr. R. Smith of Bradford University, Mr. G. Crossley of Marconi (Specialised Components), Essex, Mr. P. Hulls and Mr. A. Witt of the Electricity Council Marketing Department, for their continuing help and support over many years. We are indebted to the ex-editor of the *Journal of Microwave Power*, Dr. S. Stuchly, and many other publishers and their authors for allowing us to reproduce some of their data. Such indebtedness extends to all the authors given in the references. We would like to acknowledge the help we have received from the secretarial staff at ECRC and especially from Margaret Metaxas for typing and editing the manuscript and for her total commitment during the last stages of the preparation of the book. Finally, we would be very grateful to receive notification of any errors or amendments from any reader.

We would like to thank our publishers for their expert advice and cooperation throughout this project.

A. C. Metaxas
R. J. Meredith

Chester, England, 1982

Chapter 1

Introduction

Microwave heating as an industrial process is a technique which was originally conceived about forty years ago. The advent of the magnetron during the Second World War presented engineers and scientists in industry, universities and government establishments with a unique challenge to put such a device for generating microwaves into peaceful and profitable use. The task that lay ahead was quite formidable because of the lack of appropriate equipment and more importantly the lack of data on dielectric property of the materials which were considered as candidates for microwave heating. During the late forties and early fifties, a concerted effort was made to obtain reliable data on material properties, led by von Hippel and his co-workers at MIT. Their pioneering work on the properties of many organic and inorganic materials in the frequency region $100 < f < 10^{10}$ Hz has since formed, and still remains, a solid basis for the establishment of radio frequency and microwave energy techniques in industry (von Hippel, 1954). Their original work has since been expanded many times to fill the gaps which emerged as more and more industrial applications came up for consideration. In addition to this voluminous array of data on material properties, there have been significant developments on the design of magnetrons, power supplies and ancilliary equipment, giving greater reliability to this new technique. The engineering aspects of many applications in terms of scale-up, continuous operation, automatic control, etc., have considerably improved since the original days. More and more the various disciplines of physics, chemistry, electrical, mechanical and process engineering, thermodynamics, material science, etc., have blended according to need and priority, in order to optimise the requirement of each particular process.

Before we briefly introduce the topics that will form the basis of this book, it is important to define the frequency ranges for which the terms microwave and radio frequency will be subsequently used. At frequencies below 100 MHz, where conventional open wire circuits are used, the technique of industrial processing will be referred to as radio frequency heating. However, at microwave frequencies (above 500 MHz), wired circuits cannot be used and the power is transferred to the applicator containing the material to be processed in waveguides. This technique will be referred to as microwave heating. In between there exists a diffuse regime where the equivalent circuit representation for the two techniques blend together and become difficult to differentiate. Moreover, as regards material properties and theoretical treatises, it is often appropriate to couple the two regimes together since the arguments may be equally applicable to both techniques. For convenience therefore the term dielectric or high frequency heating will be used in this book to refer to both radio frequency and microwave heating systems. This definition combining the two frequency regimes, shown in the frequency spectrum in Fig. 1.1, will be primarily used in the early chapters when discussing dielectric loss, material properties and theoretical aspects of volumetric heating.

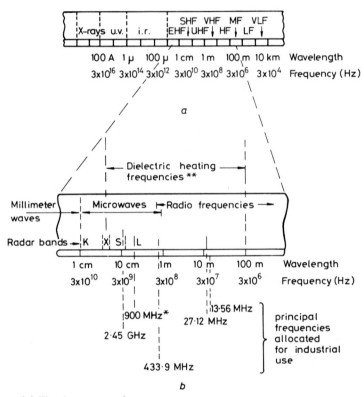

Fig. 1.1 (a) *The electromagnetic spectrum*
(b) *Definition of the various frequencies used in this book*
* Comprising 869 MHz (U.K.) and 915 MHz (U.S.A.) bands
** Referred to also throughout this book as high frequencies

Therefore, although this book is primarily concerned with industrial microwave heating, where appropriate the term dielectric or high frequency heating will be used to include the effects of radio frequency on the particular process under consideration.

Although the specific design of the various microwave systems forms the main part of this book, a subject as diverse as industrial microwave heating cannot be adequately treated solely through equivalent circuits. The chapters on applicators, microwave sources, heating circuit and applications are preceded by three chapters, starting with an elementary look at the physical concepts behind dipolar and conductivity losses which, after all, form the basis of dielectric heating. A closing paragraph on magnetic loss factor has also been included in this chapter to remind one that the magnetic field associated with an electromagnetic wave can also be used in principle to couple energy into magnetic materials. This chapter is followed by a short insight into material properties, and their interpretation. Finally the physical parameters involved in volumetric heating such as the power dissipation, penetration depth, the electric field, the specific heat, mass and heat transfer, etc., are introduced, and some of the basic equations controlling these parameters are derived. Although not a part of volumetric heating a treatise on skin depth and wall losses forms the closing paragraphs in this chapter because of their significance

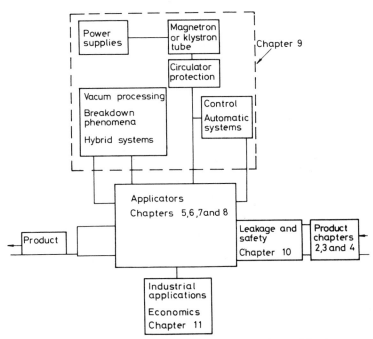

Fig. 1.2   *Block diagram of the basic microwave heating plant with the corresponding chapters discussed in the text*

in the overall utilisation efficiency of a particular piece of equipment used in high frequency processing.

More specifically, in order to appreciate the arrangement of the book consider Fig. 1.2 which shows the basic microwave heating circuit, marked with the chapter numbers which apply to each component. Essentially it consists of a microwave source, with its associated power supplies, control circuits, etc. (Chapter 9), and the applicator (Chapters 5, 6, 7 and 8). The link between the source and the applicator, usually in the form of a waveguide and a coupling structure, is vital for the efficient transfer of energy between the two (early part of Chapter 5). The properties of the products to be processed are dealt with in Chapter 3, why the product heats is considered in Chapter 2 and the physical parameters involved in volumetric heating examined in Chapter 4. Processing under vacuum, which for some applications offers distinct advantages to working at atmospheric pressures, is also considered in Chapter 9. Vacuum processing is preceded by a brief insight into high frequency breakdown phenomena and followed by outlining the principles behind hybrid systems. Radiation leakage, hazards and choking systems are considered in Chapter 10 while industrial applications and economics form the closing chapter (Chapter 11) of this book.

The book is intended for engineers and scientists working or researching in the field of microwave heating. Although there is not as yet a recognised university course which deals with the subject of dielectric heating, this book can be used as a text book in wider courses, such as electroheat, which includes processing with high frequency electromagnetic fields. For this some derivations have been treated more rigorously, the majority of which, however, has been presented in elementary form. Furthermore, much of this book can be followed quite easily without detailed knowledge of some of its previous derivations.

The book adheres for the most part to the SI system of units, although for convenience there are instances where other systems have been referred to.

Reference

VON HIPPEL, A. R., *Dielectric Materials and Applications*. MIT Press (1954).

Chapter 2

Dielectric loss

2.1 Introduction

It has long been known that an insulating material can be heated by applying energy to it in the form of high frequency electromagnetic waves. The origin of this heating lies in the ability of the electric field to polarise the charges in the material and the inability of this polarisation to follow extremely rapid reversals of the electric field. In a given frequency band, therefore, the polarisation vector, P, lags the applied electric field ensuring that the resulting current, $\partial P/\partial t$, has a component in phase with the applied electric field, which results in the dissipation of power within the insulating material. Coupled with these polarisation effects, a dielectric can be heated through direct conduction effects due to, for example, the redistribution of charge particles under the influence of the externally applied electric field forming conducting paths, particularly in mixtures of heterogeneous materials. The theory of such polarisation and conduction effects has been extensively studied in the past (Debye, 1929; Fröhlich, 1958; Daniel, 1967; Hill *et al.*, 1969; Hasted, 1973). Detailed analysis of these phenomena is beyond the scope of this book. However, this chapter will summarise the aspects which are relevant to industrial high frequency heating and formulate simple models for the processes involved which should lead to better understanding of the behaviour of many industrial materials under high frequency fields.

2.2 Polarisation

The interaction of an electric field with a dielectric has its origin in the response of charge particles to the applied field. The displacement of these charge particles from their equilibrium positions gives rise to induced dipoles which respond to the applied field. Such induced polarisation arises mainly from the displacement of electrons around the nuclei (electronic polarisation) or due to the relative displacement of atomic nuclei because of the unequal distribution of charge in molecule formation (atomic polarisation). In addition to induced dipoles some dielectrics, known as polar dielectrics, contain permanent dipoles due to the assymmetric charge distribution of unlike charge partners in a molecule which tend to reorientate under the influence of a changing electric field, thus giving rise to orientation polarisation. Finally, another source of polarisation arises from charge build-up in interfaces between components in heterogeneous systems, termed interfacial, space charge or Maxwell–Wagner polarisation. Figure 2.1 shows

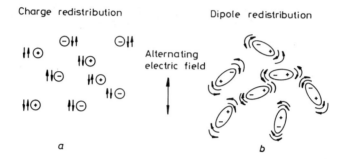

Fig. 2.1   *Interfacial (space charge) (a), and reorientation (b) polarisations*

a schematic representation of Maxwell–Wagner and orientation polarisation due to an alternating electric field. These two mechanisms, together with d.c. conductivity, are the basis of high frequency heating.

The average dipole moment of a displaced dipole is given by

$$\mu = qx_i \tag{2.1}$$

where q is the charge and x_i is the charge separation as shown in Fig. 2.2. The summation of all such dipole moments within a volume, δv, containing N dipoles, gives rise to a charge density, termed the polarisation field P, which counts bound charges of a system — such as the interface between the dielectric and the surrounding medium — and is given by

$$P = \frac{\sum_{i=1}^{N} qx_i}{\delta v} \tag{2.2}$$

The total charge of the system is given by the electric charge density vector D. The difference between the two vectors gives the remaining free charges of the system which account for the externally applied electric field E. Thus

$$D = \epsilon_0 E + P \tag{2.3}$$

where ϵ_0 is the dielectric constant of free space. However, since

$$D = \epsilon_0 \epsilon' E \tag{2.4}$$

eqn. (2.3) yields after re-arrangement

$$P = (\epsilon' - 1)\epsilon_0 E \tag{2.5}$$

where ϵ' is the relative dielectric constant.[*] The ratio of the bound to the free charges is called the electric susceptibility and is given by

$$\chi = \frac{P}{\epsilon_0 E} = (\epsilon' - 1) \tag{2.6}$$

The polarisation field P can be thought to be made up on N' individual dipole moments where

$$P = \mu N' \tag{2.7}$$

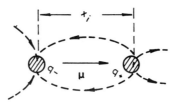

Fig. 2.2   *Electric dipole moment*

If the local field applied to the individual dipole is E', the dipole moment μ is a simple function of the field, normally assumed linear, giving

$$\mu = \alpha_t E' \tag{2.8}$$

where α_t is called the polarisability of the medium and comprises the different components of polarisation. Thus

$$\alpha_t = \alpha_e + \alpha_a + \alpha_d + \alpha_{MW} \tag{2.9}$$

where the subscripts refer to electronic, atomic, dipolar or re-orientation and inter-facial polarisations. Combining eqns. (2.7) and (2.8) with eqn. (2.5) yields

$$(\epsilon' - 1)\epsilon_0 E = \alpha_t N' E' \tag{2.10}$$

This links the macroscopic quantities ϵ' and E to the molecular parameters N' and E' in the dielectric. Equation (2.10) shows that the electric field E' experienced by a dipole differs from the applied field E. By considering the microscopic processes involved in a spherical volume which is too large compared with interatomic

---

[*] The word relative will be omitted in subsequent chapters and ϵ' will be referred to as the dielectric constant.

distances but on the other hand too small compared with interatomic inhomogen-
eities, Mosotti has derived a simple relationship between the two fields given by
(Zheludev, 1971)

$$E' = \frac{E}{3}(\epsilon' + 2) \tag{2.11}$$

which is often used as a first approximation in theoretical analysis of dielectric
phenomena for the local field for many cases but, strictly speaking, only applies
to gases and non-polar liquids.

2.3 Complex dielectric constant

Before discussing the losses arising from dipolar and other forms of polarisation,
let us examine the nature of the dielectric constant. It is well known that in a
real dielectric, to account for the losses, the dielectric constant attains a complex
form, i.e., it is given by

$$\epsilon^* = \epsilon' - j\epsilon'' \tag{2.12}$$

where the imaginary part, ϵ'', is termed the loss factor. However, to many who
possess little knowledge of dielectric theory, it would be helpful if, through some
elementary analysis, we can demonstrate the complex nature of the dielectric
constant without having to assume this premise from the start. Ampere's circuital
law in its elementary form contains all the necessary components needed for this
simple analysis. Maxwell modified Ampere's law for static fields by including a
displacement current density term caused by the rate of change of the total electric
flux passing through a surface S' bounded by l and derived the following expression
for the total current in a medium (Johnk, 1975)

$$\oint \frac{B}{\mu_a} \cdot dl = \int_{S'} J_c \cdot dS' + \frac{d}{dt}(\epsilon_0 \epsilon' E) \cdot dS' \tag{2.13}$$

where J_c is the current density due to conduction effects commonly known as
ohmic effects, B is the magnetic flux density and μ_a the permeability of the
medium. The conduction current density is a function of the electric field vector,
E:

$$J_c = \sigma E \tag{2.14}$$

where σ is the conductivity of the medium. For sinusoidal electric field variations,
$E_{max}e^{j\omega t}$, such as that usually used for high frequency heating, eqn. (2.13)
attains the following form for the total current density:

$$\nabla \times H = J = \sigma E + j\omega\epsilon_0 \epsilon' E \tag{2.15}$$

The only loss mechanism considered so far is due to conduction effects while the dielectric constant has been assumed real and contributes only to the stored energy of the system. Re-arranging eqn. (2.15) yields

$$J = j\omega\epsilon_0 (\epsilon' - j\sigma/\omega\epsilon_0)E \qquad (2.16)$$

However, in free space, where $\sigma = 0$, Maxwell's circuital law given by eqn. (2.16) becomes

$$J = j\omega\epsilon_0 \epsilon' E \qquad (2.17)$$

Comparison of eqns. (2.16) and (2.17) leads to an abridged form of eqn. (2.16), i.e.,

$$J = j\omega\epsilon_0 \epsilon_c^* E \qquad (2.16)$$

where

$$\epsilon_c^* = \epsilon' - j\sigma/\omega\epsilon_0 = \epsilon' - j\epsilon_c'' \qquad (2.18)$$

can be considered as an effective dielectric constant of the medium, where conduction effects dominate. Therefore in this treatise, the only form of loss considered, ohmic, appears as the imaginary part of the dielectric constant. By analogy any other form of loss such as dipolar, Maxwell–Wagner, etc., can be included in any subsequent analysis by considering it as part of the complex dielectric constant and forming its imaginary term. Therefore, if reorientation polarisation was the only mechanism leading to losses, its contribution could have been taken care of by writing the dielectric constant in a form similar to that given by eqn. (2.18):

$$\epsilon_d^* = \epsilon' - j\epsilon_d'' \qquad (2.18a)$$

where the subscript d refers to dipolar or polarisation loss mechanism. However, since a loss term must include all possible mechanisms, the loss factor is generally written as ϵ'' which is the form shown in eqn. (2.12). We have therefore shown that, in order to satisfy Maxwell circuital law in a real insulator, the dielectric constant must attain a complex form to account for any loss mechanisms. When all types of losses are included, eqn. (2.16) yields

$$J_e = j\omega\epsilon_0 \left[\epsilon' - j(\epsilon'' + \sigma/\omega\epsilon_0) \right] E \qquad (2.19)$$

which is shown schematically in the phasor diagram in Fig. 2.3. Since, with most dielectric measuring techniques, it is difficult to separate the losses due to conduction from those due to polarisation, all form of losses can be grouped together, thus defining an effective loss factor ϵ_{eff}'' given by

$$\begin{aligned} \epsilon_{eff}''(\omega) &= \epsilon_d''(\omega) + \epsilon_e''(\omega) + \epsilon_a''(\omega) + \epsilon_{MW}''(\omega) + \sigma/\epsilon_0\,\omega \\ &= \epsilon''(\omega) + \sigma/\epsilon_0\,\omega \end{aligned} \qquad (2.20)$$

where the subscripts d, e, a and MW refer to dipolar, electronic, atomic and Maxwell–Wagner respectively. The loss factors show a frequency dependence which often simplifies the analysis because, in a given frequency band, one or two loss mechanisms dominate over the others. This frequency dependence of the loss mechanisms has been clearly demonstrated by Hasted (1973) who has schematically represented the various contributing mechanisms to the loss factor

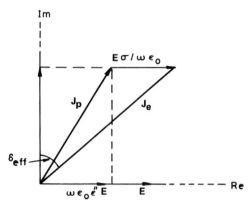

Fig. 2.3   *Current density and applied electric field vectors. J_p is only due to polarisation mechanisms, J_e is due to polarisation and d.c. conductivity effects*

of a moist material as shown in Fig. 2.4. Since industrial high frequency heating takes place in the frequency band $10^7 < f < 3 \times 10^9$ Hz, the mechanisms we should mostly be concerned with should be dipolar, represented to some extent

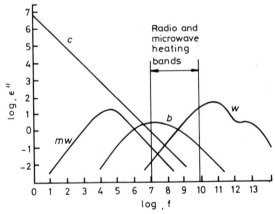

Fig. 2.4   *Loss factor of moist material as function of the frequency in Hz*
    c   d.c. conductivity, *MW* (Maxwell—Wagner) effect
    b   bound water relaxation
    w   free water relaxation
    (Extracted from *Aqueous Dielectrics*, by J.B. Hasted by permission of Chapman and Hall Ltd.)

by both *b* and *w* in Fig. 2.4, depending upon the state of water absorbed in the material, Maxwell—Wagner and d.c. conductivity. Loss mechanisms due to atomic and electronic polarisations, collectively termed distortion polarisations, occur at frequencies in the infra-red and visible part of the electromagnetic spectrum and as such play no part in high frequency heating. The complex dielectric constant follows from eqn. (2.19) and is given by

$$\epsilon^* = \epsilon' - j\epsilon''_{eff} \qquad (2.21)$$

where ϵ''_{eff} includes the loss factors which are relevant to high frequency heating. The ratio of the effective loss factor to that of the dielectric constant

$$\tan \delta_{eff} = \frac{\epsilon''_{eff}}{\epsilon'} \qquad (2.22)$$

is called the effective loss tangent, which includes the effects of d.c. conductivity, and is the angle between the total current density vector and the vertical axis in Fig. 2.3.

2.4 Dipolar loss mechanism

2.4.1 Debye equations

Of all the possible forms of loss mechanisms, dipolar or reorientation polarisation is perhaps the most significant in industrial microwave heating applications at frequencies above 1 GHz. However, it does influence the lower frequency bands as well. This is because for many polar matei.als the time constants for the establishment and decay of the polarisations occur at times comparable to the periods of oscillation of the high frequency signal. The classical approach to the treatment of permanent dipoles in liquids and in solutions of polar molecules in non-polar solvents is to consider their behaviour in alternating fields as arising from the rotation of a spherical dipole in a viscous medium dominated by friction. Debye (1929) working on electrolytes deduced the well-known equation

$$\epsilon^* = \epsilon' - j\epsilon''_d = \epsilon_\infty + \frac{\epsilon_s - \epsilon_\infty}{1 + j\omega\tau} \qquad (2.23)$$

where ϵ_s and ϵ_∞ are the dielectric constants at d.c. and very high frequencies respectively and τ is the relaxation time of the system which controls the build-up and decay of the polarisation shown in Fig. 2.5 as the external field is applied or removed respectively. P_2 and P_1 are the dipolar and distortion polarisation vectors. By assuming that the rate of increase of the dipolar polarisation is proportional to its departure from its equilibrium value, Debye established that P_2 is out of phase with the applied electric field E, thereby introducing the notion of losses. To account for this phase lag Debye had to assume a complex dielectric constant which is given by eqn. (2.23).

To put this phase lag in its proper perspective consider a high frequency electric field

$$E = E_{max} \sin \omega t \qquad (2.24)$$

applied to a dielectric. The polarisation vector then attains the following form:

$$P = P_{max} \sin (\omega t - \phi) \tag{2.25}$$

where P_{max} is a constant and ϕ accounts for the phase difference between the E and P vectors as shown in Fig. 2.6. The resulting current due to this charge redistribution is given by

$$\frac{\partial P}{\partial t} = \omega P_{max} \cos (\omega t - \phi) \tag{2.26}$$

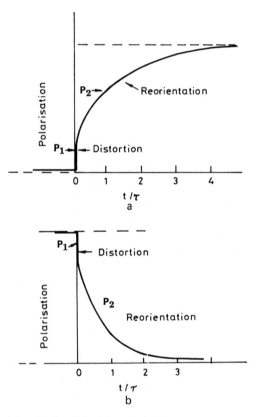

Fig. 2.5   *Build up (a) and decay (b) of the polarisation*

and the power dissipated into the dielectric is

$$\text{power} = E \frac{\partial P}{\partial t} = \tfrac{1}{2} E_{max} P_{max} \omega [\sin 2\omega t \cos \phi + (1 - \cos 2\omega t) \sin \phi] \tag{2.27}$$

Therefore, the average power dissipated is

$$P_{av} = \tfrac{1}{2}E_{max}P_{max}\omega \sin\phi = \tfrac{1}{2}E_{max}P_{max}\omega \cos\theta_0 \qquad (2.28)$$

where $\cos\theta_0$ is the power factor. It is evident that if P and E are in phase, ϕ is zero and no power will be dissipated into the dielectric.

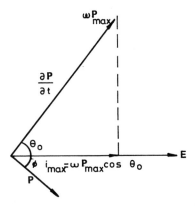

Fig. 2.6 *Phasor diagram between the external electric field vector, E, the polarisation vector, P and the resulting current vector $\partial P/\partial t$. Also $(\theta_0 + \phi) = \tfrac{1}{2}\pi$*

Separating the real and imaginary parts of the complex dielectric constant given by eqn. (2.23) we obtain

$$\epsilon' = \epsilon_\infty + \frac{\epsilon_s - \epsilon_\infty}{1 + \omega^2\tau^2} \qquad (2.29)$$

$$\epsilon_d'' = \frac{(\epsilon_s - \epsilon_\infty)\omega\tau}{1 + \omega^2\tau^2} \qquad (2.30)$$

These quantities are plotted in Fig. 2.7 as a function of f. The loss factor is a maximum when $\omega\tau = 1$. The resonance type of losses resulting from atomic and electronic polarisation, shown as P_1 in Fig. 2.5, are also included in Fig. 2.7 for comparison. For such displacement type of polarisation the real parts of ϵ^* either side of the dispersion curve should be the same. However, experimentally a difference is found between ϵ_∞ and the square of the refractive index at optical frequencies η^2, as Fig. 2.7 shows at very high frequencies, suggesting that between dipolar relaxation and optical dispersion there might be another relaxation mechanism accounting for the observed difference between ϵ_∞ and η^2 (Fröhlich, 1949).

2.4.2 Interpretation of the Debye equations
At low frequencies, near d.c., the dipoles have ample time to follow the variations of the applied field and the dielectric constant is at its maximum value; that is,

the bound charge density attains its maximum value and all the energy of the external source is stored in the material. As the frequency increases the dipoles are unable to fully restore their original positions during field reversals and as a consequence the dipolar polarisation lags behind the applied field. As the frequency increases further a point is reached where the reorientation polarisation fails to follow the applied field and contributes less to the total polarisation. The fall of the effective polarisation manifests itself as a fall in the dielectric constant and a rise of the loss factor. Energy is now drawn from the system and is dissipated as heat into the material.

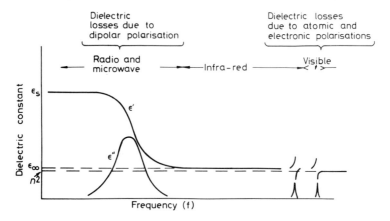

Fig. 2.7   *The dielectric constant as a function of the frequency in the region of dipolar and distortion absorption*

Debye's interpretation of this relaxation is given in terms of dipolar rotation against frictional forces in the medium. Using Stokes' theorem he derived the following expression for the relaxation time of the spherical dipole:

$$\tau = \frac{4\pi r_q^3 n_v}{k_b T} \tag{2.31}$$

where n_v is the viscosity of the medium, r_q is the radius of the rotating dipole and k_b is Boltzmann's constant. In practice, however, many liquids and solid dielectrics possess relaxation times which are much longer than those derived from eqn. (2.31), that is, the response is much flatter than that indicated in Fig. 2.7. This can be remedied by substituting the applied field used in the Debye theory by the Mosotti local field introduced earlier. The new effective relaxation time constant for the dipoles becomes

$$\tau_e = \frac{\epsilon_s + 2}{\epsilon_\infty + 2} \tag{2.32}$$

Despite this correction for the relaxation times of dipoles, the Debye interpretation is fairly inaccurate. It is difficult to imagine, particularly in solid dielectrics, the dipoles as spheres in a medium where viscosity is the dominant mechanism. Moreover, when many atoms and molecules are bonded together to form a dielectric, the dipoles are now influenced by the forces of all the neighbouring particles which must be taken into account in the theory. The Mosotti field approximation, although it is in many circumstances a very useful concept, leads to catastrophic situations when applied to polar molecules, in that the polarisation tends to infinity when the polarisability term, α, approaches a certain critical value (von Hippel, 1954). Debye (1929) recognised this shortcoming and introduced the notion of hindered motion of polar molecules in an environment of other particles exerting an influence which cannot be ignored. The concept of an activation energy, U, was introduced and is the energy required for a dipole to acquire a different equiprobable position. Of course, if $U \ll k_b T$, the dipoles would be free to reorientate and the forces of the surrounding molecular particles do not play a very important role in the ensuing polarisation mechanism, as is the case for some gaseous substances, where the interatomic forces are much smaller than $k_b T$. In solid dielectrics, however, the interatomic forces dominate, where normally $U \gg k_b T$, resulting in large deviations from the simple Debye theory to that observed in practice.

2.4.3 Potential double well

Modern methods have used the concept of an activation energy for dipole transitions between two equiprobable positions to describe the microscopic processes occurring during reorientation polarisation (Fröhlich, 1958; Anderson, 1964; Daniel, 1967). Since the essence of dielectric relaxation is the phase lag between the application of the external field and the ensuing polarisation within the material, it follows that any alternative theoretical analysis must still contain a relaxation time constant, τ_0, which determines the rise and decay of the polarisation. What is needed therefore, particularly with inhomogeneous solid dielectrics, is a link between the relaxation time constant and the energy barrier, U_a, between two dipole positions. That is

$$\tau_0 = g(U_a) \tag{2.33}$$

Such a link is indeed found in the potential double well dielectric models dealing with the transitions of dipoles between two positions separated by a potential energy, as depicted in Fig. 2.8. The external field distorts the potential energy from its original symmetrical form. The dipoles are assumed to align parallel or antiparallel with the applied field, a treatment which is analogous to transitions between states of weakly bound ions (Zheludev, 1971). Using Boltzmann statistics the number of dipole transitions from one state to the other is proportional to a factor $(1 - e^{-t/\tau_0})$. Hence, the polarisation induced by the field will change exponentially. The current density is readily determined, since $J = \partial P/\partial t$, and by using the principle of superposition of currents and integrating throughout a given volume, the total current density is obtained, comprising real and imaginary terms. By equating this to the total current density, $j\omega\epsilon^*\epsilon_0 E$, the ϵ' and ϵ'' components are easily found and shown to obey a Debye type of relaxation. However, the time constant τ_0 is now given by

$$\tau_0 = \frac{e^{U_a/k_b T}}{\nu}\left[\frac{\epsilon_s + 2}{\epsilon_\infty + 2}\right] \tag{2.34}$$

where $1/\nu$ is the time for a single oscillation in the potential well. Equation (2.34) introduces a very powerful relation between relaxation time and activation energy

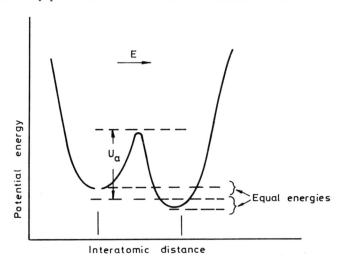

Fig. 2.8   *Potential energy diagram for two positions of a permanent dipole*

since it is now possible to extend the simple Debye theory and bring it closer to what is observed in practice in many real dielectrics. This is because, for example, in many solid dielectrics, dipoles can exhibit more than one activation energy corresponding to transitions from different potential wells giving rise to many relaxation times (Meakins, 1961). The concept of spread of relaxation times can now be easily envisaged in an inhomogeneous solid dielectric, where hindered dipole redistribution results in multi-relaxational spectra as shown qualitatively in Fig. 2.9. This is confirmed by experiments where the ϵ'' as a function of the angular frequency is much flatter and broader than the response obtained from the simple Debye equations where the time constant is viscosity dependent. To account for the broadening of the relaxation response, Cole and Cole (1941) introduced a spreading factor $\mathscr{X}$, modifying the complex dielectric constant to

$$\epsilon^* = \epsilon_\infty + \frac{\epsilon_s - \epsilon_\infty}{1 + j\omega\tau_{mean}^{(1-\mathscr{X})}} \tag{2.35}$$

where τ_{mean} is the mean of different relaxation times which correspond to transitions between the different dipole positions. Any deviations from the pure Debye theory can be readily seen if ϵ'' and ϵ' are plotted on an Argand diagram, as shown in Fig. 2.10, for a real dielectric (solid line). The factor $\mathscr{X}$ is a measure of this deviation being zero for a purely Debye response (dotted response).

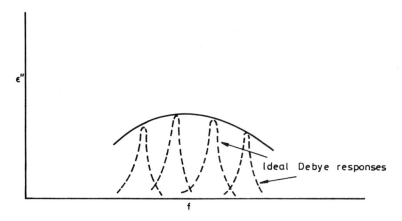

Fig. 2.9   *Ideal Debye loss factor vs. frequency responses (dotted lines) and multirelaxational response due to hindered dipole redistributions (solid line)*

2.4.4 Electric dipole moment

In a polar material there is a large number of permanent dipole moments which contribute to the total loss factor at a specific frequency. However the magnitude of the effective loss is also governed by the product $\omega\tau$. A simple formula relating the total loss factor of a mixture of polar solute into a non-polar solution is given by Debye (1929):

$$\epsilon_d''(\omega) = C_0 \frac{(\epsilon' + 2)}{k_b T} \mu^2 \frac{\omega\tau}{1 + (\omega\tau)^2} \qquad (2.36)$$

where C_0 depends on the relevant concentrations. It is evident, therefore, that the loss factor is not only proportional to the μ^2 but is also heavily influenced by the relaxation time. Such $\epsilon_d''(\omega)$ data for various polar molecules in non-polar benzene can be found in Harvey (1963). A similar expression may be valid for solid mixtures suitably modified by incorporating a mean of relaxation times to account for the observed broadening of the relaxation spectrum.

2.5 Maxwell–Wagner or interfacial loss mechanism

Interfacial or Maxwell–Wagner polarisation is very important in heterogeneous dielectrics, comprising, for example, a small fraction of conducting phase in a non-conducting medium. It relates to the build-up of charge particles at interfaces in heterogeneous dielectrics. The reason for considering it in this book is that it almost certainly influences the total polarisation of a heterogeneous dielectric at the lowest frequency band used in industrial high frequency heating, i.e., at frequencies less than 5×10^7 Hz. Wagner (1914) has shown that for the simplest model featuring this type of polarisation, consisting of conducting spheres distributed throughout a non-conducting medium, the dielectric loss factor, of volume fraction v of conducting material, is given by

$$\epsilon''_{MW} = \frac{9ve' f_{max}}{1.8 \times 10^{10}\sigma} \cdot \frac{\omega\tau}{(1 + \omega^2\tau^2)} \tag{2.37}$$

where f_{max} in Hz is the frequency of maximum losses, ϵ' is the real part of the dielectric constant of the continuous phase, σ is the conductivity in Sm^{-1} of the conductive phase and τ is the relaxation time constant.

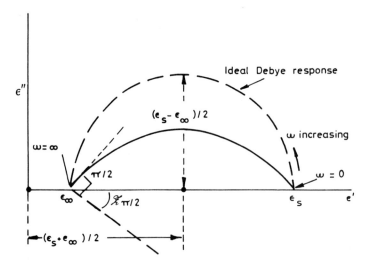

Fig. 2.10 *Qualitative representation of the Cole–Cole diagrams for an ideal Debye relaxation (dotted line) and for a real dielectric (solid line)*

It is obvious that the frequency variation of the loss factor is similar to that of dipolar relaxation, which was experimentally verified by Hamon (1953) on an artificial dielectric made up of conducting copper phthalocyanine particles in paraffin wax. Water dispersed in wax-wool shows similar Maxwell–Wagner absorption, as shown in Fig. 2.11, where the water forms small spherical particles of equal size (Dryden and Meakings, 1957). The non-symmetrical form of the relaxation is attributable to dielectric absorption in the wool-wax itself.

A more general approach to interfacial polarisation has been considered by Maxwell and Wagner in their two-layered capacitor, shown in Fig. 2.12, together with its equivalent electrical analogy. Since this treatment has been very well documented (von Hippel, 1954; Anderson, 1964; Hasted, 1973; Coelho, 1979) we will consider it very briefly. The total admittance of the combined circuits is worked out in terms of the two resistances and their corresponding time constants and equated to

$$Y = j\omega\epsilon_0\,\epsilon_{MW}^* \frac{A}{(d_1 + d_2)} \tag{2.38}$$

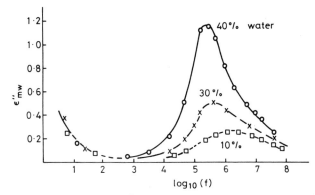

Fig. 2.11   *Maxwell—Wagner type absorption in mixtures of wool—wax and water at 20°C* (After Dryden and Meakins, 1957, Copyright The institute of Physics)

where d_1 and d_2 are the thicknesses of the two layers and A is the cross-sectional area of the capacitor. The complex permittivity is found to be equal to

$$\epsilon_{MW}^* = \epsilon_\infty + \frac{\epsilon_s - \epsilon_\infty}{1 + j\omega\tau} - j\frac{(d_1 + d_2)}{\omega A \epsilon_0 (R_1' + R_2')} \tag{2.39}$$

or

$$\epsilon_{MW}^* = \epsilon_\infty + \frac{\epsilon_s - \epsilon_\infty}{1 + j\omega\tau} - j\frac{\sigma}{\omega\epsilon_0} \tag{2.40}$$

The real part of ϵ_{MW}^* is precisely that given by the simple Debye theory. However, the loss term includes a term due to d.c., conductivity. It is evident that if either R_1' or R_2' are infinite, i.e., zero conductivity, eqn. (2.39) gives a Debye response involving a single relaxation time and with a loss component numerically equivalent to that given by eqn. (2.30). However, τ refers to the relevant $R'C$ circuit of Fig. 2.12. The contribution to the total loss due to this extra conductive part depends upon the value of the d.c. conductivity itself, as shown qualitatively in Fig. 2.13. For highly conductive dielectrics therefore, such as foodstuffs, particularly those containing large amounts of salts, there might come a point where the purely conductive term dominates the losses giving a response indicated by curve IV in Fig. 2.13. This would be very probable if the conductive phases in the two layer capacitor form paths, as shown in Fig. 2.14(b), as compared to the conducting spheres of Fig. 2.14(a) which give rise to a Debye like response depicted qualitatively by curve I in Fig. 2.13.

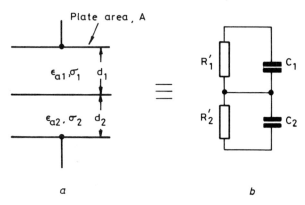

Fig. 2.12 *Maxwell-Wagner two layer capacitor (a) and its electrical model (b)*

Finally it is doubtful whether one can represent the equivalent losses due to d.c. conductivity of many heterogeneous materials by the simple expression

$$\epsilon''_{dc} = \sigma/\omega\epsilon_0 \tag{2.41}$$

as shown in eqn. (2.40). Hamon (1953) simulated such real dielectrics using copper phthalocyanine and liquid n-primary alcohols in paraffin wax and found that at low frequencies expression (2.41) was quite valid. van Beek (1967) has put forward the following empirical expression for the loss factor:

$$\epsilon''_{dc} = kf^{-k_0} \tag{2.42}$$

where k_0 approaches unity for high concentration of conducting material.

2.6 Combined effects

The processing of many industrial materials with electromagnetic energy in the frequency range $10\,\text{MHz} < f < 3\,\text{GHz}$ is a practical proposition primarily due to two distinct absorption mechanisms: dipolar and Maxwell–Wagner or interfacial relaxation. For mixtures containing very high amounts of conductive phases, interfacial polarisation is represented by the simple form of eqn. (2.41). Therefore, in an industrial heterogeneous dielectric containing some quantity of conductive phase, one cannot ignore the losses at the higher frequencies in Fig. 2.11 which combine with the dipolar losses to form an effective loss factor ϵ''_{eff}, shown in Fig. 2.15, where the rise at the lower end of the radio frequencies is attributable to effects of conductivity.

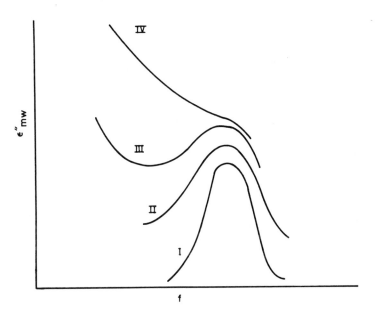

Fig. 2.13   *Influence of d.c. conductivity on the losses in interfacial polarisation — d.c. conductivity increasing between I and IV*

2.7 New physical mechanism for dielectric loss

The loss factor in the Debye type of relaxation with a distribution of relaxation times can be represented formally (Hill *et al.*, 1969) by

$$\frac{\epsilon''}{\epsilon_0} = \int_{\tau=0}^{\infty} g(\tau) \frac{\omega\tau}{1 + \omega^2\tau^2} \, d\tau \tag{2.43}$$

where $g(\tau)d\tau$ is the fraction of orientation polarisation processes with relaxation times between τ and $d\tau$. The physical interpretation of $g(\tau)$ is obscure particularly in trying to understand the physical processes (single or multiple) involved. Jonscher (1975*a*), upon examining the data for dielectric relaxation of polymeric solids has put forward an alternative physical model for the observed results.[*] Basically he suggests that the frequency dependence of the dielectric loss cannot satisfactorily be explained by the notion of distribution of relaxation times but instead it follows a universal law, ω^{n-1}, where $n < 1$ due to a mechanism for

[*] Summary by permission of McMillan Journals Ltd.

which the ratio of the energy lost per cycle to the energy stored per cycle is frequency independent. That is

$$\frac{\text{energy lost per cycle}}{\text{energy stored per cycle}} = \cotan\left(\frac{n\pi}{2}\right) = \text{constant} \tag{2.44}$$

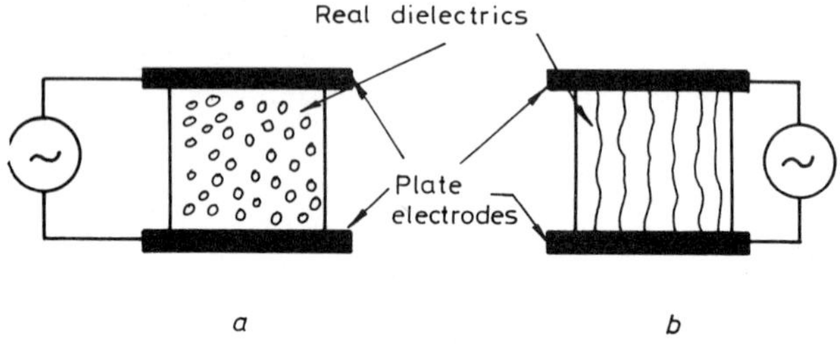

Fig. 2.14   *Transport mechanisms due to the conductive phase in a dielectric.*
a Conductive spheres
b Conductive paths

In the Debye form of relaxation this ratio is equal to $\omega\tau$. Jonscher suggests the following empirical relaxation for the loss factor:

$$\epsilon_0/\epsilon'' = \left(\frac{\omega}{\omega_2}\right)^{-m} + \left(\frac{\omega}{\omega_1}\right)^{1-n} \tag{2.45}$$

where m and n are integers less than unity. Also $(m) > (1-n)$ and $1/\omega_1$ and $1/\omega_2$ are thermally activated parameters. A physical model satisfying the energy equation and which applies to both dipolar and hopping charge systems must fulfil the following conditions. First, that charges of dipoles move in discontinuous hops or jumps between preferred positions or orientations and second, that there must be a presence of a screening charge adjusting slowly to the rapid hopping or jumping.

For many polymers and inorganic materials, where conductivity effects are also included, eqn. (2.45) reduces to

$$\epsilon''(\omega)/\epsilon_0 = B(T)\omega^{n(T)-1} \tag{2.46}$$

where $B(T)$ is a temperature dependent parameter. As eqn. (2.46) was found to be applicable to more complex dielectrics, it is proposed that it is more appropriate to use it in order to fit the experimental data rather than using the more established Debye function with a distribution of relaxation times. Note, however, the similarity between eqns. (2.46) and (2.41) the latter of which has once again been empirically formulated for dielectric materials with a conductive phase. A physical mechanism is put forward for frequencies above that where maximum losses occur, which satisfies eqn. (2.46) in a material where significant 'hopping conduction' takes place due to the large densities of localised electronic and ionic carriers. Jonscher concludes that the setting up of a given state of polarisation in a system of hopping charges must give rise to a loss of energy which is proportional to the stored energy and which does not depend on the rate at which this state of polarisation has been established.

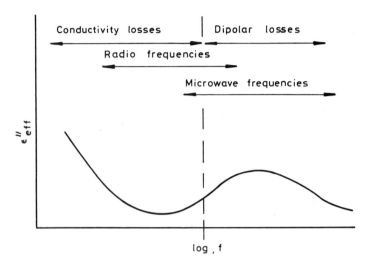

Fig. 2.15 *Effective loss factor of a hererogeneous dielectric exhibiting dipolar and tail end conductivity losses*

Jonscher (1975*b*) presents an alternative argument for the effective losses occurring for frequencies in the region of the loss peak and below. It is argued that the loss peak in solid dielectrics is due to two separate processes occurring consecutively, where, for example, the secondary process follows the main primary process, the latter's duration being thermally activated with a defined energy.

2.8 Dry, non-conducting dielectrics

Emphasis has been given so far to the conducting phase in dielectrics such as, for example, water, because high frequency heating has been applied very successfully to many drying applications. However, many applications require heating of dry, non-conducting insulators such as natural rubber or other chemical substances, in order to achieve specific reactions within the material. In this case, Maxwell–Wagner type effects, in the absence of any trace impurities, play no part at all in the heating process which is now governed solely by dipolar effects. In fact, if the insulating material is non-polar, one may deliberately introduce a controlled proportion of conducting particles, such as carbon black, to enhance the absorption of the applied high frequency energy by the material to be heated. This is discussed further in Section 3.9.

2.9 Magnetic loss factor

Although this book is primarily concerned with the behaviour of dielectric materials in electric fields, it is important to mention that a material exhibiting high magnetic losses at radio and microwave frequencies can also be effectively treated with this form of energy (Epstein, 1954). In a manner analogous to the case of a dielectric material under the influence of a high frequency electric field, a complex permeability is assigned to the following relation:

$$B = \mu_0 \mu^* H \tag{2.47}$$

where B is the magnetic flux density and H is the magnetic field. The complex permeability is represented by

$$\mu^* = (\mu' - j\mu''_{eff}) \tag{2.48}$$

where μ' is the permeability and μ''_{eff} is the effective magnetic loss factor due to relaxation and resonance processes under the influence of the alternating magnetic field. These losses are quite distinct from hysteresis or eddy current losses and are in fact associated with domain wall and electron spin resonance (Clarricoats, 1961; Tebble and Craik, 1969). The permeability spectra of sintered magnesium ferrite (Fenamic A) shows two loss factor peaks, one at about 50 MHz which is attributed to domain wall resonance arising from damped precession of the spins in the domain wall, while the other peak at about 1 GHz is due to electron spin resonance (Rado, 1953; Rado *et al.*, 1950). The permeability spectrum of polycrystalline nickel ferrite exhibits similar double magnetic loss factor peak characteristics in the radio and microwave frequency regime (Miles *et al.*, 1957). On the other hand, the magnetic losses of finely powdered Feramic A dispersed in wax (70% by weight dispersed in wax) shows only one peak at the microwave regime between 1 and 10 GHz (Rado *et al.*, 1950). This is important from the point of view of microwave processing because such lossy powders and other catalytic agents have been added to low loss materials in order to enhance the rates of energy absorption (see Section 3.9).

The parameter μ^* is obtained at microwave frequencies by impedance measurement in a transmission line. Usually two measurements are required to separate ϵ^* and μ^* which are related through the following expressions for the intrinsic impedance of the specimen Z and the propagation factor γ through the specimen

$$Z = \left(\frac{\mu_0 \mu^*}{\epsilon_0 \epsilon^*} \right)^{1/2} \tag{2.49}$$

and

$$\gamma = j\omega(\mu_0 \mu^* \epsilon_0 \epsilon^*)^{1/2} \tag{2.50}$$

respectively. Finally it must be stressed that when heating magnetic materials in single mode resonant cavities the specimens should be placed at position of magnetic field maxima for optimum absorption of the microwave energy.

2.10 References

ANDERSON, J. C., *Dielectrics*. Chapman and Hall, London (1964).

CLARRICOATS, P. J. B., *Microwave Ferrites*, pp. 87–94. Chapman and Hall, London (1961).

COELHO, R., *Physics of Dielectrics for the Engineer*, p. 77. Elsevier Publishing Co., Amsterdam (1979).

COLE, K. S. and COLE, R. H., *J. Chem. Phys.* **9**, 341 (1941).

DANIEL, V., *Dielectric Relaxation.*, Academic Press, New York (1967).

DEBYE, P., *Polar Molecules.* Chemical Catalog, New York (1929).

DRYDEN, J. S. and MEAKINS, R. J., 'Examples of the Maxwell–Wagner type of dielectric absorption using wool–wax–water mixtures', *Proc. Phys. Soc. Lond.,* **70**, 427 (1957).

EPSTEIN, D., *Permeability in Dielectric Materials and Applications* (Edited by A. von Hippel), pp. 122–134. MIT Press, (1954).

FRÖHLICH, H., *Theory of Dielectrics.* Clarendon Press, Oxford (1949).

FRÖHLICH, H., *Theory of Dielectrics,* 2nd edn., Oxford University Press, London (1958).

JOHNK, T. A., *Engineering Electromagnetic Fields and Waves* (1975).

JONSCHER, A. K., 'Physical basis of dielectric loss', *Nature* **253**, 717 (1975a).

JONSCHER, A. K., 'New interpretation of dielectric loss peaks', *Nature* **256**, 566 (1975b).

HAMON, B. V., 'Maxwell–Wagner loss and absorption currents in dielectrics', *Aust. J. Phys.* **6**, 304 (1953).

HARVEY, A. F., *Microwave Engineering*, p. 251. Academic Press, New York (1963).

HASTED, J. B., *Aqueous Dielectrics.* Chapman and Hall, London (1973).

HILL, N., VAUGHAN, W. E., PRICE, A. H., and DAVIES, M., *Dielectric Properties and Molecular Behaviour.* von Nostrand, New York (1969).

MILES, P. A., WESTPHAL, W. B., and VON HIPPEL, A., 'Dielectric Spectroscopy of Ferromagnetic Semiconductors' *Reviews of Modern Physics,* **29**, 302 (1957).

MEAKINS, R. J., 'Mechanisms of dielectric absorption in solids', *Prog. Dielectrics* **3**, 151 (1961).

RADO, T. G., 'Magnetic spectra of ferrites', *Rev. Mod. Phys.* **25**, 81 (1953).

RADO, G. T., WRIGHT, R. W., and EMERSON, W. H., 'Ferromagnetism at very high frequencies. 3. Two mechanisms of dispersion in a ferrite', *Phys. Rev.* **80**, 273 (1950).

TEBBLE, R. S. and CRAIK, D. J., *Magnetic Materials.* Wiley Interscience, John Wiley and Sons Ltd., New York (1969).

VAN BEEK, L. K. H., 'Dielectric behaviour of heterogeneous systems', *Prog. Dielectrics* **7**, 69 (1967).

VON HIPPEL, A., *Dielectrics and Waves.* MIT Press (1954).

WAGNER, K. W., *Arch. Electrotech.* **2**, 371 (1914).

ZHELUDEV, I. S., *Physics of Crystalline Dielectrics. 2, Electrical Properties.* Plenum Press, New York (1971).

Chapter 3

Dielectric properties

3.1 Introduction

In the past simple multimode oven tests have been the starting point for the evaluation of an industrial heating process. Unfortunately the apparent simplicity of this approach often leads to misleading results due to the complexity of the field distribution within the oven and the variation in dielectric properties of the material with temperature, moisture content, density and other parameters. The versatility of the microwave oven is not in dispute as many industrial systems do use the simple multi-mode principle for the cavity design. However, as will be discussed later, many applications, because of the nature of the dielectric properties and their variation during processing, require special design of applicators which calls for a detailed knowledge of dielectric data. Coupled with this requirement theoreticians have been striving for years to have a better understanding of the behaviour of heterogeneous dielectrics and mixtures of such dielectrics and new advances in measuring techniques have given way to many more accurate ways of acquiring dielectric data. Therefore, a wealth of information on dielectric properties does exist for a variety of industrial materials covering a wide range of frequency, temperature and moisture content. The purpose of this chapter is to categorise this fragmented information and extract the salient points which would be useful in helping to make a wiser decision as to the applicability of high frequency heating for various industrial processes. Emphasis has been given towards a simple presentation of data directly useful to the design engineer, rather than looking into the more fundamental aspects of such data which would be interesting to specialists of dielectric theory.

3.2 The need for dielectric property data

The knowledge of the dielectric properties of materials to be processed in the microwave regime is essential for the proper design of microwave applicators. The property which describes the behaviour of a dielectric under the influence of a high frequency field is the complex permittivity, ϵ^*, which is defined by eqn. (2.21):

$$\epsilon^* = \epsilon' - j\epsilon''_{eff} \tag{2.21}$$

where ϵ''_{eff} is the effective loss factor and includes the effects of conductivity. The subscripts c and d have been omitted since we will from now on assume that in a real dielectric both conduction and relaxational processes occur to varying degrees depending upon the operating frequency. In the previous chapter we have also defined the ratio

$$\tan \delta_{eff} = \epsilon''_{eff}/\epsilon' \tag{2.22}$$

known as the loss tangent. Both ϵ' and ϵ''_{eff} are frequency, f, and temperature, T, dependent and therefore in the past a considerable amount of data has been amassed on the variation of ϵ^* with f and T. von Hippel (1954) tabulates the dielectric properties of a wide range of inorganic materials (crystals, ceramics, glasses, water, etc.) and organic materials (crystals, simple non-crystals, plastics, elastomers, natural resins, asphalts and cements, waxes, woods, etc.) in the frequency and temperature ranges of $100 < f < 10^{10}$ Hz and $-12 < T < 200°C$ respectively. Such comprehensive data should be consulted first to determine whether a material will be readily susceptible to high frequency energy. Tinga and Nelson (1973) have also published a similar list of dielectric properties for food and biological substances.

As explained in the previous chapter, some substances contain permanent dipole moments in their structure, the magnitude of which depends on the size and symmetry of the molecule concerned. Molecules having a centre of charge symmetry such as argon, carbon tetrafluoride (CF_4), methane (CH_4), carbon tetrachloride (CCl_4), propane (C_3H_8), etc., are non-polar and exhibit zero dipole moments. On the other hand, molecules such as water, urea (CH_4N_2O), acetone ($CH_3-CO.CH_3$), polyvinyl acetate ($C_4H_6O_2$)$_n$ and particularly some protein molecules such as gelatin, haemoglobin and serum pseudoglobulin-γ, are all polar since they have no charge symmetry and exhibit strong dipole moments. It is important to stress, however, that the dipolar loss factor would also depend on the relaxation time, τ, as well as the dipole moment, a fact borne out of eqn. (2.36). Therefore, the dipole moment tables cannot give comparative values for the loss factors but should only be used in conjunction with the corresponding data for the relaxation time, τ. An interpretation of such electric dipole moment data (Wesson, 1948) in terms of other molecular parameters and the overall losses, ϵ''_{eff}, in dipolar materials can be found in Hill *et al.* (1969).

Table 3.1 shows the dielectric properties of some common industrial materials. It is perhaps evident that there is no general pattern emerging from such a table, suffice to say that the presence of water in any material is bound to increase its effective loss factor and thus render it a better candidate for processing with

Table 3.1 *Properties of some common industrial materials*△

	$T\,°C$	% Moisture dry basis	10⁷ Hz ϵ'	10⁷ Hz ϵ''_{eff}	10⁹ Hz ϵ'	10⁹ Hz ϵ''_{eff}	3 × 10⁹ Hz ϵ'	3 × 10⁹ Hz ϵ''_{eff}
Food								
Beef steak, bottom round	25	—	50	1300	50	39	40	12
frozen lean	0	—	—	—	4·4	0·72	3·95	0·3
Bacon fat, conventionally rendered	25	—	—	—	2·6	0·16	2·5	0·13
Potato raw	25	—	80	47·8	65·1	19·6	53·7	15·7
Turkey cooked 1070 kg/m³	25	—	—	—	46·0	68·0*	40·0	14·0‡
Butter	0	16·5**	—	—	—	—	4·05	0·39‡
Butter	35	—	—	—	—	—	4·15	0·44‡
Forest products								
Douglas fir	20	0	—	—	2·05	0·04	2·0	0·02
Paper, royal grey white wove	82	0	—	—	3·0	0·216*	2·94	0·235
78 g/m² E ∥ to web	22	7·0	3·1	0·25†	—	—	3·2	0·5‡
Board, box 230 g/m² E ∥ to web	22	5·0	2·8	0·3‡	—	—	2·7	0·3‡
Glass								
Fused silica	25	—	3·78	< 10⁻⁴	3·78	0·0002*	3·78	0·0002
96% SiO₂	25	—	3·85	0·0023*	3·85	0·0023*	3·84	0·0026

	°C						
Minerals and ceramics							
Ruby mica, muscovite	25	5·4	0·0016	—	—	5·4	0·0016
Marble, dry	25	9·0	0·33	—	—	9·0	0·22
Sandy soil, dry	25	2·55	0·04	2·55	0·026*	2·55	0·016
Porcelain (aluminium oxide)	25	8·95	0·0018	8·93	0·008	8·90	0·01
Barium titanate	25	1140·0	8·55	1100·0	55·0*	600·0	180·0
Oils and waxes							
Cable oil	25	2·17	0·009	2·17	0·009*	2·16	0·0043
Aviation gasoline (100 octane)	25	—	—	1·94	0·00016*	1·92	0·0027
Wax (Mitchell Rand)	25	2·36	0·0006	2·31	0·00083	2·31	0·0011
Plastics							
Araldite	—	—	—	—	—	3·14	0·076
Laminated fibreglass (69% fibreglass, 31% bakelite)	24	5·0	0·17	4·54	0·108*	4·4	0·128
Melamine formaldehyde, moulding comps.	25	5·53	0·23	4·37	0·228	4·2	0·219
Cellulose acetate	—	3·3	0·07	3·28	0·072*	3·24	0·094
Polyamide, Dupont Nylon FM10001	25	3·24	0·07	3·06	0·043	3·02	0·036

Table 3.1   continued

	$T\,°C$	% Moisture dry basis	10^7 Hz ϵ'	ϵ''_{eff}	10^9 Hz ϵ'	ϵ''_{eff}	3×10^9 Hz ϵ'	ϵ''_{eff}
Polyethylene	23	—	—	—	2·26	0·0024*	2·25	0·0026
Polythene	24	—	2·25	0·0004	—	—	2·25	0·0007
Plexiglass, perspex	27	—	2·71	0·027	2·66	0·017*	2·6	0·015
Polyester formica, field ‖ to laminate	25	—	4·08	0·23	3·53	0·138	3·41	0·106
Polytetra-fluoroethylene, (teflon)	22	—	—	—	2·1	0·0003	2·1	0·0003
Polyester resin, mylar	20	—	2·4	0·0384	2·2	0·0088	—	—
Rubber								
Natural crepe	25	—	—	—	—	—	2·15	0·0065
Cyclized, pliobond	24	—	—	—	—	—	3·76	0·28
Nitrile, natural	25	—	—	—	2·85	0·069	2·8	0·05
Silicon, 33% siloxane elastomer 33% ZnO, 33% TiO$_2$	25	—	—	—	7·9	0·245*	7·8	0·48
SBR, uncured	40	—	—	—	—	—	8·7	0·09
cured	120	—	—	—	—	—	8·4	0·3
Water								
Ice pure	− 12	—	3·7	0·07	—	—	3·2	0·003
Distilled	25	—	—	—	77·5	1·2*	76·7	12·00

Distilled	95	—	—	—	52·0	0·364*	52·0	2·44
+ 0·5 molal NaCl	25	—	—	—	69·0	269·0	67·0	41·87
Cotton 210 kg/m³	—	7	1·5	0·03†	—	—	—	—
Wool 68 kg/m³	—	20	1·2	0·01†	—	—	—	—
Wool fibre	—	7	—	—	—	—	3·9	0·29
Milk powder 800 kg/m³	30	3·3**	—	—	—	—	2·29	0·048‡
Whey powder 780 kg/m³	30	4·8**	—	—	—	—	2·04	0·025‡
Leaves Rapeseed, frost resistant	23	—	400·0	900·0	40·0	20·0	40·0	11·0

* 3×10^8 † $2·712 \times 10^7$ ‡ $2·45 \times 10^9$ ** wet basis

△Table compiled from many sources including von Hippel (1954) by permission of M.I.T. Press.

microwave energy. Before we look more closely at the effect of bound and free water to the dielectric properties of various materials, let us examine briefly the methods used for measuring ϵ^*.

3.3 Measuring techniques

The measurement of the complex dielectric constant requires specialised techniques. For frequencies below 100 MHz, bridge and Q-meter methods are employed. Above this frequency the measurement is accomplished in coaxial lines or waveguides by determining the propagation constant of the electromagnetic wave in the material. Above 10 GHz free space methods are used (Scaife, 1971; Chamberlain and Chantry, 1973).

The frequency range that we are primarily concerned with in this book is between $400 < f < 3000$ MHz since it contains the frequency bands most commonly used for industrial purposes. The bands are 433.9 ± 0.87 MHz, 896 ± 10 MHz (915 ± 13 MHz in the U.S.A.) and 2450 ± 50 MHz. Dielectric property measurements at these frequencies are usually made by setting up a waveguide in the form of coaxial conductors with the test material forming part of the dielectric medium between the conductors. The method involves the measurement of the propagation constant, γ:

$$\gamma^2 = (2\pi/\lambda_c)^2 - \omega^2 \mu_a \epsilon_a = (\alpha + j\beta)^2 \tag{3.1}$$

where λ_c is the cutoff wavelength of the waveguide. Measurement of the propagation constant of the electromagnetic wave in the material γ_2, and in the empty transmission line, γ_1, gives a direct method of obtaining ϵ^* from eqn. (3.1) which for a coaxial line ($\lambda_c = \infty$) reduces to

$$\epsilon_a = \epsilon_0 \epsilon^* = \epsilon_0 (\gamma_2/\gamma_1)^2 \tag{3.2}$$

Therefore, the measurements of ϵ^*, involves the determination of the attenuation constant, α, and phase constant, β, of the electromagnetic signal. Experimentally one of the simplest methods of measuring γ_2 is that of Roberts and von Hippel (1946) which, although it is primarily suited for use with solids, Williams (1959) and Price (1973) have adapted for the measurement of the complex dielectric constant in lossy liquids. Its major drawback is the calculation of ϵ^* from the experimental data, which involves solutions of transcendental equations. This difficulty, however, is nowadays overcome by the use of computer techniques (Nelson *et al.*, 1973; Metaxas and Driscoll, 1974). The equipment for this measurement is easily available commercially and we shall give a brief insight of the basic principles involved.

3.3.1 The Roberts and von Hippel method

This method involves the examination of standing waves in a transmission line (or waveguide) terminated by a section filled with the dielectric under investigation. Figure. 3.1 illustrates the general form of the apparatus and the standing wave pattern which is associated with it. A signal generator sends an electromagnetic wave through a slotted line to the section containing the dielectric, the far end being short-circuited. The input impedance, Z_{in}, of the dielectric filled section of the line of characteristic impedance Z_{02} is given by

$$Z_{in} = Z_{02} \tanh (\gamma_2 d) \tag{3.3}$$

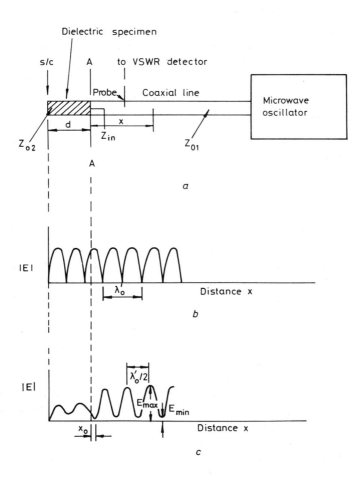

Fig. 3.1 *Method for dielectric material measurements*
   a General form of apparatus
   b Standing wave pattern with empty s/c line
   c Standing wave pattern with dielectric specimen of length *d* at the end of the s/c line

where d is the dielectric specimen length. The ratio of the characteristic impedances of the dielectric filled line Z_{02}, to that of the empty line, Z_{01}, is given by

$$\frac{Z_{02}}{Z_{01}} = (\gamma_1/\gamma_2) = (j\beta_1/\gamma_2) \tag{3.4}$$

assuming negligible losses in the empty line. Elimination of Z_{02} from eqns. (3.3) and (3.4) yields

$$Z_{in} = j\beta_1 Z_{01} [\tanh(\gamma_2 d)]/\gamma_2 \tag{3.5}$$

The combination of the forward and reflected waves give rise to a standing wave pattern which is characterised by a voltage standing wave ratio (VSWR), S. From the equations of the voltage and current at any point along the transmission line it can be readily shown that at a distance, x_0, from the interface, AA, the ratio of voltage to current is

$$Z = Z_{01}(Z_{in} + jZ_{01}\tan(\beta_1 x_0))/(Z_{01} + jZ_{in}\tan(\beta_1 x_0)) \tag{3.6}$$

If this point is a voltage minimum, then the impedance is resistive and is given by SZ_{01}. Substituting this into eqn. (3.6) yields

$$Z_{in} = Z_{01}(S - j\tan(\beta_1 x_0))/(1 - jS\tan(\beta_1 x_0)) \tag{3.7}$$

Equating the right-hand sides of eqns. (3.5) and (3.7) gives

$$(\tanh(\gamma_2 d))/\gamma_2 d = (S - j\tan(\beta_1 x_0))/(1 - jS\tan(\beta_1 x_0))j\beta_1 d \tag{3.8}$$

The determination of $\tanh(\gamma_2 d)/\gamma_2 d$ therefore entails the measurement of four parameters, namely x_0, S, d and λ'_0, where λ'_0 is the wavelength in the empty coaxial line. Equation (3.8) is solved using a computer and by substitution in eqn. (3.2), a number of possible solutions for ϵ^* can be obtained. The ambiguity of the many solutions is removed by making a second measurement at a different specimen length. Hence the correct solution is obtained by comparison of the two sets of solutions. The simple theory neglects any errors due to an imperfect short-circuit, line losses (which become important for measurement of low loss material, i.e., $(\epsilon''_{eff}/\epsilon') < 10^{-4}$), and imperfections in the dielectric specimen itself.

Moreover, additional errors could arise due to a shrinking specimen dielectric which does not entirely occupy the space between the inner and outer conductors (such as when measuring ϵ^* as a function of moisture content) and special correction factors have been employed (Metaxas and Parker, 1973).

A typical dielectric measuring set is shown in Fig. 3.2. It consists of a variable frequency oscillator with 1000 Hz internal square wave modulation (on/off 50% duty cycle) feeding a slotted line via a pad attenuator and a switch, the latter allowing connection to a digital frequency counter. A capacitive probe penetrates through a longitudinal slot of the coaxial waveguide and samples the standing wave pattern set up between the conductors. The microwave field is rectified and fed into a tuned amplifier to measure the voltage standing wave ratio (VSWR) S (see Section 5.5).

At frequencies exceeding 3000 MHz the slot in the outer conductor causes field distortion and radiation leading to considerable errors. For this, and to avoid electrical problems of the insulation supports, non-slotted lines are used. The principle of operation is similar to that of slotted lines except that now loop probes are employed to sample magnetic field distribution.

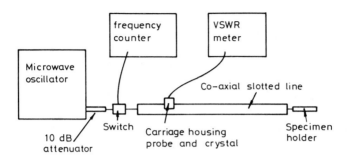

a

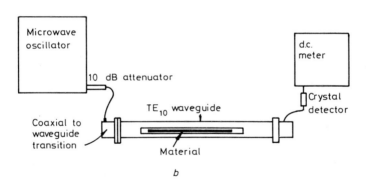

b

Fig. 3.2 (a) *Slotted line arrangement for material properties measurements.* (b) *Slotted line attenuation measuring equipment*

Precise methods of measuring ϵ' and ϵ''_{eff}, based on the von Hippel methods described above, are laborious to carry out and evaluate and the microwave heating industry generally uses a simpler but wholly adequate alternative, as shown in Fig. 3.2(*b*). It consists simply of measuring the attenuation of a launched microwave signal due to the material when it forms a thin bed inside a waveguide operating in the TE_{10} mode. This has the advantage of giving directly the property of interest to the system designer without recourse to computer evaluation.

3.3.2 X-band techniques

Before considering resonant cavities for the determination of material properties, it is important to mention the work carried out at X-band (8–10 GHz) primarily towards the design and optimisation of industrial moisture meters. Property measurements at X-band are not directly applicable for evaluating the suitability of many materials for microwave processing since most industrial heating/drying systems rarely operate above 2·45 GHz, although reference has been made to attempts at X-band heating for browning in microwave ovens (Ishitobi and Tagawa, 1978). However, measurements at X-band might indicate trends and variations in the properties of materials with some relevance to industrial processing.

Although the short-circuited technique described in the preceding paragraph together with newer techniques such as Time Delay Reflectrometry (TDR) (Iskander and Stuchly, 1972) and stripline (Kent and Steel, 1978) have been extensively used for dielectric property measurements and moisture determination, respectively, additional data on the way many materials respond to microwave energy at X-band have come about from transmission or reflection measurements in connection with the design of compact systems for moisture gauging in many materials, a technique referred to as aquametry (Kraszewski, 1980; Kalinski, 1978). The data invariably give the relationship between the attenuation α, of the electromagnetic signal as a function of the moisture content, M, in a material under test. One simple way of determining α is through a substitution measurement in a transmission type of set-up as shown in Fig. 3.3. To increase the sensitivity, bridge and other circuits have been used.

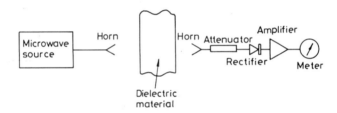

Fig. 3.3   *Block diagram of transmission type attenuation measuring equipment*

3.3.3 Cavity perturbation techniques

Methods based on the electromagnetic field perturbation of a resonant cavity by a small sample inserted in it have been used extensively for measuring the dielectric properties of low loss materials (Horner *et al.*, 1946; Harvey, 1963; Gos'Kov, 1965). This necessitates measuring the shift of the cavity resonant frequency from the original unperturbed value of ω_0' to a new perturbed value of ω_0 and the change in the cavity Q-factor values again with and without the insertion of the sample. The dimensions of the sample must be small compared with the size of the cavity resulting in a small frequency shift and ensuring the validity of the perturbation theory. Furthermore, correct positioning of the sample in the cavity is important to preserve the symmetry.

Starting from Maxwell's equations the following perturbation formula can be derived giving the cavity's complex angular frequency change due to the insertion of a small dielectric placed in the region of negligible magnetic field (Altman, 1964):

$$\frac{\hat{\omega}_0 - \hat{\omega}_0'}{\hat{\omega}_0} = -\epsilon_0 (\epsilon^* - 1) \frac{\iiint_V E^* E_0 \, dV}{4U} \tag{3.9}$$

where E and E_0 are the perturbed and unperturbed peak electric field in the region of the dielectric, U relates to the total stored energy in the cavity and V is the cavity volume. The complex angular frequency of the perturbed cavity can be written as

$$\hat{\omega}_0 = \omega_0 + j(\omega_0 / 2Q) \tag{3.10}$$

where ω_0 stands for the resonant frequency as measured on an impedance basis. Substitution of eqn. (3.10) for both $\hat{\omega}$ and $\hat{\omega}_0'$ in eqn. (3.9) yields

$$\frac{\Delta \omega}{\omega_0} + j\left(\frac{1}{2Q_0} - \frac{1}{2Q_0'}\right) = -\epsilon_0 (\epsilon^* - 1) \frac{\iiint_V E^* E_0 \, dV}{4U} \tag{3.11}$$

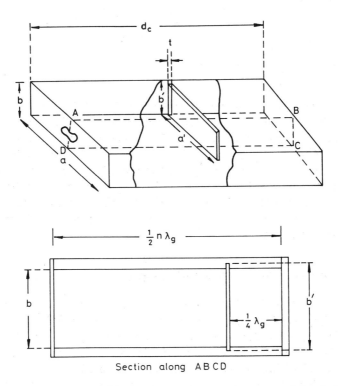

Section along A B C D

Fig. 3.4 *Resonant cavity for measuring dielectric properties via perturbation techniques*
(After Mehmet and McPhun, 1973, by permission of the National Physical Laboratory.)

where Q_0 and Q_0' are the perturbed and unperturbed unloaded Q-factors of the cavity. Here we have assumed that $\omega \simeq \omega_0$ and $|\omega| \gg |\omega/2Q_0|$ in the denominator. Thus, by measuring the shift in the resonant frequency and the change in Q, the dielectric properties ϵ' and ϵ'' can be determined. As an example, Mehmet and McPhun (1973) have applied the above theory to the case of a dielectric insertion in a rectangular TE_{10n} cavity, as shown in Fig. 3.4. By considering the relevant field distributions and substituting in eqn. (3.11), the dielectric constant, ϵ', and the loss factor, ϵ''_{eff}, are found after integration and separation of the real and imaginary parts respectively, giving

$$\epsilon' = 1 - A \left(\frac{\Delta \omega}{\omega_0} \right) \tag{3.12}$$

and

$$\epsilon''_{eff} = A \left(\frac{1}{2Q_0} - \frac{1}{2Q_0'} \right) \tag{3.13}$$

where

$$A = \frac{abd_c}{a'b't} \left[1 - \frac{a}{2\pi a'} \sin \frac{2\pi a'}{a} \right]^{-1} \tag{3.14}$$

where $\Delta \omega$ is negative. The various dimensions are illustrated in Fig. 3.4.

The equivalent expressions for cavity perturbation in cylindrical cavities have also been derived, for example for the TM_{010} mode (Terselius and Ranby, 1978) and for the TE_{011} mode (Haniotis and Günthard, 1969; Kumar and Smith, 1977). For high accuracy of measurement special techniques have been employed involving highly sophisticated circuitry (Couderc *et al.*, 1973). However, for routine measurements of industrial materials, Rzepecka (1973) described a much simpler method of measuring the width of the resonance curve while preserving the accuracy of the more sophisticated methods.

3.4 The variation of the ϵ^* with moisture content

The bulk of von Hippels (1954) data relates to materials with equilibrium moisture contents. However, since many applications involve the removal of moisture from the workload, the variation of ϵ^*, and in particular ϵ''_{eff}, with moisture content, plays an important role in the design of microwave heating/drying devices. For this a major effort was devoted to establishing the variation of ϵ' and ϵ''_{eff} with M for many industrial materials encountered in the manufacturing industry, such as paper, board, foodstuffs, leather, wood and textiles, and to interpreting the results in such a way as to be directly applicable to the designer of microwave applicators.

3.4.1 The dielectric properties of absorbed water

Liquid water is strongly polar in its structure, causing readily to absorb microwave energy and convert it into heat. When in contact with another material liquid water is referred to as absorbed water. However, the dielectric properties of absorbed water show marked differences to that of liquid water. The principle relaxation for liquid water (i.e., maximum loss factor) occurs at about 18 GHz (Hasted, 1972) with other minor relaxations taking place further up the infra-red frequency band. On the other hand the relaxation peaks of absorbed water occurs at frequencies well below the 18 GHz level. It appears, therefore, that the nature of the water absorbed in a material has a marked effect upon its dielectric properties and consequently upon the interaction of the high frequency field with that material (Hasted, 1973; De Loor, 1968; Tinga, 1969; Kraszewski, 1977).

Absorbed water in wet materials can exist in two principle states. Free water, which resides in capillaries, cavities, etc., and bound water, which is chemically combined to other molecules or physically adsorbed to the surface of the dry material. Figure 3.5 shows qualitatively, the variation of the loss factor with

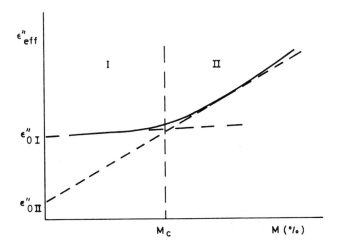

Fig. 3.5 *The effective loss factor as a function of the moisture content*

moisture content, M, of a typical wet solid. The two states of water within the original material can be related to different regions of the ϵ''_{eff} vs. M curve, these regions being characterised by the slope, $(d\epsilon''_{eff}/dM)$. The smaller slope at low moisture content is due to bound water (region I) whereas the higher slope at increased moistures is largely due to the presence of free water (region II). The water molecules 'bound' in the first uni-molecular layer at the surface of the material are less rotationally free than the water residing in capillaries and cavities. Thus the latter gives rise to much higher dielectric losses. The change of slope occurs at the critical moisture content, M_c. However, some materials exhibit a gradual change of slope making the positive identification of the two regions fairly difficult. The critical moisture content for highly hygroscopic materials occurs in the region between 10%–40% (dry basis) whereas for non-hygroscopic materials, e.g., sand, it is about 1% (Stuchly, 1970).

It is often required to express mathematically the effective loss factor ϵ''_{eff} in terms of the moisture content, M, in order to make use of the ϵ''_{eff} vs. M response in other formulae and optimise the design of particular applicators.

Assuming that the response of Fig. 3.5 can be approximated to two straight lines, as shown by the dotted lines, the effective loss factor in region I takes the form

$$\epsilon''_{eff} = \epsilon''_{0\mathrm{I}} + \left(\frac{d\epsilon''}{dM}\right)_{\mathrm{I}} M \tag{3.15}$$

while that in region II takes the form

$$\epsilon''_{eff} = \epsilon''_{0\mathrm{II}} + \left(\frac{d\epsilon''}{dM}\right)_{\mathrm{II}} M \tag{3.16}$$

where $\epsilon''_{0\mathrm{I}}$ and $\epsilon''_{0\mathrm{II}}$ are constants and $(d\epsilon''/dM)$ refer to the slopes in the two regions. Alternatively a good fit of the experimental data can be afforded by the empirical formula (for $M < M_\infty$)

$$\epsilon''_{eff} = \epsilon''_0 + \frac{AM^2}{M_\infty - M} \tag{3.17}$$

where the constants ϵ''_0, M_∞ and A are chosen to best fit the data.

3.4.2 Interpretation of the ϵ''_{eff} vs. M response

There are two important deductions that are borne out of the loss factor vs. moisture content data. First, from the ϵ''_{eff} vs. M curves obtained for different orientations of the high frequency field relative to the fibre/grain direction, the optimum field orientation to be used in an applicator can be deduced. This is quantitatively shown in Fig. 3.6(a) where the losses are higher for field orientations parallel to the grain, which is characteristic of cellulosic materials such as in paper and board (Driscoll, 1976) and wood (James and Hamill, 1965). Second, the slope of the ϵ''_{eff} vs. M curve is critical to industrial applications where moisture levelling of a material in sheet form is the prime requirement, the effect being shown qualitatively in Fig. 3.6(b). This is because moisture levelling is very effective for moisture contents above the critical value since the high $d\epsilon''/dM$ ensures that the wetter parts of the sheet material absorb more power and tends to level off the initial uneven moisture distribution. On the other hand, moisture levelling for moisture contents below the critical value is much less pronounced since ϵ''_{eff} is practically independent of the moisture content and consequently different parts of the material absorb virtually the same amount of energy from the high frequency field irrespective of the degree of wetness. In Fig. 3.6(*b*), curve III is far more effective for moisture levelling than curve II, while for a material exhibiting properties such as shown in curve I, no levelling at all will take place. A given material can exhibit properties similar to those of curves II and III at different frequencies (Metaxas and Driscoll, 1974).

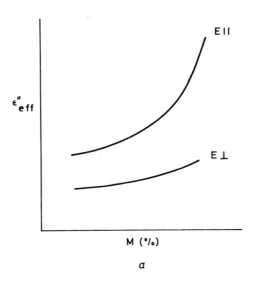

a

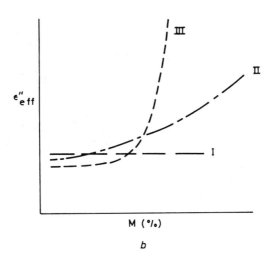

b

Fig. 3.6 *The effective loss factor as a function of the moisture content for various electric field orientations*

3.4.3 Experimental data near the industrially allocated frequency bands

A brief summary of some measurements on the variation of ϵ' and ϵ''_{eff} with moisture content will be given. Unless otherwise stated, the moisture contents, M, refers to dry basis (see Appendix I). Of all the industrial allocated microwave

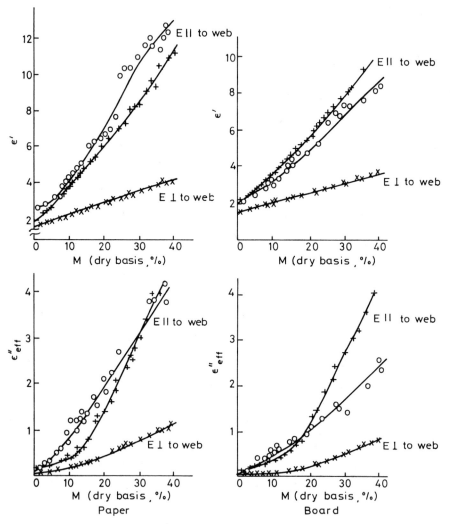

Fig. 3.7   *Dielectric properties of paper and board as a function of the moisture content for two frequencies and field orientations. Moisture content M refers to dry basis, see Appendix I*
    oooooo **2450 MHz**
    XXXXXXX **27·12 MHz**
    (After Metaxas and Driscoll, 1974, Copyright *The Journal of Microwave Power,* 1974.)

frequency bands, the one around 3 GHz (2·45 GHz) seems to have been the most popular choice for property measurements.

Metaxas and Driscoll (1974) have conducted dielectric property measurements on paper and board rings at 27·12 MHz and 2·45 GHz using the Roberts and von Hippel (1946) method, with the electric field parallel to the plane of the rings

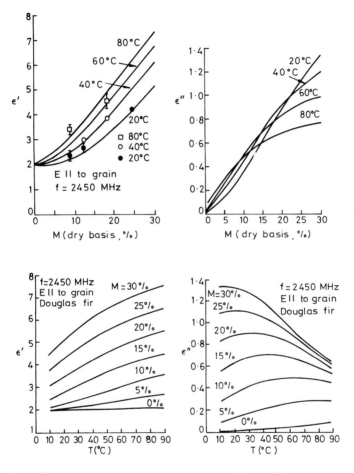

Fig. 3.8 ϵ' and ϵ'' vs. moisture content and temperature in Douglas fir
(After Tinga W.R., 1969, Copyright *The Journal of Microwave Power.*)

as shown in Fig. 3.7. The loss factor curves show a gradual change of slope, $d\epsilon''/dM$, with increasing moisture which makes the identification of M_c rather difficult. Measurements on paper with the electric field direction across the plane of the rings point to smaller ϵ' and ϵ''_{eff}. The frequency dispersion of ϵ''_{eff} for board at equilibrium moistures exhibits a maximum in the vicinity of 0·8 GHz.

Several published reports have appeared on the dielectric properties of Douglas

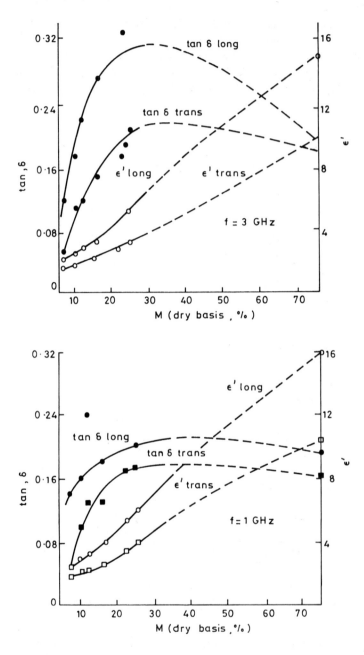

Fig. 3.9 ϵ' *and* tan δ *vs. moisture content and grain direction of Douglas fir*
(After James, W.L. and Hamill, D.W., Forest Prod., 1965.)

fir. Figure 3.8 shows the results of Tinga (1969), who studied the variation of the relaxation time with moisture content and temperature. Still on wood, Fig. 3.9 depicts the data of James and Hamill (1965) who showed the influence of the dielectric properties on grain orientation. The ϵ' and ϵ''_{eff} are higher along the grain than in a transverse direction. The loss factors at 1 and 3 GHz increase sharply with increasing moistures at low levels and exhibit maxima at some moisture near fibre saturation.

The dielectric properties of leather at 2·45 GHz, calculated from measurements on a rectangular waveguide loaded with a thin sample, are shown in Fig. 3.10

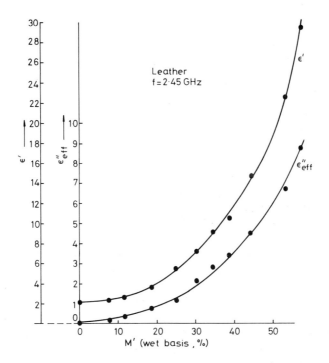

Fig. 3.10 ϵ' and ϵ''_{eff} vs. % moisture of leather. Moisture content M' refers to wet basis, see Appendix I
(After Hamid, M. A. K. *et al.*, 1972, Copyright, *The Journal of Microwave Power*, 1972.)

(Hamid *et al.*, 1972). A gradual variation of the ϵ''_{eff} with moisture content is observed in the 0–50% range of moistures (wet basis). Bhartia (1972) has interpreted the dielectric results of leather using Wiener's theory for dielectric mixtures to evaluate the degree of binding of water in the material. It has been shown that, up to volumetric moisture content of about 20%, the degree of binding increases rapidly and levels off thereafter.

A cavity method has been employed by Windle and Shaw (1954) to measure the dielectric properties of wool at 3 and 9·3 GHz and within the moisture content range, 0–14%, as shown in Fig. 3.11. The results have been analysed in terms of adsorption theory postulated by Cassie which shows that the water sorbed by wool consists of localised water which is intimately bound to polar sites in the wool and mobile (free) water which constitutes the remainder sorbed.

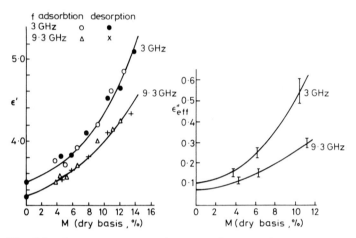

Fig. 3.11 *The dielectric constant and loss factor as a function of the moisture content for wool at two frequencies*
(After Windle, J. J. and Shaw, T. M., *Journal Chemical Physics*, 1954.)

Nelson (1978) has studied the frequency and moisture dependence of the complex dielectric constant of shelled, yellow dent field corn, in the frequency range 50 MHz to 11 GHz. As expected the dielectric constant increases with moisture content and decreases with frequency, as shown in Fig. 3.12. However, the loss factor data show some interesting deviations particularly near the two industrially allocated frequency bands, namely 1 GHz and 2·45 GHz. For example, at 2·45 GHz for moisture contents above 40%, the ϵ''_{eff} rapidly increases resulting in an overall ϵ''_{eff} vs. moisture content curve which shows good moisture levelling properties unlike the responses at 300 MHz and at 11 GHz which are much flatter. The ϵ''_{eff} vs. M response at 1 GHz shows similar trends.

Hasted and Shah (1964) have reported measurements of the complex dielectric constant at 100 mm (3 GHz), 33 mm (9·1 GHz) and 12·5 mm (24 GHz) wavelengths of specimens of concrete, mortar, hardened cement paste and brick loaded with water. The properties of three of the bricks are interpreted, a sample of which is shown in Fig. 3.13, in terms of the dielectric mixture theory of three components brick, ash and rotationally free water absorbed in cavities. The dielectric properties of cement paste were believed to have arisen from water in the same state as that which produces the low frequency loss rather than from rotationally free water.

Bengtsson and Risman (1971) have measured the dielectric properties of a range of foodstuffs, at 2·8 GHz, using a cavity perturbation technique. As far as the influence of composition on the dielectric properties, they report that both the amount of polar component (mainly water) and the addition of charge carriers, such as dissolved salts, play a paramount role. Figure 3.14 shows the properties of many foodstuffs as a function of the water content at 20°C.

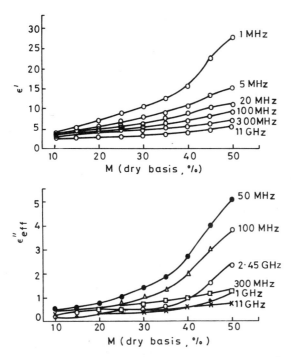

Fig. 3.12 *Dielectric properties of shelled, yellow-dent field corn at 24°C and at various frequencies*
(After Nelson, S., 1978, Copyright, *The Journal of Microwave Power*, 1978.)

Finally, dielectric property vs. moisture content data have also been published for potato starch and other organic materials (Roebuck *et al.*, 1972), for dairy products (Rzepecka and Pereira, 1974) and for tufted carpets (Gohel and Metaxas, 1978).

3.4.4 Experimental data at X-band

A considerable amount of dielectric property data have been published on the properties of many industrial materials at X-band following techniques such as Time Delay Reflectometry (Kent, 1977*b*), stripline (Kent and Steel, 1978), transmission (Kraszewski *et al.*, 1977), short-circuited (Stuchly, 1970), etc. Table 3.2 summarises some techniques and the parameters that have been measured on various materials. It must be stressed once more, however, that strict caution must be exercised when consulting such data with the view to microwave heating/drying applications since extrapolation down to the lower industrial frequency bands might lead to erroneous conclusions.

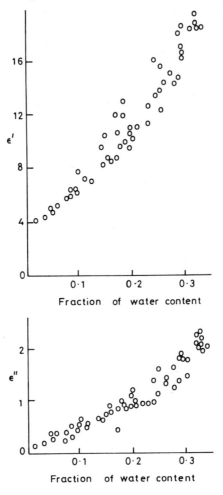

Fig. 3.13   *Dielectric properties of bricks at* $\lambda = 100\,mm$ *(3 GHz) as a function of the fraction of water content*
(After Hasted and Shah, *British J. Appl. Physics*, 1964. Extracted from *Aqueous Dielectrics* by J. B. Hasted by permission of Chapman and Hall Ltd.)

3.5 The variation of ϵ^* with temperature

3.5.1 General data

A number of investigations have been made in order to explain the temperature dependence of the loss factor in various materials. From Tinga's (1969) measurements on Douglas fir, shown in Fig. 3.8, at low moistures the loss factor increases with temperature since the physical binding reduces and the dipoles are freer to reorientate. For hydrations above 25%, the loss factor decreases with increasing temperature.

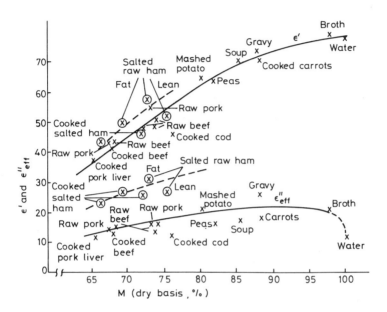

Fig. 3.14 *Relationship between water content and dielectric data at 20°C and 2·8 GHz (points for salted material are circled)* (After Bengtsson and Risman, 1971, Copyright *The Journal of Microwave Power,* 1971.)

The influence of the temperature and frequency on the ϵ^* of many foodstuffs has been thoroughly investigated by Bengtsson and Risman (1971) in the temperature range -20 to $60°C$, and by To *et al.* (1974) who have concentrated on temperatures above thawing to $65°C$. Bengtsson and Risman's (1971) work is extremely important in that it clearly showed the large differences of ϵ' and ϵ''_{eff} between ice and liquid water. Since most foodstuffs contain an appreciable amount of water, their dielectric properties follow a similar pattern to those of water in bulk when plotted as a function of the temperature, as shown in Fig. 3.15. Both the ϵ' and ϵ''_{eff} show large increases with temperature as the material thaws after which the values again decrease with increasing temperature except for salted foods which show a continuing increase. After thawing the decrease in dipolar losses with increasing temperature at 2·8 GHz is balanced by an increase of conductivity losses resulting in a practically constant ϵ''_{eff} with temperature. However, as the frequency drops to below 1000 MHz, conductivity losses dominate over dipolar losses and the effective loss factor continues to increase with increasing temperature. This is clearly demonstrated by the data of To *et al.* (1974) on cooked beef or turkey, cooked beef juice or turkey, raw beef or turkey and on aqueous non-fat dry milk,

Table 3.2 *Data acquisition at X-band on material properties*

Method	Material	f(GHz)	T °C	Measurement	Reference
Attenuation	Textiles	9·0		α vs. M	Brady (1968)
Roberts and von Hippel short circuit	Polyamide sand, mash potato powder, silica gel	9·4	−20 to +110	ϵ' vs. M / ϵ''_{eff} vs. M	Stuchly (1970)
Modulated sub-carrier New sensor	Ammonium phosphate	9·4	24	ϵ' vs. M / ϵ''_{eff} vs. M	Kraszewski et al. (1974)
Roberts and von Hippel short circuit	Wheat	9·4	25	ϵ' vs. ρ / ϵ''_{eff} vs. ρ	Nelson (1976)
Theoretical prediction	Wheat	9·4	25	ϵ' vs. M / ϵ''_{eff} vs. M	Kraszewski (1978)
	Protein	9·45	0–40	ϵ' vs. M / ϵ''_{eff} vs. M	Kent (1972a, b)
Vector null bridge	Fish meal	9·78	24	ϵ' vs. ρ / ϵ''_{eff} vs. ρ	Kent (1977b)
Two Horns-modulated sub-carrier bridge circuit	Wheat	9·4	25	ϵ' vs. M / ϵ''_{eff} vs. M	Kraszewski et al. (1977)
Attenuation TDS	Frozen fish	$10^{-2} - 35$	−30 to 0	α vs. M	Kent (1977a)
Attenuation bridge	Evaporated milk, condensed milk, cement slurry	9·4	20–40	α vs. M	Kraszewski (1974)
Stripline techniques	Fish meal, sucrose solutions, skimmed milk			α vs. M	Kent and Steel (1978)
	Bricks	9·1*		ϵ' vs. M / ϵ''_{eff} vs. M	Hasted and Shah (1964)

* Also at 3 and 24 GHz.

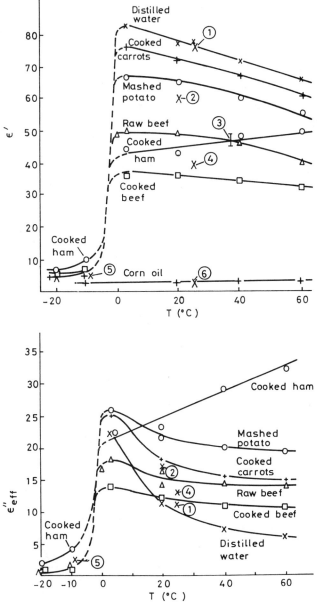

Fig. 3.15 *The dielectric properties of various foodstuffs as a function of temperature at 2·8 GHz*

1 Von Hippel, water   *4* Von Hippel, beef steak

2 De Loor, raw potato   *5* Harper *et al.*, beef steak 2 GHz

3 Püschner   *6* Pace *et al.*, corn oil

(After Bengtsson and Risman, 1971, Copyright *The Journal of Microwave Power*, 1971.)

the latter shown in Fig.s 3.16 and 3.17 respectively. A detailed study of the dielectric properties of frozen meats has been carried out by Mudgett *et al.* (1979) in the temperature range −40 to −20°C and at 300, 915 and 2450 MHz. These

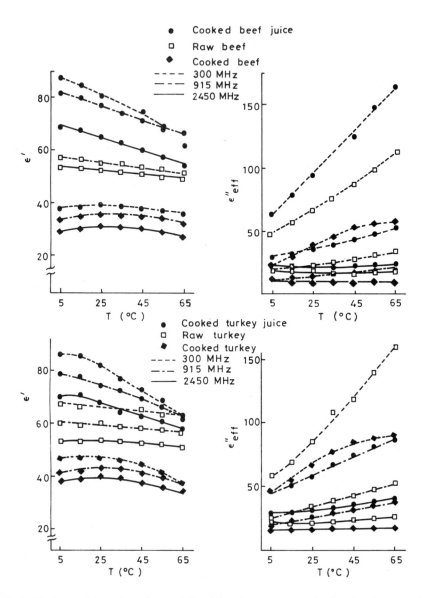

Fig. 3.16   *Temperature dependence of the dielectric properties in beef and turkey at various frequencies*
(After To *et al.*, 1974, Copyright *The Journal of Microwave Power*, 1974.)

tests suggest a significant amount of free water and dissolved ions at low temperatures. They predicted the dielectric properties of frozen meats by a physical—chemical model involving pockets of saturated aqueous sodium chloride suspended in a field of solids and ice at thermodynamic equilibrium.

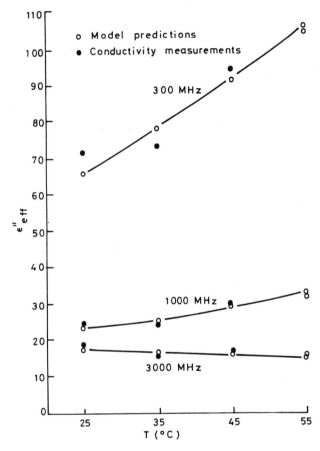

Fig. 3.17 *Effective loss factor on a function of the temperature for aqueous non-fat dry milk (After To et al., 1974, Copyright The Journal of Microwave Power, 1974.)*

Finally, the dielectric properties of some dairy products are shown in Fig. 3.18 (Rzepecka and Pereira, 1974). For both milk and whey powders the loss factor increases gradually from $-10°C$ to $70°C$ since these materials contain small amounts of water. However, in butter, where the moisture content was about 16.5% (wet basis), the effective loss factor goes through larger changes within the same temperature band since the dielectric properties of ice are significantly lower than that of liquid water. There is, however, no sudden step in ϵ''_{eff} at $0°C$ because of the salt content which depresses the freezing point progressively with concentration to saturated brine with freezing point at $-21°C$.

3.5.2 Runaway effects

Of great importance to microwave heating applications is the 'runaway effect' or the uncontrolled rise in temperature in a material brought about as a result of a positive slope, $+d\epsilon''/dT$, of the ϵ''_{eff} vs. temperature response. A typical example

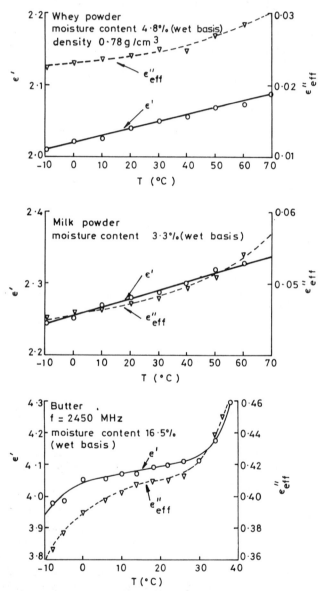

Fig. 3.18 *Dielectric properties of dairy products as a function of the temperature at 2450 MHz* (After Rzepecka and Pereira, 1974, Copyright *The Journal of Microwave Power,* 1974.)

is the case of Nylon 66 (Huang, 1976). Its properties are shown in Fig. 3.19. After an initial absorption of the microwave energy, the temperature rise causes the ϵ''_{eff} to increase which, in turn, results in a further temperature increase, and so on.

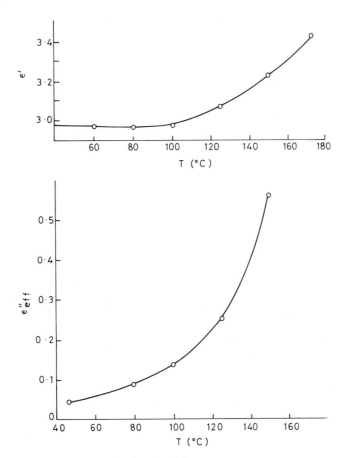

Fig. 3.19   *Dielectric properties of nylon 66 at 3 GHz*
(After Huang, 1976, Copyright *The Journal of Microwave Power,* 1976.)

Damage to the material is possible unless steps are taken to avoid this cumulative effect by either interrupting the microwave energy or removing the material from the heated zone. The data of Bengtsson and Risman (1971) on the foods show quite clearly a positive slope of the ϵ''_{eff} vs. T response around the freezing point of water, which is the cause of many difficulties with microwave defrosting applications. As soon as a few droplets of liquid water are produced, the microwave energy is preferentially dissipated in the liquid phase, causing uneven thawing with damaging effects. To avoid this, foods of high moisture content are normally thawed to a temperature of about $-2°C$, a process referred to as tempering. Other materials that show a $+ve\,d\epsilon''/dT$ are, for example, styrene butadiene rubber (SBR) in the temperature range 30–120°C (Terselius and Ranby, 1978), glass bonded mica, steatite ceramic and borosilicate glass (Couderc *et al.*, 1973). The data for SBR, shown in Fig. 3.20, show that a peak of ϵ''_{eff} is obtained at about

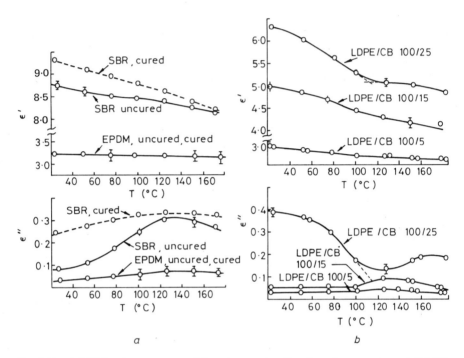

Fig. 3.20   (a) *Dielectric constant and loss factor as a function of temperature at 2·8 GHz for a polar styrene-butadiene rubber (SBR) compound and non-polar EPDM compound. (b) Dielectric constant and loss factor as a function of temperature at 2·8 GHz for different polythylene (LDPE)—carbon black (CB) mixtures*
(After Terselius and Ranby, 1978, Copyright *The Journal of Microwave Power*, 1978.)

130°C beyond which the sign of the slope changes which affords better control for microwave heating. Table 3.3 gives examples of some common materials known to exhibit positive slopes. A typical qualitative example of their response is shown in Fig. 3.21. The critical temperature T_c gives an approximate indication of the point at which the ϵ''_{eff} increases significantly. In some materials the rise of ϵ''_{eff} with T is far less pronounced making the determination of T_c much more difficult.

3.5.3 Influence of applicator characteristics on thermal runaway
The techniques in the design of microwave heating systems are aimed at avoiding the risk of thermal runaway caused by the positive value of $d\epsilon''_{eff}/dT$. Generally the rate of rise of temperature of an elemental volume, dT/dt, is proportional

Table 3.3   *Approximate temperature range for T_c*

Material	f(GHz)	$(\rho/10^6)\,\mathrm{kg/m^3}$	T_c (°C)	Reference
Aluminium oxide	3·89–3·61	3·66	800	Inglesias and Westphall (1967)
Alumina (NBS 10F2)	3·94–3·71	3·8	650–700	Inglesias and Westphall (1967)
Hot pressed boron nitride (HD 0086)	5·17–4·96	1·94	750–800	Inglesias and Westphall (1967)
Pyrolytic Boron nitride	9·21–9·04	1·23	1700	Inglesias and Westphall (1967)
Mycalex	2·45	–	450	Couderc *et al.* (1973)
Nylon	3·0	–	140–150	Huang (1976)
Steatite	2·45	–	400–450	Couderc *et al.* (1973)
SBR (uncured)	2·8	–	40	Terselius and Ranby (1978)
International pipe and ceramic (TC 30 2H)	8·52	–	400	Inglesias and Westphall (1967)
Glass ceramic	9·37	–	180	McMillan and Partridge (1972)

to its heat input, $\epsilon''_{eff}E^2f$, but heat is conducted away from that element at a rate proportional to $\alpha_t \nabla^2 T$, where α_t is the thermal diffusivity. An equilibrium temperature is established when $\epsilon''_{eff}E^2f$ is equal to the rate of heat loss and is a low value consistent with achieving an adequately fast processing time.

Multimode oven applicators can create thermal runaway as a result of their inherent standing wave characteristics, as discussed in Chapter 6. Moreover, they are generally used for processing discrete objects of large dimensions which create field distortion leading to uneven heating. A characteristic of multimode ovens is that the quality factor, Q, of the oven (given by eqn. (6.7)) is critically dependent upon the volume filling factor of the workload, v. If in this context the 'workload' is considered to be a hot spot due to non-uniform heating it will have locally a high value of $\tan \delta_{eff}$ and although of small volume can have an over-riding effect on Q particularly if the bulk material is of relatively low $\tan \delta_{eff}$. More modes may then be effective, affording better coupling of power to the hot spot and accelerating the thermal runaway.

Continuous flow applicators (described in Chapter 5) are much less prone to causing thermal runaway because the electric field is better controlled than in a multimode applicator. For example, meander and axial flow applicators at 900 MHz and 2450 MHz are extensively used in the continuous vulcanisation of rubber compounds, based on natural, EPDM, nitrile or polychloroprene rubbers all of which have positive $d\epsilon''_{eff}/dT$ in the temperature range $70°C < T < 250°C$. Butter is successfully thawed ($-15°$C to $4°$C) in horn type controlled-field applicators at 900 MHz.

3.6 The variation of ε^* with frequency

Having considered the effects of moisture and temperature on the dielectric properties of many industrial materials, we shall now briefly discuss the effects of

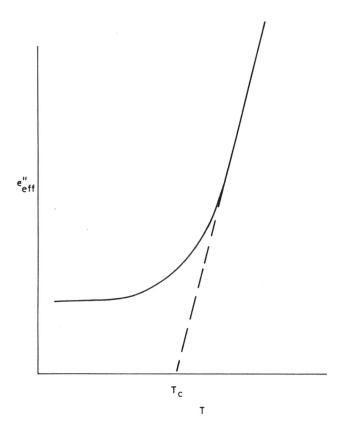

Fig. 3.21   *Qualitative representation of the loss factor as a function of the temperature*

frequency. Industrial use of microwaves calls for operation at discrete frequencies within given bands carefully selected so as not to interfere with other frequencies in use in telecommunications, defense and maritime applications. Therefore, the measurement of the relaxation response of ϵ^* in a particular industrial material, shown qualitatively in Fig. 3.22, is not in itself of particular significance other than to point out the relative values of the loss factor, ϵ''_{eff}, at the various allocated frequencies. Table 3.4 summarises some published data of dielectric relaxation in specific frequency ranges on a number of materials.

Table 3.4   *Dielectric relaxation*

Material	Frequency band (Hz)	Reference
Alkyl alcohol	$10^7 - 5 \times 10^8$	Iskander and Stuchly (1972)
Board	$10^6 - 3 \times 10^9$	Metaxas and Driscoll (1974)
Corn	$10^6 - 10^{10}$	Nelson (1978)
Glass ceramic	$10^5 - 10^{10}$	McMillan and Partridge (1972)
KLlO·5N emulsion	$10^7 - 10^{11}$	Le Petit *et al.* (1977)
n-Octonal	$5 \times 10^6 - 5 \times 10^9$	Kent (1975)
Rapeseed leaves	$6 \times 10^8 - 8 \times 10^9$	Stuchly *et al.* (1979)
Tobacco	$10^6 - 10^{10}$	Copson (1975)
Tulip tree branch leaves	$10^5 - 10^{10}$	Broadhurst (1968)
M/2 valine solution	$10^6 - 10^{11}$	Croom *et al.* (1977)
Water, pure	$5 \times 10^8 - 10^{13}$	Hasted (1973)
Water, pure	$10^8 - 10^{10}$	Suggett (1973)
Water, pure	$10^5 - 10^{10}$	von Hippel (1954)
Aqueous sodium chloride	$3 \times 10^8 - 10^{10}$	von Hippel (1954)

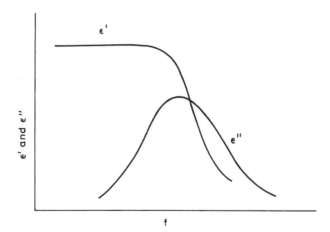

Fig. 3.22   *Dielectric relaxation of a typical polar dielectric*

Dielectric relaxation in liquid water has been studied extensively (Hasted, 1973). The dielectric properties of water in bulk are not of any major significance in industrial microwave heating since for the majority of applications the relaxation of bound water is far more important. Nevertheless it is important to briefly describe the dielectric relaxation in water and aqueous NaCl solutions, shown in Fig. 3.23(*a*), because they do highlight the importance of conductivity effects at the lower end of the industrially allocated heating frequencies. Figure 3.23(*a*) shows that the peak of the ϵ''_{eff} depends on the temperature and lies approximately between $9 < f < 30$ GHz within the temperature range $0 < T < 50°C$. The influence of salts is clearly shown where the effective loss factor reaches 100 at 900 MHz for a 0·3 molal solution. The temperature dependence of the ϵ' and ϵ''_{eff} in pure water is shown in Fig. 3.23(b) near the two most widely used microwave heating

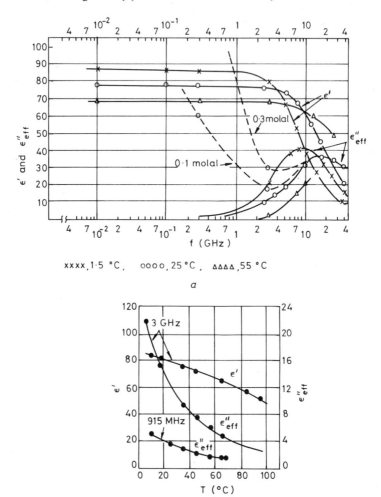

Fig. 3.23   *Temperature and frequency effects on the dielectric properties of pure water* (———) *and aqueous NaCl* (————)
(Data of pure water and aqueous NaCl from von Hippel, 1954, by permission of MIT Press. Data of loss factor at 915 MHz with temperature from To *et al.*, 1974, Copyright *The Journal of Microwave Power*.)

frequencies. As the temperature increases, the ϵ''_{eff} decreases thus offering a stabilising influence on the rate of heating. However, as was seen earlier, any additions of salts, etc., would have the opposite effect and increase the likelihood of thermal runaway.

3.7 *Q*-factor versus moisture content response

The perturbation theory for the determination of material properties applies to cases where the losses introduced into the measuring cavity are small. However, as shall be described later in Chapter 7, resonant cavities can be used as applicators for drying or heating processes. To assess the suitability of a particular cavity/material combination for drying with high frequencies, it is important to examine the changes of the cavity's *Q*-factor with moisture content. To trace the origin of the link between *Q* and *M*, consider a resonant cavity completely filled with a homogeneous dielectric, in isolation from any external circuit. The unloaded *Q*-factor, Q_0, with the dielectric inserted in it is defined as (see Section 7.2)

$$Q_0 = 2\pi \frac{\text{energy stored}}{\text{energy dissipated/cycle}} = \frac{\omega U}{P_w + P_s} \tag{3.18}$$

where P_w and P_s are the powers dissipated in the dielectric and cavity walls respectively.

Re-arrangement of eqn. (3.18) and substituting for P_w and U using eqns. (4.10) and (6.3), respectively, yields

$$\frac{1}{Q_0} = \frac{P_s}{\omega U} + \frac{\int_V \frac{1}{2} \epsilon_0 \epsilon''_{eff} \hat{E}^2 \, dV}{\int_V \frac{1}{2} \epsilon_0 \epsilon' \hat{E}^2 \, dV} \tag{3.19}$$

or

$$\frac{1}{Q_0} - \frac{1}{Q'_0} = \epsilon''_{eff}/\epsilon' \tag{3.20}$$

where Q'_0 is the *Q*-factor of the cavity without any dielectric inserted into it, and $\hat{E}$ is the peak electric field established in the cavity.

It is usual to assume that $Q_0 \ll Q'_0$ reducing eqn. (3.20) to

$$Q_0 = \epsilon'/\epsilon''_{eff} \tag{3.21}$$

Since the loaded *Q*-factor Q_L is directly related to Q_0 (see Section 7.2), this too becomes a function of the ratio $\epsilon'/\epsilon''_{eff}$. As we have shown earlier in this chapter the ϵ' and ϵ''_{eff} are strongly dependent upon the moisture content and therefore the Q_L becomes also strongly dependent on *M*. Figure 3.24 shows a typical variation of the Q_L vs. *M* for tufted carpets. The advantage of such data is that they are made directly at the industrially allocated frequencies and as such are directly applicable for assessing the materials' suitability for microwave processing. Similar *Q* vs. *M* data have been obtained at radio frequencies in assessing textile and paper dryers at about 27 MHz (Morrow, 1973) and correlated with dielectric property data (Driscoll, 1973, 1974, 1976). The link between *Q* and *M* at radio frequencies is far simpler to derive because the former quantity for the circuit shown in Fig. 3.25 is

$$Q = \omega C_p R'_p \tag{3.22}$$

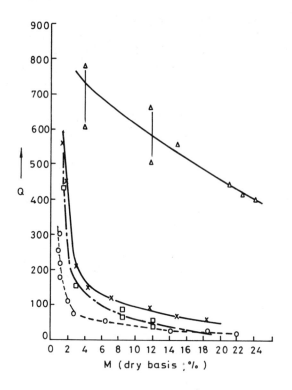

Fig. 3.24   *Q-factor as a function of the moisture content for various tufted carpet structures at the 896/915 MHz frequency band*
△△△, primary backed carpet ~ 0·5 kg/m²
xxx, foam backed (cured, rewetted) ~ 2 kg/m²
□□□, precoat backing ~ 0·5 kg/m²
ooo, uncured foam latex ~ 1 kg/m² on polypropylene backing

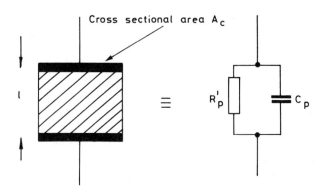

Fig. 3.25   *A capacitive load and its equivalent circuit*

where R'_p is the resistance of a dielectric between two parallel plates of cross-sectional area, A_c, and separation, l. Substituting $C_p = \epsilon_0 \epsilon' A_c/l$ and $R'_p = l/\sigma A_c$ in eqn. (3.22) gives

$$Q = \frac{\omega \epsilon_0 \epsilon' A_c}{l} \times \frac{1}{\sigma} \frac{l}{A_c} = \frac{\epsilon'}{\epsilon''_{eff}} = \frac{1}{\tan \delta_{eff}} \tag{3.23}$$

having made use of eqn. (2.41) under the assumption that at radio frequencies, i.e., below 50 MHz, conductive losses dominate over dipolar losses, i.e., $\sigma = \omega \epsilon_0 \epsilon''_{eff}$. Since $\tan \delta_{eff}$ varies with moisture content we can write the generic expression

$$Q_L = g(M) \tag{3.24}$$

where $g(M)$ is a function of the moisture content, and includes any loading effects by connecting the radio frequency or microwave applicator to the external circuit (see Sections 7.2 and 7.4.5).

3.8 Discussion on published data

Primarily, dielectric property data are important because they give values of the loss factor, ϵ''_{eff}, which, as shall be seen in the following chapter, controls, along with other parameters such as the electric field and the frequency, the power that can be dissipated in a given material volume. Alternatively, for a given power dissipation, ϵ''_{eff} controls the rate of rise of temperature. The larger the loss factor the easier the material absorbs the incident microwave energy. As a general practical rule, loss factors less the 10^{-2} require very high electric field strengths in order to ensure a reasonable rate of rise of temperature in the material; such low loss factor would almost certainly require fundamental mode resonant applicators (Chapter 7) as distinct from travelling wave (Chapter 5) or multimode (Chapter 6) applicators. On the other hand, loss factors of greater than 5 might present depth of penetration problems, in that because the material is highly absorptive to microwave radiation, most of the incident energy is absorbed within the first few mm, leaving the internal parts little affected. This causes non-uniformities of heating which for in-depth heating are totally unacceptable. Therefore, loss factors between the limits $10^{-2} < \epsilon''_{eff} < 5$, and bearing in mind that those limits are somewhat flexible, would present materials which, in general, are good candidates for microwave heating applications.

In materials where there is a definite fibre or grain orientation, such as in paper, board, textile, wood, etc., the dielectric properties do strongly depend on the relative orientations of the exciting electric field to that of the grain of the material. Furthermore, the influence of the electric field orientation on the dielectric properties is important for drying materials in sheet form, thin webs, etc., where the moisture needs to be levelled across the section of the web.

For general drying applications it is crucial to determine the loss factor as a function of the moisture content for basically three reasons. First, an overall

Table 3.5   *Determination of critical moisture content* (M_c)

Material	Frequency GHz	Temperature °C	Critical moisture content (M_c)
Board 230 g/m²	0·027	22	14% dry basis
$E \parallel$ fibre	2·45	22	14% dry basis
Butter	2·45		9·5% wet basis
Douglas fir	2·45	20	10% dry basis
Field corn	0·05	24	35% dry basis
	0·1	24	35% dry basis
	2·45	24	40% dry basis
Leather	2·45	20	20% wet basis
Mash potato powder	9·4	20	29% wet basis
Milk powder	2·45	20	4% wet basis
Paper 78·5 g/m²	0·027	22	12% dry basis
$E \parallel$ fibres	2·45	22	9% dry basis
Polyamide	9·4	20	17% wet basis
Protein WF M	9·45	23	8% dry basis
HB	9·45	40	12% dry basis
Sand	9·4	20	1–2% wet basis
Silica gel	9·4	20	28% wet basis
Tufted carpet Nylon 183 kg/m³	0·896	22	3% dry basis
$E \parallel$ fibre	2·45	22	3% dry basis
Whey powder	2·45	–	3% wet basis
Wool	3·0	–	4% dry basis
Wool 43·9–78·5 kg/m³	0·027	23	≈ 25% dry basis
$E \perp$ fibre	0.027	40	30–35% dry basis
Rayon $E \perp$ fibre 128–134 kg/m³	0·027	21–24	16% dry basis
Cotton 210 kg/m³ $E \perp$ fibre	0·027	22	15% dry basis
Acrylic 92 kg/m³	0·027	23	1–2% dry basis

indication of the absolute values of the loss factor is given. Second, the best choice of frequency and electric field orientation is obtained. Third, and perhaps the most important since it reflects on capital costs, we can determine through such data the critical moisture content, M_c, which, although loosely marks the boundary between bound and free water, gives the designer of high frequency equipment an indication of the moisture content required where conventional heating systems such as hot air, etc., become inefficient since it involves the evaporation of tightly bound water. A possible upper limit for the microwave set-up is indicated, since operation well above the M_c level would mean evaporation of water bound in cavities and capillaries which, in general, can be efficiently removed with conventional low cost systems. In view of its importance in drying applications, Table 3.5 shows the value of M_c for many industrial materials at various frequencies and at near room temperatures. As can be seen M_c lies within wide limits depending upon the particular material under consideration.

Temperature effects on the dielectric properties give us a better understanding of microwave processes such as defrosting, meat tempering, following dielectric data on frozen materials and highlighting the difference between the effective loss factors of ice and liquid water. Since many foodstuffs contain over 90% water, their properties follow more closely the dielectric properties of water and here comparisons can be made more readily with liquid water.

It has long been known that as the frequency drops below 1 GHz conductive losses, in many aqueous dielectrics such as foodstuffs, increase and for frequencies below 100 MHz may dominate over dipolar losses. Dielectric property data on foodstuffs (To *et al.*, 1974) and corn (Nelson, 1978) have confirmed these trends. Although the effective loss factors are larger at lower frequencies, that in itself is not an advantage since the data show that a positive $d\epsilon''/dT$ is often obtained at radio frequencies which might give rise to thermal runaway effects. Indeed thermal runaway effects have been observed within a wide frequency range and has been a major problem to manufacturers of equipment for thawing frozen materials.

3.9 Catalysts and agents

Many materials in their natural state are virtually transparent to high frequency energy, exhibiting loss factors below 10^{-3}, and are therefore extremely difficult to heat with microwaves. However, some attention has been given to the possibility of making such materials more absorptive of the microwave energy by introducing controlled lossy 'impurities' or additives. One such application successfully operating in industry for many years, has been the additions of carbon black such as ISAF, HAF, and/or FEF grades to natural, SBR, butyl or EPDM rubbers (Meredith, 1976; Costemalle *et al.*, 1975), to enhance their pre-treatment with microwave energy. Bulky substituents such as $-(CH_2)_2CF_3$ or $-(CH_2)C$, in silicon rubbers as well as additions of carbon black or Fe_2O_3 in dimethylsiloxanes are also reported to improve the speed and uniformity of curing these materials with microwave energy (Lee, 1979).

The possibility of adding a lossy ferrimagnetic powder such as magnetite (Fe_3O_4) to plaster moulds and core mixes has been reported (Dench and Freedman, 1977). The powder which is lossy in the microwave frequency regime, is used as a catalyst to remove small percentages of residual water from the mix by controlling the absorption of microwaves around its Curie temperature. This is because above the Curie temperature the loss factor of magnetite falls sharply. Similarly, microwave energy could effectively harden moulding materials containing thermosetting resins (phenol-formaldehyde), silica and sand and acid salt (Kuroki Akihiro, 1974). Also the curing of sodium silicate bonded cores with additions of ZnO and Cr_2O_3 to alter the core properties have been studied by Cole *et al.* (1979). Graphite powder has also been used in moulds as a catalyst.

Wheat gluten bakery products cannot be effectively heated by microwave energy. However, additions of egg white, casein starch or leavening agents to the wheat have resulted in a mixture which has been readily and most effectively heated with microwave energy (Tsnyuki Hideo, 1977).

Urea has been extensively considered in microwave heating applications, for example, as an additive in resin adhesive used in the production of wooden laminates (Matsuda Kenishi *et al.*, 1975) or in dye bath solutions to improve the subsequent dye fixation by microwave energy (Evans and Skelly, 1972).

Finally, the susceptibility to microwave energy of refractory Al_2O_3 containing (H_3PO_4) was improved by additions of NiO or C_2O_3 to it, these acting as strong and weak coupling agents respectively for the electromagnetic energy (Johnson *et al.*, 1980).

3.10 References

ALTMAN, J. L., *Microwave Circuits.* Van Nostrand, New York (1964).

BENGTSSON, N. E., and RISMAN, P. D., 'Dielectric properties of foods at 3 GHz as determined by a cavity perturbation technique. II. Measurements in food materials', *J. Microwave Power* 6(2), 107 (1971).

BHARTIA, P., 'On the degree of binding of water absorbed in leather', *J. Microwave Power* 7(1), 51 (1972).

BRADY, P., 'Loss measurements of wet textiles at 9 GHz', *J. Microwave Power* 3(4), 194 (1968).

BROADHURST, M. G., National Bureau of Standards Report, Washington DC (1968).

CHAMBERLAIN, I., and CHANTRY, G. W., *High Frequency Dielectric Measurements.* IPC Ltd., New York (1973).

COLE, G. S., NOWICKI, R. M., and OWUSU, Y. A., *Trans. Am. Foundrymen's Soc.* 87, 605 (1979).

COPSON, D. A., *Microwave Heating,* Second Edition. AVI Publishing Co. Inc. (1975).

COSTEMALLE, B., MOREE, J., and TZIDON, B., *Rev. Gen. Caoutch. Plast.* 52(9), 593 (in French), (1975).

COUDERC, D., GIROUX, M., and BOSISIO, R. G., 'Dynamic high temperature microwave complex permittivity measurements on samples heated via microwave absorption', *J. Microwave Power* 8(1), 69 (1973).

CROOM, E. J., SHACK, R., SHEPHERD, J. C. W., and SHEPHERD, R. J., 'Dielectric dispersion of DL-α-valine in aqueous solution', *J. Microwave Power* 12/2, 111 (1977).

DE LOOR, G. P., Dielectric properties of heterogeneous mixture containing water, *J. Microwave Power* 3(2), 67 (1968).

DENCH, E.C. and FREEDMAN, G., 'Drying of casting molds by microwave energy', British Patent No. 1481356 (1977).

DRISCOLL, J. L., 'Measurements on the dielectric properties of wood and other textile fibres at radio frequencies', The Electricity Research Centre Memorandum, ECRC/M659, October, Capenhurst, Chester (1973).

DRISCOLL, J. L., 'The dielectric properties of beech and cedar woods at radio frequencies', The Electricy Council Research Centre Memorandum, ECRC/M683, February, Capenhurst, Chester (1974).

DRISCOLL, J. L., 'The dielectric properties of paper and board and moisture profile correction at radio frequencies', *Paper Technology and Industry*, April, 17(2), 71 (1976).

EVANS, D. G., and SKELLY, K., 'Application of microwave heating in dye fixation', *J. Soc. Dyers and Colourists*, 88(12), 429 (1972).

GOHEL, H. P., and METAXAS, A. C., 'Microwave drying of nylon tufted carpets. I. Dielectric property and Q measurements', *Proc. IMPI Symposium*, Ottawa, Canada (1978).

GOS'KOV, P. I., 'Measurement of ϵ' and tan δ by the perturbation method in a rectangular cavity', *Sov. Phys. J.* 8(4), 1 (1965).

HAMID, M. A. K., STUCHLY, S. S., BHARTIA, P., and MOSTOWY, N., 'Microwave drying of leather', *J. Microwave Power* 7(1), 43 (1972).

HANIOTIS, Z., and GÜNTHARD, H. H., *Z. Angew. Math. Phys.* 20, 771 (1969).

HARVEY, A. F., *Microwave Engineering*. Academic Press, New York (1963).

HASTED, J. B., in *'Water' A Comprehensive Treatise* (Edited by F. Franks), Vol. 1, pp. 255–305. Plenum Press, New York (1972).

HASTED, J. B., *Aqueous Dielectrics*. Chapman and Hall, London (1973).

HASTED, J. B., and SHAH, M. A., 'Microwave absorption by water in building materials', *Brit. J. Appl. Phys.* 15, 825 (1964).

HILL, N., VAUGHAN, W. E., PRICE, A. H., and DAVIES, M., *Dielectric Properties and Molecular Behaviour*. van Nostrand, New York (1969).

HORNER, F., TAYLOR, T. A., DUNSMUIR, R., LAMB, J., and JACKSON, W., 'Dielectric measurements at centimeter wavelengths', *J. Inst. Elec. Engrs.* 93, Pt. III, 53 (1946).

HUANG, H. F., 'Temperature control method in a microwave resonant cavity system for rapid heating of nylon monofilament', *J. Microwave Power* 11(4), 305 (1976).

INGLESIAS, J., and WESTPHALL, W. B., 'Supplementary dielectric constant and loss measurements on high temperature materials', Technical Report 203, MIT, January (1967).

ISHITOBI, Y., and TOGAWA, M., Browning of foodstuff by heating with 10 GHz microwaves, *Proc. IMPI Symposium*, p. 8, Ottawa, Canada (1978).

ISKANDER, M. F. S., and STUCHLY, S. S., 'A time domain technique for measurement of the dielectric properties of biological substances'. *IEEE J.* 1M-21, No. 4, 425 (1972).

JAMES, W. L., and HAMILL, D. W., 'The dielectric properties of Douglas Fir', *Forest Prod. J.* 15, 51 (1965).

JOHNSON, W. E., and SUTTON, W. H., Special Metals Co-op., US Pat. 4219361, 26 August (1980).

KALINSKI, J., 'An industrial microwave attenuation monitor (MAM) and its application for continuous moisture content measurement', *J. Microwave Power* 13(3), 275 (1978).

KENT, M., 'Microwave dielectric properties of fish meal', *J. Microwave Power* 7(2), 109 (1972a).

KENT, M., 'Complex permittivity of protein powders at 9·4 GHz as a function of temperature and hydration', *J. Appl. Phys.* D5, 394 (1972b).

KENT, M., 'Time domain measurements of the dielectric properties of frozen fish', *J. Microwave Power* 10(1), 37 (1975).

KENT, M., 'Microwave attenuation by frozen fish', *J. Microwave Power* 12(1), 101 (1977a).

KENT, M., 'Complex permittivity of fish meal: a general discussion of temperature, density and moisture dependence', *J. Microwave Power* 12(4), 341 (1977b).

KENT, M., and STEEL, D. J., 'Microwave stripline techniques applied to moisture measurement in food materials', *Proc. IMPI Conf.*, pp. 31–36, Ottawa, Ontario, Canada (1978).

KRASZEWSKI, A., 'Determination of the strength of water suspension using microwave bridge technique', *J. Microwave Power* 9(4), 295 (1974).

KRASZEWSKI, A., 'Prediction of the dielectric properties of two phase mixtures', *J. Microwave Power* 12(3), 215 (1977).

KRASZEWSKI, A., 'A model for the dielectric properties of wheat', *J. Microwave Power* 13(4), 293 (1978).

KRASZEWSKI, A., 'Microwave aquametry – a review', *J. Microwave Power* 15(4), 211 (1980).

KRASZEWSKI, A., KULINSKI, S., and CHECINSKI, K., 'Measurement of moisture content in granular amonium phosphate by microwave method', *J. Microwave Power* 9(4), 361 (1974).

KRASZEWSKI, A., KULINSKI, S., and STOSIO, Z., 'A preliminary study on monitoring of moisture content in wheat by microwave methods', *J. Microwave Power* 12(3), 241 (1977).

KUMAR, A., and SMITH, D. G., *IEEE Trans. on Instrumentation and Measurement* IM-26, 95 (1977).

KUROKI, AKHIRO, and ICHINOMIYA NABUSHIGE, Asahi Organic Chemical Co. Ltd., Patent 7428562 (July, 1974).

LEE, C. L., ACS Symposium, Serv, 1978, 107 (Energy Conservation Text Polym Processes), pp. 45–50 (1979).

LePETIT, J. P., DELBOS, G., BOTTREAU, A. M., DUTUIT, Y., MORZAT, C., and CABANAS, R., 'Dielectric relaxation of emulsions of saline aqueous solutions', *J. Microwave Power* 12(4), 335 (1977).

MATSUDA KENISHI, KAMBARA MORIMINE, and KAGOSHIMA DAIGATO, *Kyoi Kuga Rubu Kenkyu Kiyo Shizen Kagaku Hen* 27, 59 (1975).

McMILLAN, P. W., and PARTRIDGE, G., 'The dielectric properties of certain $ZnO-Al_2^{'}O_3-S_iO_2$ glass ceramics', *J. Mat. Sci.* 7, 847 (1972).

MEHMET, K., and McPHUN, M. K., *Measurement of the Properties of Rectangular Phases using a Rectangular Cavity. High Frequency Dielectric Measurements* (Edited by I. Chamberlain and G. W. Chantry), pp. 69–72. IPC Science and Technology Press (1973).

MEREDITH, R. J., 'Microwave energy for high speed efficient vulcanisation of extruded rubber', *J. Elastomers and Plastics* 8, 191 (April, 1976).

METAXAS, A. C., and DRISCOLL, J. D., 'Comparison of the dielectric properties of paper and board at microwave and radio frequencies', *J. Microwave Power* 9(2), 79 (1974).

METAXAS, A. C., and PARKER, I., 'The complex dielectric constant of moist paper and board in the microwave region', Electricity Council Research Centre, Memorandum ECRC/M557 (1973).

MORROW, R., 'The Q of r.f. applications, its measurements and use', The Electricity Council Research Centre, ECRC/M654, December, Capenhurst, Chester (1973).

MUDGETT, R. E., MUDGETT, D. R., GOLDBLITH, S. A., WANG, D. I. C., and Westphall, W. B., 'Dielectric properties of frozen meats', *J. Microwave Power* 14(3), 209 (1979).

NELSON, S. O., 'Microwave dielectric properties of insects and grain kernels', *J. Microwave Power* 11(4) 299 (1976).

NELSON, S. O., 'Radio frequency and microwave dielectric properties of shelled field corn', *J. Microwave Power* 13(2), 213 (1978).

NELSON, S. O., SCHLOPHOFF, C. W., and STETSON, L. E., 'A computer program for short-circuited waveguide dielectric properties measurements on high or low loss materials', *J. Microwave Power* 8(1), 13 (1973).

OHLSSON, T., BENGTSSON, N. E., and RISMAN, P. O., 'The frequency and temperature dependence of dielectric food data as determined by a cavity perturbation technique', *J. Microwave Power* 9(2), 129 (1974).

PRICE, A. H., *Dielectric Measurements on Liquids in the Frequency Range 250 MHz to 140 GHz in High Frequency Dielectric Measurements* (Edited by J. Chamberlain and G. W. Chantry), pp. 28–38. IPC Science and Technology Press Ltd. London (1973).

ROBERTS, S., and VON HIPPEL, A. R., *J. Appl. Phys.* 17, 610 (1946).

ROEBUCK, B. D., and GOLDBLITH, S. A., and WESTPHALL, W. B., 'Dielectric properties of carbohydrate water mixtures at microwave frequencies', *J. Food Sci.* 37(2), 199 (1972).

RZEPECKA, M., 'A cavity perturbation method of routine permittivity measurements', *J. Microwave Power* 8(1), 3 (1973).

RZEPECKA, M., and PEREIRA, R. R., 'Permittivity of some dairy products at 2450 MHz', *J. Microwave Power* 9(4), 277 (1974).

SCAIFE, B. K. P., *Complex Permittivity.* The English University Press (1971).

STUCHLY, S. S., 'Dielectric properties of some granular solids containing water', *J. Microwave Power* 5(2), 62–68 (1970).

STUCHLY, S. S., MLADEK, J., STUCHLY, M. A., and PARISIEN, B., 'A method for measurement of the permittivity of thin samples', *J. Microwave Power* 14(1), 7 (1979).

SUGGETT, A., *Time Domain Spectroscopic Measurements in High Frequency Dielectric Measurements* (Edited by J. Chamberlain and G. W. Chantry), pp. 96–103. IPC Science and Technology Press Ltd. (1973).

TERSELIUS, B., and RANBY, B., 'Cavity perturbation measurements of the dielectric properties of vulcanising rubber and polyethylene compounds', *J. Microwave Power* 13(4), 327 (1978).

TINGA, W., 'Multiphase dielectric theory applied to cellulose mixtures', PhD Thesis, University of Alberta, Edmonton, Canada (1969).

TINGA, W., and NELSON, S. O., 'Dielectric properties of materials for microwave processing – tabulated', *J. Microwave Power* 8(1), 23 (1973).

TO, E. C., MUDGETT, R. E., WANG, D. I. C., GOLDBLITH, S. A., and DECAREAU, R. V., 'Dielectric properties of food materials', *J. Microwave Power* 9(4), 303 (1974).

TSNYUKI HIDEO, *New Food Ind.* 19(8), 23 (1977).

VON HIPPEL, A. R., *Dielectric Materials and Applications.* MIT Press (1954).

WESSON, L. G., *Tables of Electric Dipole Moments.* MIT Press (1948).

WILLIAMS, G., *J. Phys. Chem.* 63, 534 (1959).

WINDLE, J. J., and SHAW, T. M., 'Dielectric properties of wool–water mixtures', *J. Chem. Phys.* 22, 1752 (1954).

Chapter 4

Theoretical aspects of volumetric heating

4.1 Introduction

In this chapter we shall describe the physical parameters which play an important role in the interaction of a dielectric with microwave energy. The power dissipated within the material will be derived from first principles and leads to a simple expression involving the electric field established in the material E, which is linked to the field developed within the microwave device containing the material and known as the applicator. The power will be attenuated as the electromagnetic fields penetrate the dielectric, an effect depending upon the dielectric properties. Some aspects of the electric field strength within microwave applicators and its determination through various methods are described. A simplified form of the equations controlling the internal transport processes such as heat, mass and total pressure are presented. This is followed by a qualitative discussion of the physical principles of high frequency drying, stressing the importance of total internal pressure on the drying characteristics of a material in the presence of a volumetric heat source, particularly when the temperature of the liquid phase attains its boiling point. A simple formulation of moisture profiling or end drying is presented. Finally the skin depth' for metals, which is akin to the penetration depth in insulators, is derived and discussed in terms of the best choice of metallic surface to be used for the internal wall of applicators to minimise losses and therefore maximise their efficiency.

4.2 Power dissipated

4.2.1 Derivation from Maxwell's equations

Microwave heating involves the conversion of electromagnetic energy into heat. Energy is transported through space or any medium by means of electromagnetic waves. The power flow through a closed surface can be calculated from the integration of the Poynting vector

$$\mathbf{\rho} = E \times H \quad \text{W/m}^2 \tag{4.1}$$

Therefore, we seek (Johnk, 1975)

$$\int_{S'} (E \times H^*) \cdot dS' \tag{4.2}$$

Starting from Maxwell's current law,

$$\nabla \times H = J + j\omega\epsilon_0\epsilon^* E \tag{4.3}$$

and substituting $J = \sigma E$ and $\epsilon^* = \epsilon' - j\epsilon''$, eqn. (4.3) yields

$$\nabla \times H = \sigma E + (\omega\epsilon_0\epsilon'' + j\omega\epsilon_0\epsilon')E = \omega\epsilon_0\epsilon''_{eff}E + j\omega\epsilon_0\epsilon'E \tag{4.4}$$

where

$$\epsilon''_{eff} = \epsilon'' + \sigma/\omega\epsilon_0 \tag{2.20}$$

and ϵ'' in eqn. (2.20) includes all losses apart from those due to conductivity. Also

$$(\nabla \times H^*) \cdot E = \omega\epsilon_0\epsilon''_{eff}E \cdot E^* - j\omega\epsilon_0\epsilon'E^* \cdot E \tag{4.5}$$

However, using the Maxwell third law for $(\nabla \times E)$ and finding its dot product with H^* gives

$$(\nabla \times E) \cdot H^* = -j\omega\mu_0\mu' H \cdot H^* \tag{4.6}$$

Subtracting eqn. (4.5) from eqn. (4.6) yields

$$(\nabla \times E) \cdot H^* - (\nabla \times H^*) \cdot E = -j\omega\mu_0\mu' H \cdot H^*$$
$$+ j\omega\epsilon_0\epsilon' E \cdot E^* - \omega\epsilon_0\epsilon''_{eff}E \cdot E^* \tag{4.7}$$

Integrating eqn. (4.7) over volume V and using the divergence theorem yields:

$$\int_V \nabla \cdot (E \times H^*)dV = \int_{S'} (E \times H^*) \cdot dS' = -j\omega$$
$$\int_V (\mu_0\mu' H^* \cdot H - \epsilon_0\epsilon' E \cdot E^*)dV \tag{4.8}$$
$$- \int_V \omega\epsilon_0\epsilon''_{eff}E \cdot E^* dV$$

However, by definition the average power is

$$P_{av} = -\tfrac{1}{2}\int_S \text{Real}\,(E \times H^*) \cdot dS' \tag{4.9}$$

Therefore,

$$P_{av} = \tfrac{1}{2}\omega\epsilon_0\epsilon''_{eff}\int_V (E^* \cdot E)dV \tag{4.10}$$

The electric field is left in integral form to remind us that in most cases E is not a constant quantity but varies in space within the microwave applicators. In special cases, where the electric field can be assumed constant, eqn. (4.10) attains the more familiar form, using $E \cdot E^* = E^2$

$$P_{av} = \omega \epsilon_0 \, \epsilon''_{eff} E^2_{rms} \, V \tag{4.11}$$

Substituting $\epsilon_0 = 8 \cdot 8 \times 10^{-12}$ F/m in eqn. (4.11) yields with $\omega = 2 \pi f$

$$P_{av} = 0 \cdot 556 \times 10^{-10} f \epsilon''_{eff} E^2_{rms} \, V \, \text{W} \tag{4.11a}$$

where E is in V/m, f in Hz and V is in m^3. If the material exhibits magnetic losses as well (see Section 2.9) the permeability attains the complex form shown by eqn. (2.48). The term $- \omega \mu_0 \mu''_{eff} H \cdot H^*$ should now be added to the right-hand side of eqn. (4.7) which will lead to a modified version of eqn. (4.11) to include magnetic wall domain and electron spin losses, given by

$$P_{av} = \omega \epsilon_0 \, \epsilon''_{eff} E^2_{rms} \, V + \omega \mu_0 \, \mu''_{eff} H^2_{rms} \, V \tag{4.11b}$$

4.2.2 *Lossy capacitor*

A simplified, but equally valid, method of deriving eqn. (4.11) is to consider the current flow through a lossy capacitor connected across a sinusoidal voltage as shown in Fig. 4.1:

$$V = \hat{V} e^{j\omega t} \tag{4.12}$$

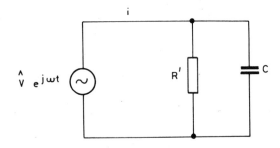

Fig. 4.1   *Capacitive load across a sinusoidal voltage*

where $\hat{V}$ is the peak value. The total current flow is given by

$$i = i_R + i_c = \frac{V}{R'} + C \frac{dV}{dt} \tag{4.13}$$

However, differentiating eqn. (4.12) gives

$$\frac{dV}{dt} = j \omega V \tag{4.14}$$

Substitution in eqn. (4.13) yields after rearrangement

$$i = j\omega\epsilon_0 \frac{A_c}{d}\left(\epsilon' - j\frac{\epsilon'}{R'\omega C}\right)V \tag{4.15}$$

since

$$C = \frac{\epsilon_0 \epsilon' A_c}{d} \tag{4.16}$$

where A_c is the cross-sectional area of the capacitor plates and d their separation. With $Q = \omega C R' = \epsilon'/\epsilon''_{eff}$ the maximum current density through the condenser is

$$\hat{J} = j\omega\epsilon_0 \epsilon^* \hat{E} \tag{4.17}$$

where $\hat{E} = \hat{V}/d$. However, the power dissipation per unit volume is given by $\frac{1}{2}$ Re $(J^* \cdot E)$, therefore making use of eqn. (4.17) we obtain

$$\text{power/unit volume} = \omega\epsilon_0 \epsilon''_{eff} \frac{\hat{E}^2}{2} = \omega\epsilon_0 \epsilon''_{eff} E^2_{rms} \tag{4.11}$$

which is the same expression derived earlier through Maxwell's equations. In order to get an appreciation of the quantities involved in microwave heating applications the electric field established within a dielectric has been calculated as a function of the dissipated power density for a range of ϵ''_{eff} values and at various industrially allocated frequencies, namely 27·12 MHz, 433·9 MHz, 896/915 MHz and 2450 MHz, the plots shown in Figs. 4.2, 4.3, 4.4 and 4.5 respectively. For a power dissipation of 10^7 W/m^3 and $\epsilon''_{eff} = 0\cdot1$ the required electric fields at 27·12 MHz and 2450 MHz are 257 kV/m and 27 kV/m respectively. Thus, the higher the frequency the smaller the electric field required for a specific power dissipation, assuming the loss factors are of similar magnitude.

Equation (4.11a) shows that the product $(f\epsilon''_{eff})$ is crucial in determining the electric field required to establish a given power density within a material. For example, consider the dielectric properties of tobacco shown in Fig. 4.6. In this case the effective loss factor is made up of both conductive and dipolar elements, the latter being more pronounced at 2450 MHz whereas the former dominate at radio frequencies. The temperature or the packing density are very important parameters because they control the form of the ϵ''_{eff} vs. f response and therefore the absolute values of ϵ''_{eff} at specific frequencies. Figure 4.6 shows that although at a packing density of 150 kg/m^3 the effective loss factors at 27·12 MHz and 2450 MHz are approximately equal, at an elevated packing density of 250 kg/m^3 the losses are four times higher at 2450 MHz which would depress even more the value of the electric field to that required at the lower packing density for a given power dissipation.

On the other hand, some materials exhibit very high losses at radio frequencies due to d.c. conductivity, for example in aqueous NaCl solutions, as shown in

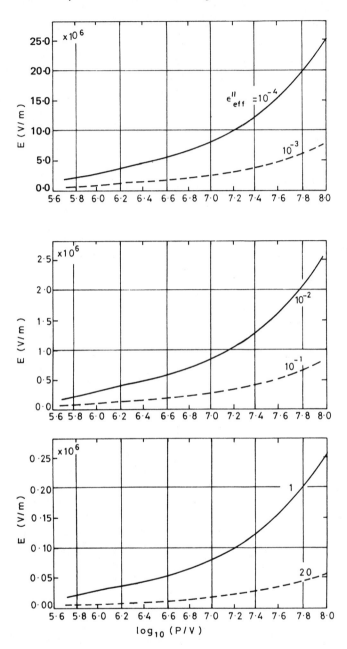

Fig. 4.2   *The r.m.s. electric field strength (V/m) as a function of the power density (W/m³) for various ϵ''_{eff} at 27·12 MHz*

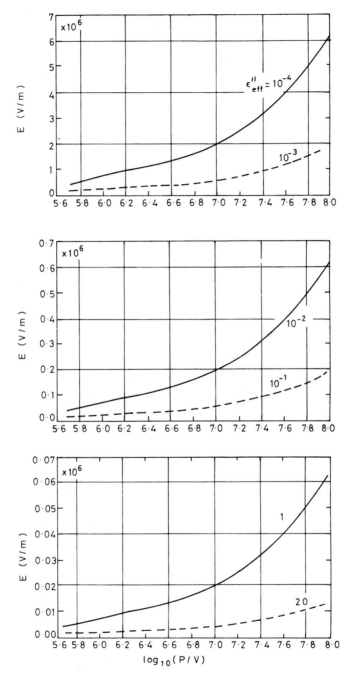

Fig. 4.3   *The r.m.s. electric field strength (V/m) as a function of the power density (W/m³) for various ϵ''_{eff} at 433·9 MHz*

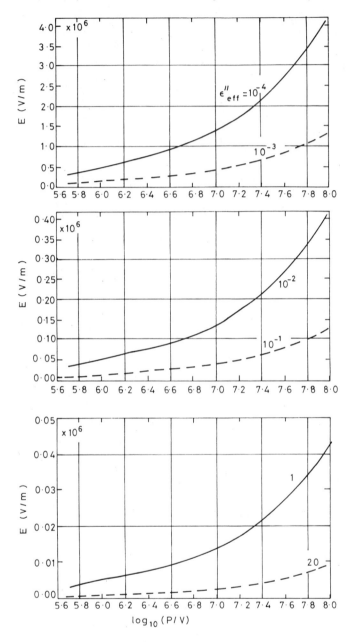

Fig. 4.4   *The r.m.s. electric field strength (V/m) as a function of the power density (W/m³) for various ϵ''_{eff} at 896/915 MHz frequency band*

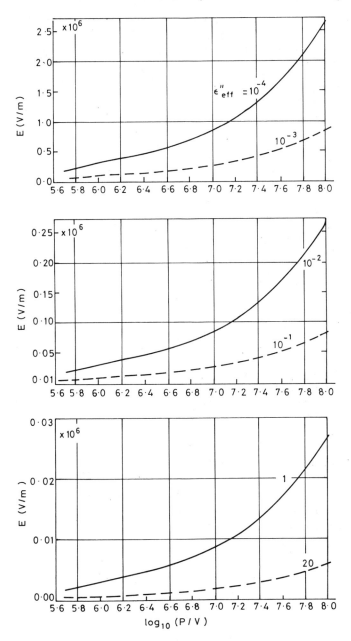

Fig. 4.5 *The r.m.s. electric field strength (V/m) as a function of the power density (W/m³) for various ϵ''_{eff} at 2·45 GHz*

Fig. 3.23(*a*). This has the effect of depressing quite considerably the electric field developed within the material at radio frequencies. The situation is now the reverse to that shown in Fig. 4.6, in that the ϵ''_{eff} at 27·12 MHz for a 0·1 molal solution is some ten times higher than the value at 2450 MHz, which cancels the frequency difference and makes the electric field required for a given power dissipation the same at the two frequencies.

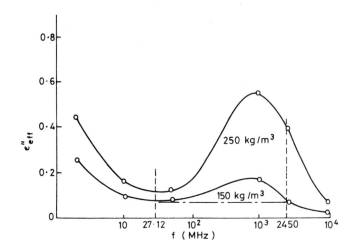

Fig. 4.6   *The effective loss factor as a function of the frequency for tobacco (Hi-Lite) at a moisture content of 13·2% (wet basis)*
(Courtesy of Toshiba Coorporation, data presented at 1969 Microwave Power Symposium.)

4.3 Propagation factor and penetration depth

Many problems in microwave engineering involve the use of Maxwell's equations through which one can derive the following wave equations of the electromagnetic field in the z direction (von Hippel, 1954):

$$\frac{\partial^2 E}{\partial z^2} = \epsilon_0 \epsilon^* \mu_0 \mu^* \frac{\partial^2 E}{\partial t^2} \tag{4.18}$$

and

$$\frac{\partial^2 H}{\partial z^2} = \epsilon_0 \epsilon^* \mu_0 \mu^* \frac{\partial^2 H}{\partial t^2} \tag{4.19}$$

The solution to be considered here is that of a plane wave (see Section 5.2), which for the electric field attains the form

$$E = E_{max} e^{j\omega t - \gamma z} \tag{4.20}$$

This is a periodic field travelling in the z direction with a complex propagating factor γ, given by

$$\gamma = j\omega(\epsilon_0\,\epsilon^*\mu^*\mu_0)^{1/2} = \alpha + j\beta \tag{4.21}$$

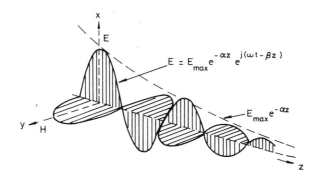

Fig. 4.7 *Propagation of a plane wave in a lossy medium*

where α is the attenuation factor and β is the phase factor. Figure 4.7 shows the essential features of such a propagation. The wave is attenuated as it traverses the medium and therefore the power dissipated, which is a function of E^2, reduces to an even larger extent. To derive an expression for the attenuation of the incident power, we equate the real and imaginary parts of eqn. (4.21), yielding after solving for α and β and assuming $\mu^* = \mu'$

$$\alpha = \omega\left(\frac{\mu_0\,\mu'\,\epsilon'\,\epsilon_0}{2}\right)^{1/2}[(1 + (\epsilon''_{eff}/\epsilon')^2)^{1/2} - 1]^{1/2} \text{ Np/m} \tag{4.22}$$

and

$$\beta = \omega\left(\frac{\mu_0\,\mu'\,\epsilon'\,\epsilon_0}{2}\right)^{1/2}[(1 + (\epsilon''_{eff}/\epsilon')^2)^{1/2} + 1]^{1/2} \text{ rad/m} \tag{4.23}$$

The expression for the attenuation factor can be simplified as follows:

(*a*) For a highly lossy medium, where $(\epsilon''_{eff}/\epsilon') \gg 1$, eqn. (4.22) reduces to

$$\alpha = \left(\frac{\omega^2\,\mu'\,\mu_0\,\epsilon''_{eff}\,\epsilon_0}{2}\right)^{1/2} \text{ Np/m} \tag{4.24}$$

which is the case for conducting materials, since from eqn. (2.20) with $\epsilon'' = 0$, $\sigma = \omega\epsilon_0\,\epsilon''_{eff}$. Therefore eqn. (4.24) yields $\alpha = 1/\delta_s$ where δ_s is the skin depth (see Section 4.8).

(b) For a low loss medium, where $(\epsilon''_{eff}/\epsilon') \ll 1$, eqn. (4.22) after substitution of $\omega = 2\pi f = 2\pi c/\lambda'_0$ reduces to

$$\alpha = \frac{\omega}{2} \left(\frac{\mu' \epsilon_0 \mu_0}{\epsilon'}\right)^{1/2} \epsilon''_{eff} = \frac{\pi \epsilon''_{eff}}{\lambda'_0 (\epsilon')^{1/2}} \ \text{Np/m} \tag{4.25}$$

where λ'_0 is the free space wavelength and the free space velocity c has been equated to $(\mu_0 \epsilon_0)^{-1/2}$ and $\mu' = 1$.

Substitution of eqn. (4.21) into eqn. (4.20) yields

$$E = E_{max} e^{-\alpha z} e^{j(\omega t - \beta z)} \tag{4.26}$$

The first exponential term gives the attenuation of the electric field and therefore the dissipated power follows the form

$$P \propto e^{-2\alpha z} \tag{4.27}$$

The penetration depth is defined as the distance from the surface of the material at which the power drops to e^{-1} from its value at the surface, that is

$$D_p = \frac{1}{2\alpha} \tag{4.28}$$

Substitution of eqn. (4.22) into eqn. (4.28) yields the general expression for the penetration depth

$$D_p = \frac{1}{2\omega} \left(\frac{2}{\mu' \mu_0 \epsilon_0 \epsilon'}\right)^{1/2} [(1 + (\epsilon''_{eff}/\epsilon')^2)^{1/2} - 1]^{-1/2} \tag{4.29}$$

In terms of the free space wavelength eqn. (4.29) reduces with $\mu' = 1$ to

$$D_p = \frac{\lambda'_0}{2\pi(2\epsilon')^{1/2}} [(1 + (\epsilon''_{eff}/\epsilon')^2)^{1/2} - 1]^{-1/2} \tag{4.30}$$

For low loss dielectrics $(\epsilon''_{eff}/\epsilon') \ll 1$ and the penetration depth approximates to

$$D_p = \frac{\lambda'_0 (\epsilon')^{1/2}}{2\pi \epsilon''_{eff}} \tag{4.31}$$

Equations (4.30) and (4.31) show that the power penetration depth increases with larger wavelengths or in other words with decreasing frequencies. In general the penetration depths at frequencies below 100 MHz are of the order of metres and presents little problem as far as power penetration unless the loss factors are exceedingly high. At frequencies near the microwave heating regime the penetration depths are correspondingly smaller and often the size of the material to be treated, particularly when it is very wet, is many times larger than D_p and microwave heating could result in unacceptable non-uniformities in the temperature distribution. Ohlsson *et al.* (1974) have calculated the variation of the penetration depths of foodstuffs with temperature near the three industrially allocated frequency bands. Their results are shown in Fig. 4.8. The general trends, as expected, are higher penetration depths at lower temperatures and frequencies, since as we have seen the effective losses for ice are much less than those for liquid

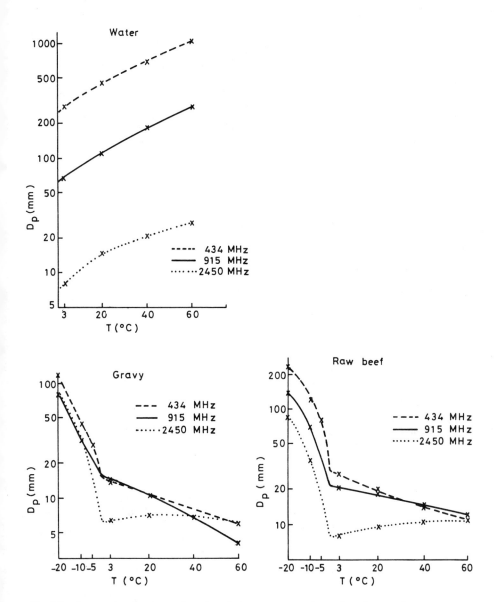

Fig. 4.8 *Penetration depth as a function of the temperature for various frequencies* (After Ohlsson *et al.*, 1974, Copyright *The Journal of Microwave Power*, 1974.)

water. For raw beef, depths of penetration of about 150 mm at sub-zero temperatures reduce to about 20 mm at room temperature, whereas in gravy the penetration depths are correspondingly smaller since the effective losses are primarily
conductive and larger.

Perhaps, surprisingly, gravy would be more uniformly heated to temperatures
above 40°C at 2·45 GHz than at the other two lower frequencies. In contrast, in
pure water the penetration depths increase with increasing temperature highlighting
the differences in the dielectric properties of foodstuffs with large salt content.
More depth of penetration data have been published by Bengtsson and Risman
(1971) on a variety of foodstuffs at 2·8 GHz and in cod at various temperatures
from − 20°C to + 60°C at 2·45 GHz, as shown in Fig. 4.9. It transpires that there
could be limitations in the treatment of lossy dielectrics if the dimensions are
comparable with, or exceed, the penetration depth. Web materials on the other
hand do not usually suffer from such limitations as their thicknesses are small.

4.4 Specific heat

The internal energy of a system U in terms of its pressure p, volume V and external
heat supplied to it is given in differential form by (Tabor, 1969)

$$dU = dQ_h - pdV \tag{4.32}$$

Moreover the system enthalpy is given by

$$H_h = U + pV \tag{4.33}$$

Differentiating eqn. (4.33) and using eqn. (4.32) yields

$$dH_h = dQ_h + Vdp \tag{4.34}$$

The specific heat of a material in SI units, c, is the amount of heat required to
raise a Kg by 1°C. We can observe two definitions of specific heat which follow
from the above equation. First, differentiating eqn. (4.32) with respect to temperature at constant volume gives

$$c_v = \left(\frac{\partial U}{\partial T}\right)_v = \frac{\partial Q_h}{\partial T} \tag{4.35}$$

which defines the specific heat at constant volume. Also, differentiating eqn. (4.34)
with respect to temperature T at constant pressure gives

$$c_p = \left(\frac{\partial H_h}{\partial T}\right)_p = \frac{\partial Q_h}{\partial T} \tag{4.36}$$

which defines the specific heat at a constant pressure. In gases there is a significant
difference between c_p and c_v which does not apply to liquids and solids. In fact, in

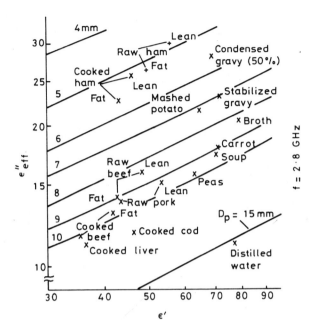

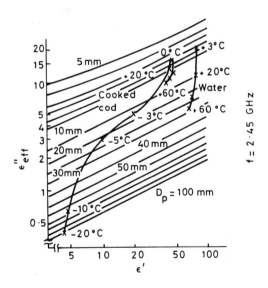

Fig. 4.9 *Penetration depth data for various foodstuffs*
(After Bengtsson and Risman, 1971, Copyright *The Journal of Microwave Power*, 1971.)

the latter, the difference is extremely small and can normally be neglected. Furthermore, the specific heat of most materials can be taken as constant with temperature up to well below the freezing point and starts to decrease at significantly lower temperatures. Measurement of the specific heat in solids is normally made at constant pressure. Table 4.1 shows the values of c_p for some common industrial materials.

Table 4.1 *Specific heat of some common industrial materials*

Material	Specific heat c_p	
	Cal/(g°C)	kJ/(kg°C)
Acetone	0·51	2·13
Alcohol ethyl	0·55	2·31
Asbestos	0·2	0·84
Asphalt	0·4	1·67
Bakelite	0·3–0·4	1·26–1·67
Beeswax	0·82	3·43
Brick common	0·22	0·92
hard	0·24	1·0
Cellulose	0·32	1·34
Charcoal, wood	0·24	1·0
Clay	0·23	0·96
Coal anthracite	0·3	1·26
Coal bituminous	0·33	1·38
Coal tar oils	0·34	1·42
Coke	0·27	1·13
Concrete (stone)	0·17	0·71
Cork board	0·45	1·88
granulated rolled	0·49	2·05
Earth (dry)	0·3	1·26
Fibre board (light)	0·6	2·51
Fibre hard board	0·5	2·09
Glass crown	0·16–0·2	0·67–0·84
flint	0·12	0·5
pyrex	0·2	0·84
silicate	0·19	0·79
wool	0·16	0·67
Graphite powder	0·16	0·67
Gypsum board	0·26	1·09
Ice (0°C)	0·49	2·05
India rubber	0·48	2·0
Leather (dry)	0·36	1·5
Limestone	0·22	0·92

Table 4.1   *continued*

Material	Specific heat c_p	
	Cal/(g°C)	kJ/(kg°C)
Marble	0·21	0·88
Mica	0·11	0·46
Mineral wool blanket	0·2	0·84
Oils caster	0·44	1·84
olive	0·47	1·97
Paper	0·33	1·38
Paraffin wax	0·7	2·89
Plaster light	0·24	1·0
sand	0·22	0·92
Porcelain	0·22	0·92
Plastics foamed	0·3	1·25
solid	0·4	1·67
Sand	0·19	0·79
Sandstone	0·22	0·92
Sawdust	0·21	0·88
Silica aerogel	0·2	0·84
Sodium chloride brine + 10 part H_2O	0·8	3·34
+ 200 part H_2O	0·98	4·00
Turpentine	0·411	1·72
Water	1·00	4·18
Wood fir	0·65	2·12
oak	0·5	2·09
Pine	0·67	2·8
Wool felt	0·33	1·38
loose	0·3	1·26

4.5 Rate of rise of temperature

As the microwave energy is absorbed in the material its temperature increases at a rate depending upon a number of distinct parameters. The power required to raise the temperature of a mass M_a kg of material from T_0°C to T°C in t seconds is given by extending eqn. (4.36):

$$P = \frac{Q_h}{t} = M_a c_p (T - T_0)/t \qquad (4.37)$$

Substituting P using eqn. (4.11a), eqn. (4.37) yields

$$(T - T_0)/t = \frac{0\cdot556 \times 10^{-10} \epsilon''_{eff} f E^2_{rms}}{\rho c_p} \text{ °C s}^{-1} \qquad (4.38)$$

where ρ is the density of the material in kg/m^3 and the specific heat is given in J/kg °C. For a given material heated by high frequency energy at a given f, the rate of rise of T depends on $(\epsilon''_{eff} E^2_{rms})$, which is usually a function of the temperature (due to the variation of ϵ''_{eff} with T).

4.6 The electric field strength

The electric field is the prime parameter in microwave heating; it offers an intangible link between the electromagnetic energy and the material to be treated. It is often difficult to predict the magnitude of the electric field developed in the material since its introduction into the cavity or applicator alters the absolute value of the field. Perturbation techniques are often used to determine the field distribution. However, the perturbation must be small in order for the theory to be valid. Moreover, as we have seen dielectric properties play a very important role in field calculations, and these properties are not always easily available for a given set of operating experimental conditions.

One way to determine the absolute value of the effective electric field is through calorimetry, making use of eqn. (4.38) which after re-arrangement gives

$$E_{rms} = \left(\frac{\rho\, c_p (T - T_0)/t}{0 \cdot 556 \times 10^{-10} f \epsilon''_{eff}} \right)^{1/2} \text{V/m} \tag{4.39}$$

It is apparent that if the field distribution in the material is not constant, eqn. (4.39) is not applicable. The form of the electric field distribution should be substituted in eqn. (4.10) and a similar expression to eqn. (4.39) is obtained. For example, the axial electric field in a TM_{010} cavity is given by the Bessel form (see Section 7.7.2)

$$E_z(r) = E_{max} J_0(Kr) \tag{4.40}$$

where E_{max} is the axial electric field, r is the radius and K is a constant. Substitution of eqn. (4.40) in eqn. (4.10), with $V = \pi r^2 l_e$, where l_e ($l_e < h$, see Fig. 4.10) is the effective length of the cavity and $dV = 2\pi r l_e dr$ yields (Metaxas, 1976)

$$P = 0 \cdot 556 \times 10^{-10} f E^2_{max} \epsilon''_{eff} V (J^2_0(KR_w) + J'^2_0(KR_w)) \tag{4.41}$$

where R_w is the radius of the dielectric. The maximum value of the field E_{max} can now be calculated through a measure of the rate of rise of T, which upon substitution for P in eqn. (4.41) yields a similar expression to eqn. (4.39).

The electric field is maximum at the centre and reduces to zero at the walls as shown in Fig. 4.10. The difficulty with eqns. (4.39) and (4.41) is that a knowledge of the variation of ϵ''_{eff} with T is required to complete the calculation. For example, in water in the temperature range $25 < T < 75$°C and at 2·45 GHz, ϵ''_{eff} varies as $(320/T)$. This enables the electric field to be estimated at various power levels.

The electric field distribution of the cylindrical TM_{010} mode cavity described above is one of a special category of fundamental mode cavities (see Chapter 7) where the electric field in the air space surrounding the dielectric must be smaller

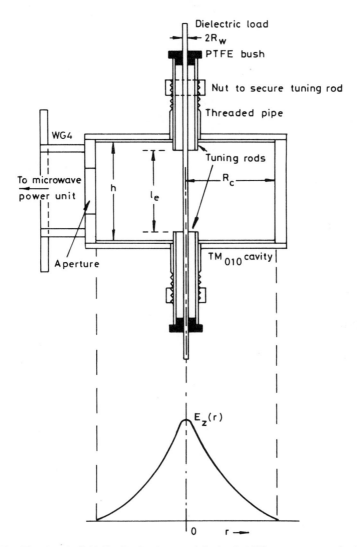

Fig. 4.10   *The electric field distribution in a partially loaded TM_{010} resonant cavity $R_c \gg R_w$*

than the field in the dielectric, if the boundary conditions are to be satisfied. However, this is not always the case. Take for example the case of a spherical water load placed in a multimode oven (see Chapter 6). Calculation of the electric field in such ovens through simple calorimetry is now subject to some fundamental limitations because the electric field developed within the spherical load, E_i, differs substantially to that established in the surrounding air, E_{ext}, and whose value is now sought (White, 1970).

Attempts have been made in the past to predict E_{ext} through a simple relation of electrostatics which for a spherical water load yields (Bleaney and Bleaney, 1957)

$$E_i = E_{ext} \left(\frac{3}{\epsilon' + 2} \right) \tag{4.42}$$

Substitution of eqn. (4.42) into eqn. (4.39) yields an expression for the electric field strength, E_{ext} in terms of the rate of rise of temperature. By assuming a series of values for E_{ext} for a particular case, say a volume of water placed in a microwave field, a family of curves can be obtained which can be combined by experimental data to yield a value for E_{ext} as shown in Fig. 4.11 (Metaxas, 1973). However,

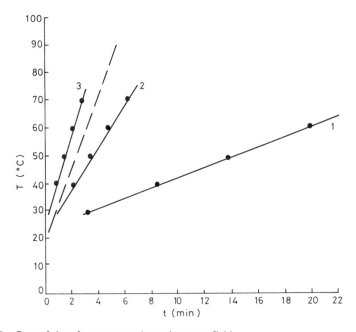

Fig. 4.11   *Rate of rise of temperature in a microwave field*
  ●━●━● Theory 1. $E_{ext} = 10\,\mathrm{kV/m}$
            2. $E_{ext} = 20\,\mathrm{kV/m}$
            3. $E_{ext} = 30\,\mathrm{kV/m}$
  ——— Experiment mass $= 1\cdot38\,\mathrm{kg}$, $P = 1500\,\mathrm{W}$

recent data suggest that the value derived through such a simplified assumption leads to much higher values than actually do exist (MacLatchy and Clements, 1980). It is argued that to use eqn. (4.42), which is based on a scalar potential, is invalid in this case when clearly the electric field is a product of both scalar and vector potentials. A new approach is used where a spherical load is assumed to absorb energy from a surrounding radiation field. A significant fraction of the incident energy is reflected, the latter being equal to $(1 - \tau')$ where τ' is the fraction reaching the load, out of which an amount A is absorbed and produces a heating effect. The fraction A is found equal to

$$A = (1 - e^{-2r/\delta_s}) \tag{4.43}$$

where δ_s is an effective skin depth and r is radius of the spherical load. Typically about 90% of the power that enters the load is absorbed. The value of τ' is found to lie between $0.1 < \tau' < 0.4$ by assuming that the plane waves to be incident from one direction or from all directions respectively. The unperturbed electric field E_{ext} is then related to the calorimetrically absorbed power (P/V) by

$$E_{ext}^2 = \frac{(P/V)r}{3c\epsilon_0\tau'A} \tag{4.44}$$

This theory has been tested using a novel gas-breakdown technique to measure the unperturbed field strengths, whereby a small glass cell is evacuated and filled with helium containing a drop of mercury. Using well-documented data for the effective breakdown field strength E_e, which is related to the E_{ext} (in r.m.s.) value of the applied microwave field by (Francis, 1960)

$$E_e^2 = E_{ext}^2 \left(\frac{v_c^2}{v_c^2 + \omega^2}\right) \tag{4.45}$$

where ω is the angular frequency of the microwave signal and v_c is the electron–neutral particle collision frequency, the above theory was tested, showing that the assumptions made for τ' and A were essentially correct.

4.7 Heat and mass transfer phenomena

4.7.1 Introduction
Conventional drying of relatively thick materials is a slow process relying on heat conduction from the outer layers towards the interior. The thicker the material the slower the evaporation process becomes. High frequency electromagnetic energy, with its vastly superior penetration, offers a unique opportunity for enhancing the rate of evaporation and optimising the overall drying process. As long as the liquid phase of the water and other solvent present is capable of large electromagnetic energy absorption at radio and microwave frequency then this energy can be readily and efficiently transferred to the wet solid. Unlike conventional drying, this energy is absorbed throughout the volume of the wet solid and thus gives rise to a volumetric heat source in the material. This volumetric absorption of high frequency electromagnetic energy can, under suitable conditions, result in the temperature of the wet solid reaching the boiling point of

the liquid. The accompanying generation of vapour due to evaporation of moisture within the pores of the solid results in an increase in the internal gas pressure which can rapidly drive the moisture from the interior of the solid.

When evaporation takes place at the surface of the material lower temperatures are experienced due to evaporative cooling. Temperature gradients can be formed at the surface due to the difference in surface and interior temperatures. It has been suggested that at lower temperatures, these temperature gradients assist the moisture migration to the surface. However, excessive energy dissipation may be detrimental to drying high density, non-porous and brittle dielectrics, due to viscous resistive forces preventing easy migration of moistures towards the surface. In extreme cases internal boiling can occur generating very high internal pressure sufficient to rupture the material. Most of the following analysis is restricted to non-hygroscopic and weakly hygroscopic porous wet dielectrics.

4.7.2 Physical processes

Before presenting a theoretical formulation of the heat and mass transfer equations which govern the removal of moisture from a wet solid, let us describe the physical processes involved. After an initial heating up period, where the material temperature rises towards the wet bulb temperature of the liquid, the rate of moisture evaporation from the wet solid takes place under two distinct periods shown qualitatively in Fig. 4.12 (Perry and Chilton, 1973; Jones, 1975; Stuchly and

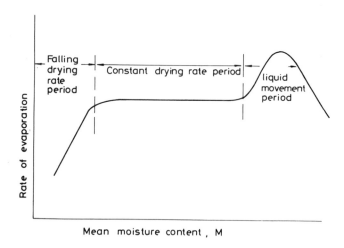

Fig. 4.12   *Rate of evaporation as a function of the mean moisture content*

Hamid, 1972). During the 'constant drying rate period' the moisture content is very high and evaporation will occur from the surface at a constant rate so long as the ambient conditions remain constant. As the moisture content reduces below a critical level, which depends on the type of material, the rate of evaporation becomes progressively less with decreasing moisture, being limited by the reduction of water migration from the interior of the solid towards its surface. This stage is termed the 'falling drying rate period' during which the menisci of the water in the pores fall below the evaporating surface thus impeding drying. A qualitative representation of the rate of moisture reduction during the constant and falling drying rate periods is shown in Fig. 4.13. The continuous supply of

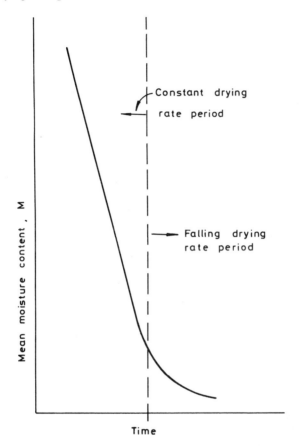

Fig. 4.13 *Decay of moisture content during drying*

moisture through capillary forces from the interior layers towards the surface to replace the water being evaporated from the hot surface is a pre-requisite of the constant drying rate period. Eventually a point is reached where, because of the falling moisture content, it is not possible to maintain the hot surface saturated with water, resulting in the evaporating zone moving further within the interior of the solid. During the falling drying rate period the moisture moves towards the outer surfaces by mass flow of the liquid and vapour phases. As the solid dries out the network of capillaries ceases to be continuous introducing pockets of air which impedes the liquid migration towards the outer layers. Vapour flow is now the dominant flow mechanism. Bories and Pourhiet (1979) have analysed the fundamental mechanisms of moisture migration from porous media towards the evaporating surface by capillary forces. Finally Fig. 4.14 shows

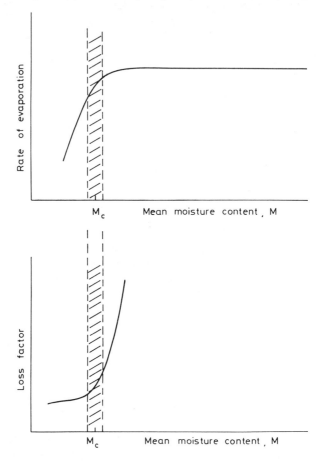

Fig. 4.14   *Rate of evaporation and loss factor as a function of the mean moisture content for a web material*

qualitatively the relation between the critical moisture below which the evaporative efficiency starts to fall and where for the same web material the loss factors are relatively flat and represent bound water absorption effects. The shaded area shows the range of moistures where these transitions are most likely to take place.

4.7.3 High frequency drying

As was explained earlier, radio frequency and microwave heating does not depend on the transfer of heat through a surface. The existence of a volumetric heat source affords the rapid transfer of energy throughout the body of the wet solid and alters the physical characteristics of drying. Considering the temperature changes which occur during drying, shown in Fig. 4.15, Perkin (1979) describes three categories for radio frequency and microwave drying.

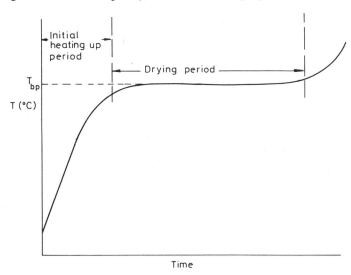

Fig. 4.15   *Rate of rise of temperature during high frequency drying. T_{bp} is the temperature at the boiling point*

(*a*) The solid is rapidly heated to the moisture's boiling point temperature, thereafter $\partial T/\partial t \simeq 0$. Applicable to preheated materials which are not temperature sensitive or heat sensitive materials dried under vacuum.

(*b*) The temperature increases throughout the drying operation – this may occur when the throughput of the material is very fast and the boiling point of the liquid is not attained.

(*c*) The solid is heated to a critical temperature below the boiling point temperature of the liquid phase and the material is force cooled. Applicable to heat sensitive materials.

The temperature of the solid is not limited to the wet bulb temperature. When the temperature of the wet solid approaches the boiling point of the liquid then internal evaporation and an increase in the total pressure can occur within the pores. Mass transfer is now primarily governed by the total pressure gradients due to rapid establishment of the vapour phase in the wet solid. When the solid material contains very high proportions of water, some moisture may be removed from the solids as liquid, due to filtrational flow driven by the total pressure gradient. This phase of drying precedes the constant drying rate period and is termed 'liquid movement period' (Lyons *et al.*, 1972). During this period the rate of moisture removal increases beyond the constant drying rate limit shown in Fig. 4.12.

Usually radio frequency and microwave techniques are rarely used during the initial heating up period. However, Lefeuvre *et al.* (1978) put forward a simulating model of equations governing heat, mass and waves where the electromagnetic energy supplies the sensible heat to increase the internal migration of liquid to the surface and thus act as a water pump.

The rate of drying, which is still carried out by conventional means, is now enhanced. The duration of the constant rate period can be extended provided the values of the effective mass diffusion $\alpha_m \delta_T$ and thermal diffusivity α_T coefficients are such that the increased moisture flow to the surface is greater than the additional moisture evaporated at the surface due to the heat conducted from the wet solid (Perkin, 1979) (see eqns. (4.46) and (4.47)).

The uniqueness of radio frequency and microwave techniques becomes very apparent when considering the falling drying rate period. Energy can now effectively be transferred to the remaining free or loosely bound water which rapidly increases the temperature of the solid. When the solid can be heated to the boiling point of the liquid phase, the pores are filled with vapour and the moisture flow is now governed by the total pressure equation (see eqn. (4.47)) where $\epsilon_v = 1$. The important result is that the constant drying rate period now extends beyond its previous limit at the critical moisture content, which was set by the inability to maintain an adequate moisture flow to the evaporating surface. With radio frequency and microwave drying the limit seems to be the maximum pressure which can be tolerated within the material (Perkin, 1980). If the solid matrix is not rigid, as assumed in the continuity equations described below, the excess pressure can cause the material to expand thereby increasing the pore size and the pressure diffusivity.

The removal of tightly bound or crystallization water with radio frequency and microwave methods is technically difficult because of the low absorption of the energy by the residual liquid phase in the solid. Furthermore, because of the small amounts of moisture, the stabilization of temperature so pronounced at the higher moisture levels, does not now occur with the risk of overheating and charring of the solid.

4.7.4 Transport equations

The rigorous treatment of the internal transport processes taking place within the solid material is extremely complicated and is outside the scope of this book. However in order to appreciate the physical processes discussed above, it is necessary to present a mathematical formulation of the continuity equations, based on a number of assumptions. A simplified form of the equations which govern the mass, heat and total pressure in a porous media can be written as (Perkin, 1979; Luikov, 1966)

$$\frac{\partial M}{\partial t} = \alpha_m \nabla^2 M + \alpha_m \delta_T \nabla^2 T + \alpha_m \delta_p \nabla^2 p \qquad (4.46)$$

$$\frac{\partial T}{\partial t} = \alpha_T \nabla^2 T + \frac{\epsilon_v}{c_p} L_h \frac{\partial M_l}{\partial t} + \frac{\delta P}{\rho c_p} \tag{4.47}$$

and

$$\frac{\partial p}{\partial t} = \alpha_p \nabla_p^2 - \frac{\epsilon_v}{c_a} \frac{\partial M_l}{\partial t} \tag{4.48}$$

where δP is the localised power density, p is the total pressure, α_m, α_T and α_p are the mass, temperature and pressure diffusivities, respectively, δ_T and δ_p are the thermal and pressure gradient coefficients respectively, c_a is the specific moisture capacity of vapour phase. The moisture content equals $M = M_l + M_v$ where M_l and M_v are the mass contents of liquid and vapour phases respectively and ϵ_v is the ratio of vapour flow to total moisture flow. The above equations assume that the solid matrix does not alter as the solid dries and that the diffusivities are constant. The temperature diffusivity is the ratio (thermal conductivity/specific heat × density), while α_p and α_m are the equivalent diffusion coefficients describing the changes of pressure and moisture within the material (Luikov, 1966).

In conventional convective drying the pressure term can be ignored and the volumetric heat source is zero. During the constant rate drying period the internal transport equations simplify to

$$\frac{\partial T}{\partial t} \simeq 0 \qquad \nabla T \simeq 0 \qquad T_s = T_w \tag{4.49}$$

and

$$h_T(T_g - T_s) = -L_h \rho \alpha_m \frac{\partial M_l}{\partial x} \simeq \text{constant} \tag{4.50}$$

where T_g, T_w and T_s are the dry bulb, wet bulb and materials surface temperatures, h_T is a heat transfer coefficient and α_m is the apparent diffusion coefficient representing liquid movement by capillary forces. During the falling rate period in conventional convective drying, the moisture flow is given by

$$\frac{\partial M}{\partial t} = \nabla(\alpha_{mv} \nabla M_v) \tag{4.51}$$

where α_{mv} is the diffusion coefficient of vapour flow.

4.7.5 A simple drying formulation

Having discussed in qualitative terms the internal transport processes taking place during radio frequency and microwave drying, let us consider the theoretical model of one of the most common uses of high frequencies, i.e., end drying or moisture profiling of web materials (Jones and Lawton, 1974). In this process radio frequency or microwaves supply the energy in the last stage while sensible as well as most of the evaporative energy needed is usually supplied by conventional means.

The localised power per unit volume required for evaporation is contained in eqn. (4.47) where $\partial T/\partial t = 0$, $\nabla T = 0$, $\epsilon_v = 1$ and $M \simeq M_l$:

$$\rho L_h \frac{dM}{dt} + \delta P = 0 \tag{4.52}$$

Changing time to co-ordinate distance in the dryer, i.e., $v_w\, dt = dz$, where v_w is the velocity of the moving web eqn. (4.52) gives

$$-v_w\, \rho L_h \frac{dM}{dz} = +\delta P \tag{4.53}$$

Equating this to the microwave power per unit volume available for drying given by eqn. (4.11) we obtain

$$-v_w\, \rho L_h \frac{dM}{dz} = \omega \epsilon_0\, \epsilon_{eff}''\, E_{rms}^2\, g'(M) \tag{4.54}$$

where $g'(M)$ is a term which accounts for energy convected away by the ventilating air and is a function of the moisture content. Integration of eqn. (4.54) yields, after rearrangement,

$$-\int_{M_i}^{M_f} \frac{1}{\epsilon_{eff}''}\, dM = \frac{\omega \epsilon_0\, g'(M)}{\rho L_h v_w} \int_0^z E_{rms}^2\, dz \tag{4.55}$$

where M_i and M_f are the initial and final moisture contents, where $M_i > M_f$ and z is the length of the dryer. By considering an approximate linear functional relationship between ϵ_{eff}'' and M given by eqns. (3.15) or (3.16), assuming a known variation of the electric field with the distance in the dryer and a constant $g'(M)$ function, eqn. (4.55) gives

$$\log_e \left[M\left(\frac{d\epsilon_{eff}''}{dM}\right) + \epsilon_0'' \right]_{M_i}^{M_f} = \text{constant} \tag{4.56}$$

or

$$\frac{M_f(d\epsilon_{eff}''/dM) + \epsilon_0''}{M_i(d\epsilon_{eff}''/dM) + \epsilon_0''} = \text{constant} \tag{4.56a}$$

The above equation can be used to supply data on moisture levelling. For example in paper at $27 \cdot 12$ MHz the effective loss factor varies approximately as $(16 \cdot 1\,M - 1 \cdot 6)$ in the moisture range $10 < M < 30\%$ (dry basis) (Metaxas and Driscoll, 1974). Assuming that the input and output moistures from the high frequency dryer under a given set of conditions are 20% and 12·6% respectively, substitution in eqn. (4.56a) establishes the value of the constant, which in this case is about 0·26. For any given input moisture we can now calculate the output moisture level using eqn. (4.56a), where, for example, with an input moisture of 30% this equation gives an output moisture of 15%. This result shows clearly the levelling effect of the high frequency energy where a difference of 10% in the input moisture level has been reduced to a difference of only 2·4% in the output moisture level. Moisture levelling has also been observed at other frequencies within the radio and microwave frequency heating band. A more rigorous analysis, taking into account the variations of $g'(M)$, is considered by Jones, *et al.* (1974).

4.7.6 Temperature distribution

Heat transfer phenomena, in the absence of any significant mass transfer, have been extensively studied in the past in order to calculate the distribution of temperature across a dielectric, heated volumetrically by radio frequency or microwave energy (Giles *et al.*, 1970; Decareau, 1965). Assuming no evaporation, the heat flow, eqn. (4.47), reduces in one dimension to

$$\frac{\partial T}{\partial t} = \alpha_T \frac{\partial^2 T}{\partial x^2} + \frac{\delta P}{\rho c_p} \tag{4.57}$$

As the electromagnetic energy travels through the material to be heated, it attenuates exponentially by a factor depending upon the dielectric properties. Kruhl *et al.* (1978) have divided the material in homogeneous slabs each having a given set of parameters representing variations in their dielectric properties. The above simplified heat flow equation was solved numerically to obtain the temperature distribution in an inhomogeneous dielectric formed by three homogeneous slabs. Heat transfer by conduction will take place along with volumetric heating due to the electromagnetic energy from the plane wave. Losses from the heated body could be accounted for, first due to radiation, by the expression

$$q_{rad} = \sigma_T \epsilon_i A_c T^4 \text{ watts} \tag{4.58}$$

where T is the absolute temperature, σ_T is the Stephan–Boltzmann constant, ϵ_i is the material emissivity and A_c is the area. Second, convective losses from the surface can be taken into consideration by the use of the expression

$$q_{conv} = \bar{h}_T A_c (T - T_s) \tag{4.59}$$

where T is the gas stream temperature at the surface, T_s is the surface temperature and $\bar{h}_T$ is the mean heat transfer coefficient. Figure 4.16 shows the temperature profiles for the three slabs for an input power density of 5×10^3 W/m^2 (0·5 W/cm^2) and at a frequency of 2450 MHz.

4.8 Skin depth

Most applications in microwave heating employ applicators which consist of metallic enclosures confining the electromagnetic radiation within it. Reflection of the energy on the walls is very significant particularly in resonant applicators. The interaction of the electromagnetic wave at the metallic surface gives rise to a flow of currents and therefore power dissipation in it which manifests itself as conductor losses. An analysis of the depth of penetration of the electromagnetic signal into the metallic surface can be made in a treatise similar to that in Section

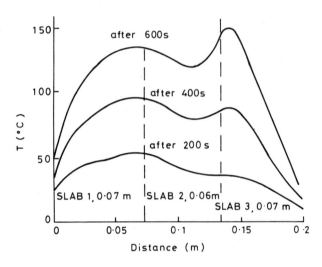

Fig. 4.16   *Temperature profiles for a different heating time*
SLAB 1 $\epsilon' = 5$, tan δ = 0·05
SLAB 2 $\epsilon' = 4$, tan δ = 0·05 $(1 + e^{-0.008\,T})$
SLAB 3 $\epsilon' = 3$, tan δ = 0·05 − 0·4 $e^{-0.008\,T}$
(After Kruhl *et al.*, 1978, Copyright *The Journal of Microwave Power*, 1978.)

4.3. However, we assume that no displacement currents occur as conductive current density, σE, dominates in this case. This gives the following form of Maxwell equations for the fields in the conductor (Jackson, 1962):

$$\nabla \times E_s = -\mu_a \frac{\partial H_s}{\partial t} \tag{4.60}$$

and

$$\nabla \times H_s = \sigma E_s \tag{4.61}$$

where E_s and H_s are the electric and magnetic field strengths in the conductor respectively. Taking the curl of eqn. (4.60), using eqn. (4.61) and assuming a solution for the electric field within the conductor of the form

$$E_s = E_t\, e^{j\omega t - \gamma_c z} \tag{4.62}$$

yields

$$\nabla^2 E_s - \gamma_c^2 E_s = 0 \tag{4.63}$$

with γ_c given by

$$\gamma_c = (j\omega\sigma\mu_a)^{1/2} = (\sigma\omega\mu_a/2)^{1/2}(1+j) \qquad (4.64)$$

where E_t is the tangential electric field at the surface. Substituting γ_c into the field, eqn. (4.62), and defining a skin depth, δ_s, as that where the field is attenuated by $1/e$ of its value at the surface, we obtain

$$\delta_s = \left(\frac{2}{\sigma\omega\mu_a}\right)^{1/2} \qquad (4.65)$$

Substitution of eqn. (4.65) into eqn. (4.64) yields

$$\gamma_c = (1+j)/\delta_s \qquad (4.66)$$

Equation (4.65) is the inverse of eqn. (4.24) which was derived under the assumption $(\epsilon''_{eff}/\epsilon') \gg 1$, that is, for sinusoidal varying fields the imaginary part of the total current density in eqn. (2.15) can be ignored which is equivalent to a system comprising only of conductive currents. Using eqn. (4.65) the dependence of the skin depth on the conductivity and frequency is shown in Fig. 4.17. Alternatively,

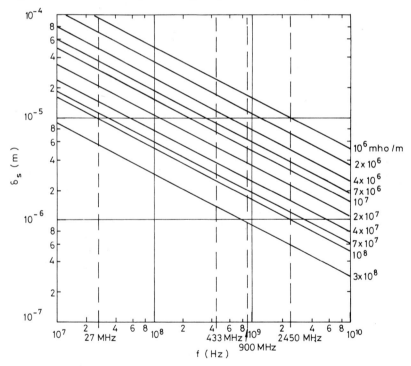

Fig. 4.17 *Skin depth, δ_s, as a function of the frequency for various conductivities, σ*

the skin depth has been plotted as a function of the conductivity as shown in Fig. 4.18 for a number of industrially allocated frequencies including, for comparison, the 27·12 MHz band at radio frequency. Table 4.2 shows the d.c. conductivity of some common metals often used as the internal wall material. Although stainless steel exhibits larger losses compared to most metals it is widely used as the internal wall of applicators on account of its resistance to corrosion and surface hardness. It is particularly useful in industrial applicators where the

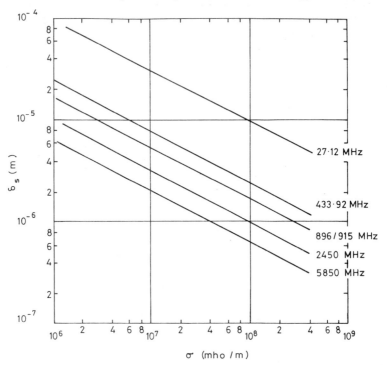

Fig. 4.18 *Skin depth, δ_s, as a function of the conductivity at the industrially allocated frequencies for high frequency heating*

Table 4.2   *Conductivity of some metals*

Metal	Conductivity mho/m
Aluminium 100%	$34 \cdot 3 \; \times \; 10^6$
Brass (90% copper)	$24 \cdot 1 \; \times \; 10^6$
Brass (70% copper)	$14 \cdot 5 \; \times \; 10^6$
Copper (100%)	$58 \cdot 0 \; \times \; 10^6$
Silver (100%)	$61 \cdot 0 \; \times \; 10^6$
Stainless steel type 304	$1 \cdot 39 \times 10^6$

Q-factor with the dielectric inserted in it, Q_0 (see Sections 6.4, 6.7 and 7.2), is relatively low, say about 100, since in such a case the loss in the processed material dominates over wall losses. Also domestic multimode ovens use stainless steel quite extensively, the relatively high loss giving enough equivalent resistance to afford an adequate 'empty oven' match to avoid magnetron damage when switched without any dielectric workload. However, stainless steel should not be used in high Q-factor applications and although silver and gold present the least losses aluminium offers a good compromise of low cost and small skin depth.

4.9 Wall loss

An appreciation of the influence of the material choice to the operation of an industrial cavity as a heater can be obtained from deriving an expression for the power density dissipated in the walls. Figure 4.19 shows the electromagnetic

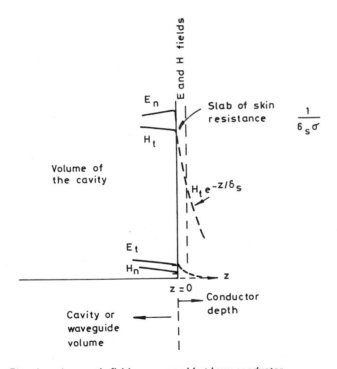

Fig. 4.19 *Electric and magnetic fields near a good but lossy conductor*

fields at the walls of a near perfect conductor, i.e., one exhibiting some loss due to eddy current flow. As before, let the electric and magnetic fields in the conductor be E_s and H_s. At the surface, where $z = 0$ we have larger E_n and H_t fields. Neglecting any displacement currents, the magnetic field in the conductor H_s, in terms of its tangential component at the surface H_t, is computed once again using eqns. (4.60) and (4.61) and attains the form

$$H_s = H_t e^{j\omega t - \gamma_c z} \tag{4.67}$$

Substitution of eqns. (4.67) and (4.62) into Maxwell's equation (4.61) yields the interrelation between the tangential components of the electric and magnetic fields

$$E_t = -(\gamma_c/\sigma)H_t \tag{4.68}$$

The power dissipated at the surface of the conductor follows using Poynting's vector (eqn. 4.2):

$$P_s/\text{unit area} = \frac{1}{2}\frac{1}{\delta_s\sigma}|H_t|^2 \tag{4.69}$$

having also made use of eqn. (4.66). The product $1/\delta_s\sigma$ is known as the skin resistance. Substitution for δ_s from eqn. (4.65), eqn. (4.69) yields

$$P_s/\text{unit area} = \frac{1}{2}\left[\frac{\mu_a\omega}{2\sigma}\right]^{1/2}|H_t|^2 = \omega^{1/2}F \tag{4.70}$$

where F is a frequency independent term assuming that, in the frequency range considered, σ remains substantially constant.

4.10 References

BENGTSSON, N. E., and RISMAN, P. O., 'Dielectric properties of foods at 3 GHz as determined by a cavity perturbation technique. II. Measurements in food materials', *J. Microwave Power* **6**(2), 107 (1971).

BLEANEY, B. I., and BLEANEY, B., *Electricity and Magnetism*. Clarendon Press, Oxford (1957).

BORIES, S., and LE POURHIET, A., 'Heat and mass transfer in porous materials, modelling, simulation, experimentation', *Proc. IMPI Conf. Monaco*, pp. 105–108 (1979).

COPSON, D. A., *Microwave Heating*. AVI Publishing Co. Inc., New York (1975).

DECAREAU, R. V., 'For microwave heating tune to 915 MHz or 2450 MHz', *Food Engineering* No. 6 (1965).

FRANCIS, G., Ionisation Phenomena in Gases. Butterworths Scientific Publications (1960).

GILES, P. G., MOORE, E. E., and BOUNDS, L., 'Investigation of heat penetration of food sample at various frequencies', *J. Microwave Power* **5**(1), 40 (1970).

JACKSON, J. D., *Classical Electrodynamics*. John Wiley & Sons, New York (1962).

JOHNK, T. A., *Engineering Electromagnetic Fields and Waves*. John Wiley & Sons, New York (1975).

JONES, P. L., 'RF Paper Drying – its theory and practice', Paper Technology and Industry, June, pp. 187–193 (1975).

JONES, P. L., and LAWTON, J., 'Comparison of microwave and radio frequency drying of paper and board', *J. Microwave Power* **9**(2), 109 (1974).

JONES, P. L., LAWTON, J., and PARKER, I. M., 'High frequency paper drying – paper drying in radio and microwave frequency fields', *Trans. Instn. Chem. Engrs.* **52**, 121 (1974).

KRUHL, L., ATTEMA, E. P. W., and DE HAAN, C. D., 'Modelling of microwave heating processes', *Proc. IMPI Conf.* Ottawa, pp. 77–79 (1978).

LEFEUVRE, S., PARESI, A., MANGIN, B., and REZUAN, Y., 'Industrial material drying by microwave and hot air', *Proc. IMPI Conf.* Ottawa, pp. 65–67 (1978).

LUIKOV, A. V., *Heat and Mass Transfer in Capillary Porous Bodies*. Pergamon Press, New York (1966).

LYONS, D. W., HATCHER, J. D., and SUNDERLAND, J. E., 'Drying of a porous medium with internal heat generation', *Int. J. Heat and Mass Transfer* **15**, 897 (1972).

McLATCHY, C. S., and CLEMENTS, R. M., 'A simple technique for measuring high microwave electric field strengths', *J. Microwave Power* **15**(1), 7 (1980).

METAXAS, A. C., 'Effect of real and imaginary parts of the dielectric constant and material size on the performance of a microwave oven', Electricity Council Research Centre memorandum, ECRC/M609 (April, 1973).

METAXAS, A. C., 'Rapid heating of liquid foodstuffs at 896 MHz', *J. Microwave Power* **11**(2), 105 (1976).

METAXAS, A. C., and DRISCOLL, J. L., 'A comparison of the dielectric properties of paper and board at microwave and radio frequencies', *J. Microwave Power* **9**(2), 79 (1974).

OHLSSON, T., BENGTSSON, N. E., and RISMAN, P. O., 'The frequency and temperature dependence of dielectric food data as determined by a cavity perturbation technique', *J. Microwave Power* **9**(2), 129 (1974).

PERKIN, R. M., 'Prospects of drying with radio frequency and microwave electromagnetic fields', *J. Separation Process Technology* **1**(2), 14 (1979).

PERKIN, R. M., 'The heat and mass transfer characteristics of boiling point drying using radio frequency and microwave electromagnetic fields', *Inst. J. Heat and Mass Transfer* **23**, 687 (1980).

PERRY, R. H., and CHILTON, C. H., *Chemical Engineers Handbook*, 5th Edition. McGraw-Hill, New York (1973).

STUCHLY, S. S., and HAMID, M. A. K., 'Physical parameters in microwave heating processes', *J. Microwave Power* **7**(2), 117 (1972).

TABOR, D., *Gases, Liquids and Solids*. Penguin Press (1969).

VON HIPPEL, A., *Dielectrics and Waves*. MIT Press (1954).

WHITE, J. R., 'Measuring the strength of the microwave field in a cavity', *J. Microwave Power* **5**(2), 145 (1970).

Chapter 5

Travelling wave applicators

5.1 Introduction

Previous chapters have examined the physical principles involved in high frequency heating; this and subsequent chapters consider the design principles of various heating applications in industrial practice. The most common applicator is undoubtedly the multimode oven which paradoxically is the most difficult to analyse with precision and which in practice is the subject of much experimental design.

This chapter, devoted to travelling wave applicators, is presented first because these are relatively simple to analyse and the methods used illustrate well the basic principles. A preliminary section is therefore devoted to developing a qualitative description of guided waves from the superposition of two, plane waves, leading to an introduction to reflected waves, standing waves and impedance matching. These terms are frequently used in subsequent chapters and familiarity with them is important to the understanding of all microwave heating devices. For a more rigorous approach to microwave theory the reader is referred to the extensive theory on the subject (Moreno, 1948; Ginzton, 1958; Sucher and Fox, 1963; Harvey, 1963; Montgomery *et al.*, 1965; Ramo *et al.*, 1965; Baden Fuller, 1979).

It will be appreciated from preceding chapters that heating by microwave energy is essentially an interaction between the molecules of the material to be heated and the electric field component of the microwave energy. For this reason the electric field component receives special scrutiny in this chapter.

5.2 Plane waves

Electromagnetic (EM) energy propagates through free unbounded space at 3×10^8 metres/second irrespective of frequency, and has two principal and inseparable components:

(*a*) An electric field E, measured in volts/metre.
(*b*) A magnetic field H, measured in amperes/metre.

Both these components have magnitude and direction in space and are therefore vectors, and they are always mutually orthogonal in space as shown in Fig. 5.1. They also oscillate sinusoidally with time and this can be represented by rotating time vectors familiar to electrical engineers. The power density transmitted is the product (volts/metre × amps/metre) = watts/metre^2. Basic to all sinusoidal wave propagation is the well known relation

$$c = f\lambda \tag{5.1}$$

where c is the velocity of propagation, f is the frequency of oscillation and λ is the distance along the direction of propagation of two similar points in phase, i.e., the wavelength.

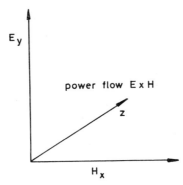

Fig. 5.1 *Electric and magnetic field vectors and power flow of a plane wave*

The simplest way in which electromagnetic energy can propagate is as a plane wave, which, by way of example, consists of a sinusoidally varying electric field, E, vector vertically polarised, together with a horizontally polarised magnetic field, H, sinusoidally varying in phase with the electric field. The two components E and H lie in one plane, and the direction of propagation of the field is at right angles to the plane. By definition the plane is infinite requiring a generator of infinite power and in practice only an approximation to a plane wave is possible.

Practical devices carrying electromagnetic fields employ metal conductors for transmission from generator to load and for the envelope of a microwave heating tunnel or oven. In these cases the fields are more complicated than a plane wave, but it is always possible to synthesise the actual field by superposition of two or more plane waves with different directions of propagation. This technique is valuable in illustrating the fundamental properties of guided waves by geometry.

5.3 Guided waves

It is well-known that electromagnetic waves will propagate inside hollow metal tubes and may resonate inside a metal box. In this section a pair of plane waves is shown to synthesise the propagation of EM waves in a metal tube of rectangular cross-section.

First consider the effect of a perfectly conducting plane in the path of a plane wave. By definition the electric field intensity tangential to the surface of the conducting plane is zero because its resistance is zero. However, the electric field component normal to the plane is unimpeded by the presence of the plane. This property is of fundamental importance in all microwave structures, namely, that the electric field component grazing the surface of a conductor is zero, but the component normal to the surface can attain a finite value. For the magnetic field the converse applies, and the grazing magnetic field induces currents to flow in the surface of the conducting plane.

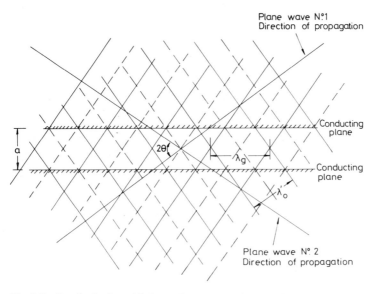

Fig. 5.2   *Synthesis of a guided wave between conducting planes by two coherent plane waves*

Figure 5.2 shows a diagram of a pair of plane waves of equal amplitude and frequency propagating along paths which converge with angle 2θ. Their electric fields are polarised normal to the diagram. Solid lines show the position at a time of maximum positive E field and broken lines the position at a time of maximum negative E field at an instant in time; in between these times the amplitude of the fields varies sinusoidally. Obviously at any point where a solid line intersects a broken line the E field resultant is zero, but wherever a solid line intersects another solid line the E field has twice the maximum E field value of the individual plane waves. Further inspection of Fig. 5.2 reveals that if the two waves are considered at some later time their maximum E fields will have travelled forward along their respective axes of propagation but that the points of resultant zero field will move along a direction bisecting the angle of convergence of the two waves. It may be seen that there are several such loci of zero resultant field, and two are shown spaced apart but parallel to each other. Since the E field intensity is zero along these loci it is possible to insert conducting planes vertical to the diagram without affecting the fields. The E field between an adjacent pair of planes now varies sinusoidally in a transverse direction between them having a maximum value at the midpoint, and sinusoidally in the direction parallel to them.

The field outside the two planes can now be eliminated and the resultant condition is a pair of plane waves synthesising a single wave propagating between a pair of parallel conducting plates, that is, a guided wave.

An important feature may be seen from Fig. 5.2, in that the wavelength of this guided wave λ_g is longer than that of the plane waves forming it, and by geometry the wave length is $\lambda'_0/\cos\theta$. This wavelength is the waveguide wavelength λ_g. Notice that if the two planes are set at this fixed distance apart and the wavelength is increased, the zero E field condition only remains satisfied if the angle of convergence 2θ is increased, and as a consequence the λ_g increases even more rapidly than λ'_0. Ultimately $\theta = 90°$ and λ_g becomes infinite. This condition occurs when the free-space wavelength λ'_0 is equal to twice the separation distance between the planes, and this critical wavelength is the cutoff wavelength λ_c. At wavelengths greater than λ_c the wave cannot propagate between the planes without high attenuation.

It is now possible to introduce a second pair of planes at right angles to the first set to form a closed tube with axis coincident with the direction of propagation. This pair of planes is now at right angles to the electric field vector and satisfies the condition that E can have a finite value normal to a conducting surface. The separation between this second pair of planes does not affect the ability of the wave to propagate since the planes remain at right angles to the E vector irrespective of their separation. The tube now formed is a rectangular waveguide and the field supported inside it, as synthesised above, is the simplest 'mode' of propagation, as shown in Fig. 5.3.

This 'mode' of propagation is the simplest and has the lowest frequency that can propagate freely in a rectangular waveguide. It is characterised by an electric field vector (E) which has one component only, linearly polarised parallel to the narrow faces of the waveguide with a half sinusoid variation between them, with a maximum value at the midpoint.

In the direction between the broad faces the E field is of constant value. The accompanying magnetic field has two components forming closed loops. One component is in the direction of the waveguide axis (H_z) the other at right angles to the axis and parallel to the broad faces (H_x). All the components have a sinusoidal variation with time, and also with position along the waveguide. Note that E_y and H_x have their maximum values at the same position along the axis of the waveguide and are in phase in time, whereas the component H_z has its maximum value displaced a quarter wavelength away along the axis.

As already indicated other field patterns can also exist as the wavelength is reduced, having more than one half sinusoid of variation of E field between the narrow faces. Other modes have a sinusoidal variation of E in a direction normal to the broad faces. All these modes are designated H_{lm} or TE_{lm} (transverse electric) where the suffixes l and m indicate the number of half sinusoids of variation of E field along the x- and y-axes respectively. The lowest mode is obtained when $l = 1$ and $m = 0$ giving the H_{10} or TE_{10} mode.

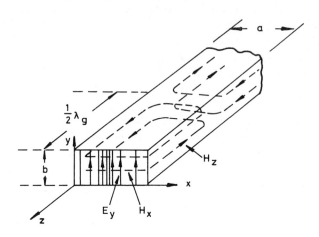

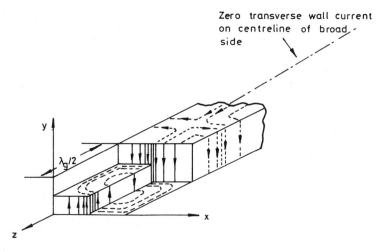

Fig. 5.3   *Fields and wall currents in a TE_{10} mode rectangular waveguide*
——— Electric field
– – – magnetic field
. . . . . . wall current.

For microwave heating applications two features of waveguide propagation are particularly important.

(*a*) In magnitude of the electric field vector, since this is the component involved in heating.

(*b*) The distribution of current in the walls of the waveguide as this determines where slots can be made without causing excessive power to leak out which can result in waste of energy or a hazard.

The relation between power flow and peak field stress in a rectangular waveguide operating in the H_{10} mode is given by (Harvey, 1963; Montgomery *et al.*, 1965)

$$E = \left(4P \frac{\lambda_g}{\lambda_0'} \frac{1}{ab} \left(\frac{\mu_a}{\epsilon_a} \right)^{1/2} \right)^{1/2} \quad (\text{V/m}) \tag{5.2}$$

where P is the power flow in watts and a and b are the dimensions of the broad and narrow faces. Figure 5.4 gives the electric field stress in rectangular waveguides used at the two principal microwave heating frequencies.

The values of electric field strength shown in Fig. 5.4 relate to the peak of the sinusoidal space-variation across the width of the waveguide and are therefore at the centre-plane through the broad faces. They also correspond to an empty waveguide having a dielectric constant of unity, i.e., with no dielectric filling.

When a workload is introduced the E field amplitude falls for three reasons. First, because the dielectric constant of the workload is greater than unity, the load thereby shunting the field capacitively and, second, because the load has a resistive component to its impedance representing power absorption which also shunts the E field. Third, as power is absorbed by the workload the field will fall in the direction of travel of energy along the waveguide away from the generator. The reduction of E field due to the load is therefore a function of its electrical properties (ϵ', $\tan \delta_{eff}$) and its thickness.

5.4 Impedance matching

In an energy-efficient heating system transfer of energy from generator to workload must clearly be achieved with minimum loss. A microwave heating system has an unusually high transfer efficiency, limited by two factors.

First, energy is absorbed in the walls of the transmission waveguide by $I^2 R'$ heating. In most cases this is a very small proportion of the transmitted power and is negligible.

Second, power is reflected from the load if 'optimum matching' has not been achieved. The concept of optimum matching for maximum power transfer from a source to a load is well known and the load resistance R_L' must equal the generator internal resistance R_G'. If the load resistance is less than R_G' more current flows and more power is dissipated in the generator perhaps causing damage if excessive. If the load resistance is higher than R_G' less current flows so that less power is dissipated in the load (the power in the load is $I^2 R_L'$ so that although the resistance has increased the fall in current is the greater effect due to the square law). The same argument applies with complex impedances leading to the result that for optimum power transfer the load impedance should be the complex conjugate of the generator impedance.

However, at very high frequencies there is an added complication in that the

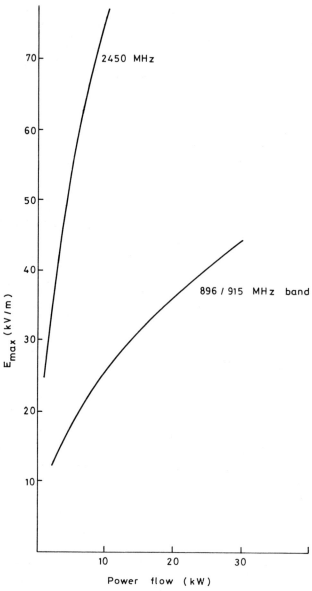

Fig. 5.4   *Maximum electric field strength in a TE_{10} rectangular waveguide as a function of the power flow*

generator and load are necessarily connected via a transmission line or waveguide, as is shown in Fig. 5.5, in which electrical length transforms the apparent impedance of the load seen by the generator at its terminals (Adam, 1969). At microwave frequencies, therefore, it is not practicable to compare the reactive and resistive components of the load and generator impedances and it is easier to consider impedance matching in terms of forward and reflected waves which are readily measured, the presence of any reflected wave indicating a loss of efficiency.

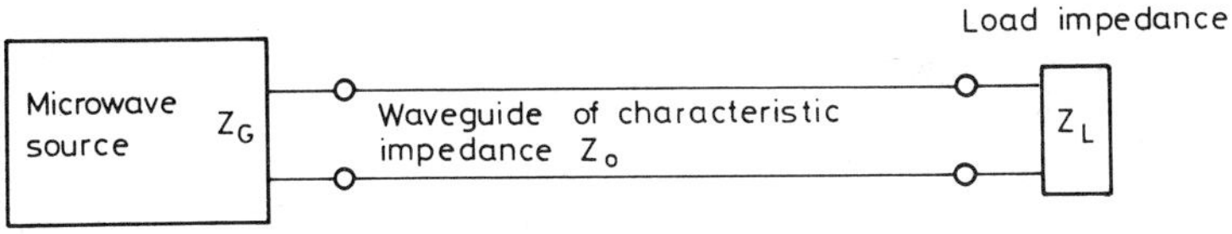

Load impedance

Microwave source Z_G Waveguide of characteristic impedance Z_o Z_L

Fig. 5.5   *Load impedance connected to a source via a waveguide*

5.5 Voltage standing wave ratio (S)

Referring back to Fig. 5.5 by definition the generator would be matched to the waveguide if $R'_G = Z_0$. Similarly the load would be matched to the waveguide if $Z_L = Z_0$ and all the available power from the source will be transmitted to the load. However, if $Z_L \neq Z_0$ there is now a reflected wave in addition to the forward wave. Let their respective amplitudes be V_- and V_+.

The reflection coefficient ρ, a vector quantity, is defined as

$$\rho = \frac{V_-}{V_+} \tag{5.3}$$

Since the power is proportional to the square of the voltage the reflected power is equal to $|\rho|^2 P_0$, where P_0 is the forward power incident at the termination. The transmitted power is therefore given by $P_0 (1 - |\rho|^2)$ (see also Section 7.9).

The reflection coefficient can readily be measured because the forward and reflected waves add vectorially along the transmission line giving rise to a cyclical variation of the field intensity along the line, creating a standing wave. Figure 5.6(a) is a series of vector diagrams all at the same instant but at points (1/8) wavelength apart along the line beginning at point A (Fig. 5.6(a)) where the standing wave is at its maximum value. The amplitudes of the forward wave, V_+, and the reflected wave, V_-, are constant but as the point of observation moves along the transmission line towards the load, the phase angle of the forward wave is retarded whilst that of the reflected wave is advanced by a similar amount. The forward and reflected wave vectors therefore rotate in opposite directions as the point of observation moves along the transmission line, coming in-phase and phase opposition at points half a wavelength apart along the line. The resultant voltage therefore varies from a maximum equal to the arithmetic sum, to a minimum equal to the arithmetic difference as shown in Fig. 5.6(b). The ratio of maximum to minimum voltage is the voltage standing wave ratio, S (VSWR), which

is defined as

$$S = \frac{|V_+| + |V_-|}{|V_+| - |V_-|} = \frac{1 + |\rho|}{1 - |\rho|} \tag{5.4}$$

VSWR ranges from 1, for zero reflected power, to infinity for 100% reflected power. Sometimes the reciprocal, $1/S$, is quoted having the range 1 to 0. Instruments for measuring VSWR are available and called standing wave detectors, or slotted lines. Also directional couplers are available which separate the forward from the reflected waves as small samples coupled from the waveguide, enabling the ratio of forward to reflected power to be measured. Figure 5.7 shows the fractions of the reflected and transmitted powers and the VSWR as a function of the reflection coefficient, ρ.

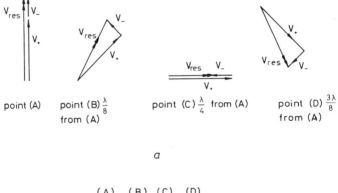

a

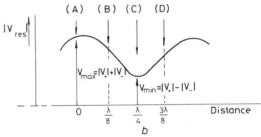

b

Fig. 5.6 *Resultant amplitude of the standing wave, V_{res}, due to the superposition of the forward and reflected waves*
$V_+ =$ Amplitude of forward wave
$V_- =$ Amplitude of reflected wave

In practice a microwave generator will change its efficiency with load VSWR (and phase angle of reflection coefficient) so that more forward power may be obtained under certain 'mismatched' loading conditions. The data of Fig. 5.7, whilst strictly accurate, should not be taken in isolation in estimating the overall transmission efficiency. Usually an industrial microwave applicator is designed to have an input VSWR smaller than 2 over the operating range of frequencies of the generator, giving a maximum of about 10% reflected power and stable operation of the generator.

5.6 Travelling wave applicators

Microwave heating applicators in which power, fed into a chamber from the generator, is substantially absorbed by the workload with the residue being dissipated in an absorbing terminating load are described as travelling wave applicators. They are characterised by a good input VSWR irrespective of the amount of workload within them and can be operated empty without risk to the generator. Their efficiency depends on the ϵ' and tan δ values of the workload and its cross-sectional area. Invariably they are used on a continuous flow basis, usually with a conveyor belt. Often they have balanced symmetrical configuration such that the entry and exit slots for the workload are at points of minimum wall current thereby giving inherently low levels of leakage energy.

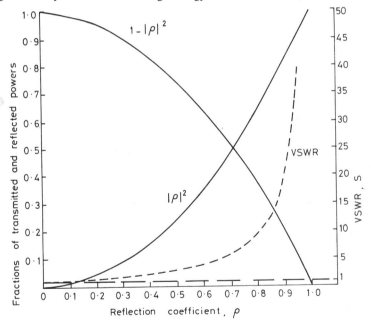

Fig. 5.7 *The fractions of transmitted* $(1 - |\rho|^2)$, *reflected* $|\rho|^2$ *powers and the VSWR, S, as a function of the reflection coefficient,*

Travelling wave applicators are not suitable for materials of low loss because they become inconveniently long. However, all travelling wave applicators can, in principle, be made into resonant applicators by replacing the terminal load by an adjustable short circuit, and fitting an aperture plate at the generator end, as detailed in Chapter 7. Under these conditions low-loss materials can be heated very efficiently.

In some cases where the workload has a very high loss factor, the travelling wave applicator is terminated in a short circuit because there is only negligible residual power. However, precautions must then be taken to avoid operation in the absence of an adequate volume of workload in the chamber to avoid risk of damage to the generator, such as fitting a ferrite circulator and a high power load. This combination is referred to as an iso-circulator (see Section 9.2.7).

Several travelling wave applicator types are used in practice as will be described in Chapter 8 and many more have been described in the literature (Dunn, 1967; Püschner, 1966; Heenan, 1968; Williams and Warner, 1968). This section surveys the principal features of the commonest types.

5.6.1 Axial travelling wave applicators

This is the simplest travelling wave applicator and in its elementary form is a waveguide operating in the TE_{10} mode with E field horizontal in which the work-load travels axially usually, but not always, in contraflow to the microwave energy.

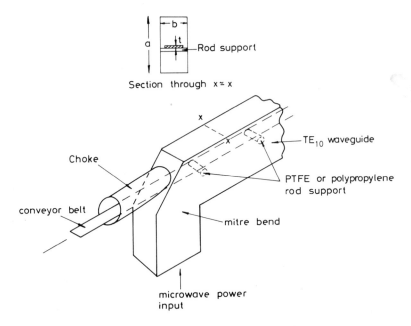

Fig. 5.8 *Material and conveyor belt feed into an axial travelling wave applicator using a mitre bend*

The conveyor belt is supported at intervals so that the mid-depth plane of the workload is coincident with the mid-points of the broad faces of the waveguide as illustrated in Fig. 5.8.

At each end a mitre bend (usually $90°E$-plane) permits connection to the generator and terminating load. The mitre plates of the bends have holes with cutoff waveguide chokes to permit the belt and workload to enter and leave the applicator.

This applicator has the advantage of simplicity and low cost. It is suitable for materials where the thickness and loss factor are such as to yield an attenuation per unit length high enough to provide the required microwave efficiency within a practical length, as indicated by eqn. (5.12). Its limitation is the narrow width available for the workload of about $\frac{3}{8}\lambda_0'$. This is sometimes increased to $\frac{3}{4}\lambda_0'$ at the risk of exciting the cross-polarised H_{10} mode. A major feature of this applicator is excellent transverse heating uniformity.

In another form of waveguide applicator a ridged waveguide is used to increase the value of E field in the vicinity of the workload, as shown in Fig. 5.9. This applicator is used for continuous flow heating of extruded rubber and is considered further in Chapter 8.

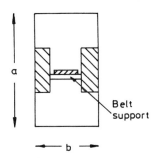

Fig. 5.9 *Ridged axial travelling wave applicator*

All applicators with a plane-polarised E field apply heat to discrete objects moving on a conveyor band in a non-uniform way due to field distortion at the boundaries of the objects, as shown in Fig. 5.10. This form of non-uniformity gives rise to excessive heating at the ends of the objects, compared with the centres. By applying an axially polarised E field the heating profile is reversed and good compensation can be achieved by adjusting the relative values of the axial and transverse E fields. An axial E field is created by corrugating the broad faces of the waveguides to form slow-wave structures, as shown in Fig. 5.11. Various structural variations are possible with this arrangement to simplify fabrication, which are based on creating a finite non-zero surface impedance across which axial currents flow creating, thereby, an axial electric field.

5.6.2 Meander (or serpentine) travelling wave applicators
To overcome the limit to the width of material which can be processed in an axial travelling wave applicator, the planar material can still be passed through

the centre of the waveguide but move at right angles to its length, as shown in Fig. 5.12.

For this the waveguide is slotted at the centre of its two broad faces, so that the material lies in the maximum electric field to couple effectively with the flowing microwave energy. Such a slot is inherently non-radiating since, as shown in Fig. 5.13, it does not cut through the flow of currents on the waveguide walls. Various slots are illustrated in Fig. 5.13 and of these only 1 and 2 are non-radiating, whereas the rest behave as small aerials radiating to a degree which depends on the extent to which their orientation interferes with current flow.

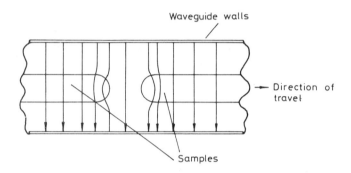

Fig. 5.10   *Field distortion due to discrete objects*

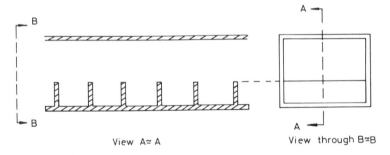

View A≃A   View through B≃B

Fig. 5.11   *Corrugated waveguide generating axially polarised E-field*

Although in principle any width of material can now be accommodated, such an arrangement poses new problems in processing sheet dielectrics, not least the fact that the electric field in the z-direction falls exponentially, as illustrated qualitatively in Fig. 5.12(b). The decay follows the simple relation $e^{-\alpha z}$, where α is the attenuation per unit length and depends upon the effective loss factor of the processed planar material. Such a single pass applicator can therefore lead to either high non-uniformity of heating or insignificant heating depending upon the value of $t\epsilon''_{eff}$, where t is the thickness of the material. A practical solution to this is offered by passing the material through a number of waveguide passes leading to the development of the meander type travelling wave applicator typically

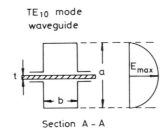

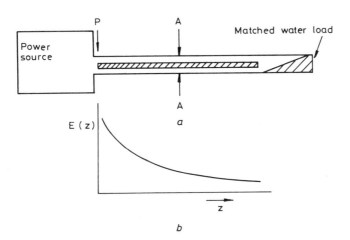

Fig. 5.12 *Electric field decay along the length of the waveguide*

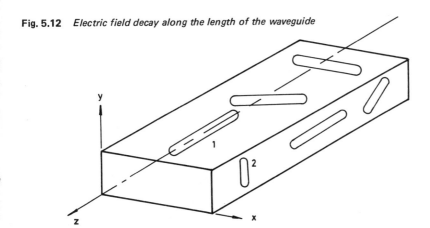

Fig. 5.13 *A number of slots in a TE$_{10}$ mode rectangular waveguide. Slots 1 and 2 are inherently non-radiating*

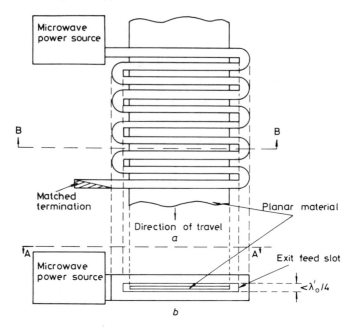

Fig. 5.14   *Meander applicator*
a Plan view *A–A*
b Cross-sectional view *B–B*

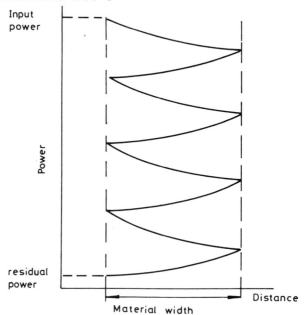

Fig. 5.15 *Exponential decay* $(e^{-2\alpha z})$ *of power in an eight pass meander applicator (electric field decays as* $e^{-\alpha z}$)

illustrated in Fig. 5.14. It consists of a number of rectangular waveguides mounted with their broad faces side-by-side electrically connected in series, each waveguide having slots in the centres of its broad faces through which the workload travels at right angles to the waveguide. A proportion of the microwave energy is absorbed by the workload in each waveguide pass, typically 10—40%, giving an attenuation ideally in the range of 0·4—1·5 dB. A sufficient number of waveguides (or 'passes') are provided to yield a high efficiency (e.g., 10 passes of 1 dB would give 10 dB total attenuation, or 10% residual power corresponding to 90% absorption efficiency). The electric field strength in the meander drops gradually from the source to the matched termination, as shown in Fig. 5.15 for an eight pass, resulting in a more even power distribution across the width.

The meander or serpentine (as it is otherwise known) applicator is not primarily limited in width of process slot, such limit being determined by the attenuation per pass and constructional arrangement.

The leakage of microwave energy is inherently small because of the symmetry of the process slot, with respect to the fields in the waveguide. However, the introduction of a workload disturbs this symmetry and can increase the leakage. It is therefore normal practice to provide a choking system and power absorber at each end of the process slot to reduce leakage to a low level. The height of the process slot is determined by cross-coupling between waveguide passes, and by microwave leakage. These effects are much influenced by assymmetry of loading and constructional tolerances, and are difficult to estimate. Experience has shown that with a process slot not more than $\frac{1}{4}\lambda_0'$ high as shown in Fig. 5.14(b), there is little difficulty with cross-coupling, but in excess of this amount the effect may be significant. However, the height is not critical and there is no special characteristic in a process slot height of $\frac{1}{4}\lambda_0'$.

Whilst the meander applicator is fundamentally simple in concept it is expensive to manufacture if 180° waveguide bends, shown in Fig. 5.16, are used to join the passes. An alternative strap-coupled structure is shown in Fig. 5.17(a) which is widely used.

5.6.3 Standing waves in meander applicators
Uniformity of heating is affected by the attenuation per pass and by the presence of standing waves. Considering the latter first, they arise from numerous small reflections, principally at the edges of the workload in each pass, residual reflection from the 180° bends or coupling straps, and unwanted cross-coupling between passes. All these reflections add vectorially to give a standing wave in each pass, giving a sinusoidal variation in heating intensity across the process slot, with a distance between the peaks of half a guide wave length, $(\frac{1}{2}\lambda_g)$.

It is possible to minimise standing waves by choice of the overall width of the applicator to give an optimum broadband match for the generator and compensation of standing waves within the applicator in the direction of travel of the workload. Consider the geometry of Fig. 5.17 which shows three passes of a multipass serpentine with an overall electrical width per pass, d_0, and with an

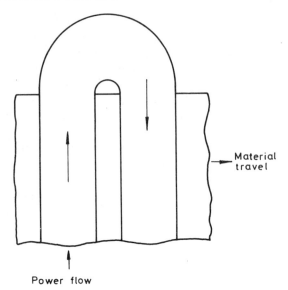

Material travel

Power flow

Fig. 5.17   *A 180° waveguide bend*

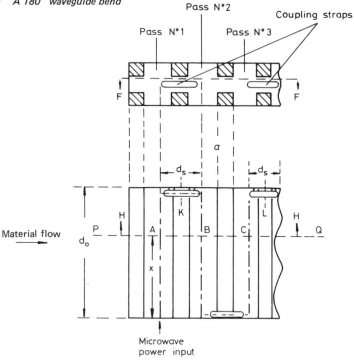

Pass N°1   Pass N°2   Pass N°3   Coupling straps

d_s d_s

a

Material flow d_o P   A   K   B   C   L   H   Q

x

Microwave power input

b

Fig. 5.17   *Strap-coupled meander applicator*
a Section *H–H*
b Plan view *F–F*

effective interpass or strap electrical length, d_s. A web of the material to be heated passes through the applicator, the line PQ representing one of its edges at a distance x from one side. By inspection we can deduce the following relationships for the various lengths:

$$l_{AB} = d_s + 2(d_0 - x) \tag{5.5a}$$

$$l_{BC} = d_s + 2x \tag{5.5b}$$

$$l_{AC} = 2(d_s + d_0) \tag{5.5c}$$

If we assume the reflections at the line PQ are small, the reflection at C can be made to cancel that at A by making

$$l_{AC} = (2m + 1)\frac{\lambda}{4} = 2(d_s + d_0) \tag{5.6}$$

where m is an integer. Note that this condition is independent of x, so by super-position, all lines PQ are matched if the condition is satisfied. Further alternate passes will be mutually self-matching throughout the applicator. The value to ascribe to d_0 in this relation requires care since the transmission path $A-C$ includes waveguide, waveguide with dielectric loading (viz. the workload and conveyor), and for coupling-strap interpass coupling, a section with TEM propagation. In effect the electrical length from A to C must be an odd integral number of quarter wavelengths.

By similar reasoning if L and K (Fig. 5.17) are the electrical mid-points of the coupling devices, a match will be obtained if l_{KL} is an odd number of quarter wavelengths.

$$l_{KL} = 2(d_0 + d_s) = (2n + 1)\frac{\lambda}{4} \tag{5.7}$$

where n is an integer. Subtracting eqn. (5.6) from eqn. (5.7) yields

$$d_s = (n - m)\frac{\lambda}{2} \tag{5.8}$$

Thus for optimum matching the coupling means should be a half wavelength long or multiples thereof.

These considerations give guidance only in the design of meander applicators for best uniformity and matching. Attenuation of the workload obviously violates the assumption that the amplitude of the reflection coefficient ρ_C, referred to A, is equal $|\rho_A|$, and similarly for ρ_L referred to K is $|\rho_K|$.

Nevertheless, attenuation will reduce the amplitude of the reflection coefficient elements when referred to the generator input. Moreover, variations in workload thickness, width and ϵ' (e.g., due to moisture changes) affect the wavelength in the loaded region of the applicator. The dimensional design of a meander applicator requires some care to optimise performance, but it should be noted that even with the most inappropriate dimensions the performance may well still be acceptable particularly for drying.

5.6.4 Attenuation in travelling wave applicators

Additional to standing wave effects, the uniformity of heating of planar dielectric materials processed in travelling wave applicators depends on the attenuation per metre, α. Consider the arrangement of Fig. 5.12, where P is the output power from the source incident on a rectangular waveguide operated in its fundamental TE_{10} mode. Since, from eqn. (4.20), the electric field strength decays exponentially as $E e^{-\alpha z}$, the power at a point distance z away from the source is given by

$$P(z) = Pe^{-2\alpha z} \tag{5.9}$$

Differentiation yields

$$\frac{dP(z)}{dz} = -2\alpha Pe^{-2\alpha z} = -2\alpha P(z) \tag{5.10}$$

The fractional loss of power per metre, $-1/P(z)\,(dP(z)/dz)$, is therefore given by 2α.

The attenuation coefficient, α, relates to the partially loaded rectangular waveguide which for the general case of a dielectric slab of thickness t and height h placed off-centre in a rectangular waveguide, as shown in Fig. 5.18(a), and using

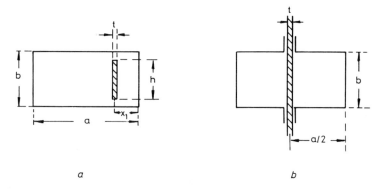

a

b

Fig. 5.18 *Partially loaded TE_{10} waveguide with a dielectric slab*
a General case
b Travelling wave applicator

cavity perturbation formulae, can be shown to be equal to (Altman, 1964)

$$\alpha = \alpha_0 + 2\pi\epsilon''_{eff}\,\frac{th}{ab}\,\frac{\lambda_g}{\lambda_0'^2}\,\sin\left(\frac{\pi x_1}{a}\right) \text{Np/m} \tag{5.11}$$

where α_0 is the corresponding attenuation of the microwave power due to radiative and wall losses and x_1 is the distance of the slab from the narrow waveguide face. In travelling wave applicators the computation of α is simplified due to the symmetrical placing of the planar load (Fig. 5.18(b)) and by ignoring the α_0 term except in cases where the dielectric is extremely low loss. Therefore, substituting $h = b, x_1 = a/2$ and bearing in mind that 1 neper $\equiv 8\cdot686$ dB, eqn. (5.11) yields

$$\alpha = 17\cdot37 \, \pi \, \epsilon_{eff}'' \, \frac{t}{a} \, \frac{\lambda_g}{\lambda_0'^2} \ \text{dB/m} \tag{5.12}$$

Substituting further for values corresponding to the principal frequencies 896 MHz, 915 MHz and 2450 MHz with the appropriate waveguide sizes (viz. WG4, RG204U, WR975 247·5 × 123·8 mm and WG9A, RG112U, WR340 86·36 × 43·18 mm) (see Appendix VII), eqn. (5.12) becomes

$$\text{For 896 MHz} \qquad \alpha = 891 \, t \, \epsilon_{eff}'' \ \text{dB/m} \tag{5.13a}$$

$$\text{For 915 MHz} \qquad \alpha = 894 \, t \, \epsilon_{eff}'' \ \text{dB/m} \tag{5.13b}$$

$$\text{For 2450 MHz} \qquad \alpha = 7\cdot35 \times 10^3 \, t \, \epsilon_{eff}'' \ \text{dB/m} \tag{5.13c}$$

It will be noted that the attenuation is some eight times greater at 2450 MHz than at the lower frequencies for the same value of $t \, \epsilon_{eff}''$ which arises from two components, the fundamental increase in frequency and the increased 'filling factor' of the waveguide. Also the mechanical length of the meander along the axis of the workload is less at the higher frequency by the square of the frequency ratio giving greater compactness. The attenuation per pass is proportional to the cross-section area of the workload in the direction of travel which can usually be adjusted for powders and particulate material for optimum compromise between uniformity and efficiency.

The above expressions relate the attenuation coefficient per unit length to the thickness and the effective loss factor of the dielectric slab. The corresponding expressions for the fractional loss per meander pass for a slab of 1 mm thickness are

$$\text{At 896 MHz} \qquad 2\alpha l = 0\cdot891 \, (2l) \, \epsilon_{eff}'' \ \text{dB} \tag{5.14a}$$

$$\text{At 915 MHz} \qquad 2\alpha l = 0\cdot894 \, (2l) \, \epsilon_{eff}'' \ \text{dB} \tag{5.14b}$$

$$\text{At 2450 MHz} \qquad 2\alpha l = 7\cdot35 \, (2l) \, \epsilon_{eff}'' \ \text{dB} \tag{5.14c}$$

where l is the width of the processed material, in metres. For example, for a material with $\epsilon_{eff}'' = 1$, eqn. (5.14b) gives, with $l = \frac{1}{2} m$, $2\alpha l = 0\cdot894$ dB. This represents a reduction of about 19% of the input power in one pass. In this case six passes would be required to dissipate most of the available power into the material. Alternatively, if the effective loss factor of the material is only, say, 0·1, substitution in eqn. (5.14b) now yields a fractional loss per pass of 0·089 dB which results in barely more than 2% dissipation of the input power. For such dielectrics meander applicators are not practical and resonant applicators (Chapter 7) are more suitable.

Table 5.1   *Non-uniformity of heating in a meander due to attenuation. No. of passes = 10. Attenuation per pass = 1·0 dB*

Pass No.	Input power/Residual power	Residual power/Input power
1	10·00	7·94
2	6·31	7·94
3	6·31	5·01
4	3·98	5·01
5	3·98	3·16
6	2·51	3·16
7	2·51	1·99
8	1·58	1·99
9	1·58	1·26
10	1·00	1·26
Σ	39·76	38·72
Ratio	1·000	0·974

Table 5.2   *Heating uniformity and number of passes to achieve > 90% efficiency vs. attenuation per pass of meander applicators*

Attenuation per pass dB	No. of passes for 90% efficiency	Uniformity (power ratio) gen. feed side : non-feed side
0·25	40	1 : > 0·99
0·5	20	1 : > 0·99
0·75	14	1 : 0·985
1·0	10	1 : 0·974
2·0	5	1 : 0·882
3·0	4	1 : 0·800
6·0	2	1 : 0·444
10·0	1	1 : 0·100

5.6.5 Uniformity of heating

Having dealt with the effects of standing waves in Section 5.6.3, and with the attenuation per pass in Section 5.6.4 we discuss in this section the overall effect of attenuation on heating uniformity. Such effects are readily calculable using eqns. (5.14) and are demonstrated in Table 5.1 in which a ten pass applicator of 1 dB per pass attenuation is considered. Residual power at the nth pass is fed to the $(n + 1)$th pass and it is assumed that the power absorbed by the workload is a fixed proportion of the incident power in each pass, i.e., the ϵ''_{eff} remains substantially constant during the processing period. By adding the powers at each end of the passes as in the table the sums are proportional to the powers absorbed at the two edges of the workload.

It will be seen that excellent compensation is achieved for attenuation of workload, giving uniform heating within 2·6%. Furthermore, Table 5.2 shows the uniformity ratio and number of passes for 90% efficiency as a function of the attenuation.

In simple heating applications, meander applicators usually have at least five passes, but not more than twenty because if the attenuation per pass falls below 0·5 dB the wall losses become a significant factor in overall efficiency. Moreover, the very large number of reflection components cause difficulty in obtaining a good broad-band impedance match to the generator. In drying applications where the dry matter has a low attenuation it is possible to have less than five passes as the attenuation per pass is then a function of moisture content, and falls with evaporation.

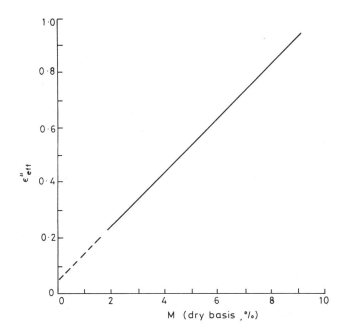

Fig. 5.19 *Effective loss factor as a function of the moisture content in the range 0·9 < f < 3 GHz*

The data of Tables 5.1 and 5.2 have assumed that the effective loss factors of the processed materials remain substantially constant during exposure to microwave energy. In drying operations this is hardly true as the data in Chapter 3 have shown. Let us therefore calculate the effects of a variable loss factor on the attenuation per pass in a meander applicator processing a material of width $\frac{1}{2}$ m and 1 mm in thickness possessing a linear variation of ϵ''_{eff} vs. moisture of the form indicated by eqn. (3.16). Such a variation is shown in Fig. 5.19 for a planar material where it has been assumed that the effective loss factor varies as $(0\cdot05 + 0\cdot1\,M)$, where M is the moisture content (dry basis, see Appendix I). Assuming further that the material shows very little change in its properties between 915 MHz and 2450 MHz

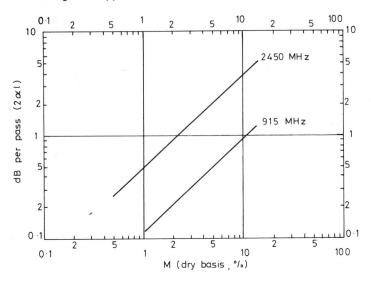

Fig. 5.20 *Attenuation per pass as a function of the moisture content at two frequencies*
$t = 1$ mm, $l = \frac{1}{2}$ m and $\epsilon''_{eff} = 0.05 + 0.1$ M

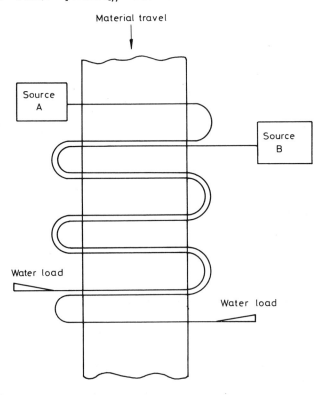

Fig. 5.21 *Two opposite fed sources for more uniform heating in a meander applicator*

the attenuation per pass has been calculated using eqns. (5.14*b*) and (5.14*c*) and is shown in Fig. 5.20. As expected the attenuation per pass reduces as the material dries, an effect which must be taken into account in the design of meander applicators. When the attenuation per pass is so high, for example at 2450 MHz in Fig. 5.20, as to give unacceptable heating non-uniformity it is possible to have two meanders fed from opposite sides of the conveyor as shown in Fig. 5.21. By symmetry, and neglecting non-linear effects, the two edges of the workload then have equal heat energy input, but the centre-line becomes overheated.

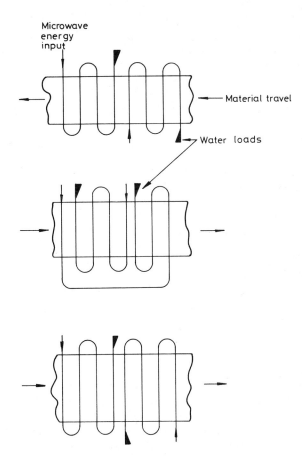

Fig. 5.22   *Various meander applicator configurations*

Alternatively the flow of the material may be reversed through a single meander applicator such that the wet parts enter near the termination where they experience lower electric fields than the dry parts which are, in turn, subjected to higher electric fields at the source end. Various combinations of meander units, some of which are shown in Fig. 5.22, are used to obtain better uniformity of heating

Table 5.3   *Comparison of construction materials*

	Stainless steel	Aluminium	Mild steel aluminium coated	Copper	Brass
Wall loss	High	Low	Moderate	Low	Moderate
Hygiene	Good	Moderate	Poor	Moderate	Moderate
Corrosion	Good	Moderate	Fair	Moderate	Moderate
Thermal conductivity	Poor	High	Moderate	High	Moderate
High temperature °C	Up to 600	80 max	250 max	300	300
Thermal expansion	Moderate	High	Moderate		
Construction method	Fab. weldg.	Fab. weldg. or casting	Fab. weldg.	Fab. and brazing or TIG* welding	As copper

* TIG: Tungsten inert gas.

for processing a variety of planar and other materials of different dielectric properties.

5.6.6 Construction

Although meander applicators appear technically superior at 2450 MHz when compared with those in the 915/896 MHz band, in practice they are significantly more expensive to build because fabrication techniques adequate for the lower frequencies (tolerance ± 1 mm) lack the accuracy required for a 2450 MHz serpentine (tolerance ± 0·3 mm).

Because of its symmetry about the centre plane at the half-height of the process slot, across which no wall currents flow, the serpentine structure can be constructed in two halves with this plane as the joint face. The two halves may be hinged together so that the applicator may be easily opened for access.

Choice of material of construction for a meander applicator is a compromise of wall losses, corrosion, hygiene and surface working temperature. Table 5.3 summarises the principal features of the commonly used materials.

For reasons of safety and efficiency the amount of energy permitted to radiate from the processing aperture of a travelling wave and other applicators must be limited in terms of total power and radiated power density, the levels for safety being outlined in Chapter 10 which reviews recommended and proposed limits on leakage power density. Leakage of energy can be controlled through structures generally referred to as chocking systems, the details of which are also discussed in Chapter 10.

5.7 References

ADAM, S. F., *Microwave Theory and Applications*. Prentice Hall, New York (1969).

ALTMAN, J. L., *Microwave Circuits*. van Nostrand, New York (1964).

BADEN FULLER, A. J., *Microwaves*, 2nd Edition. Pergamon Press, New York (1979).

DUNN, D. A., 'Slow wave couplers for microwave dielectric heating systems', *J. Microwave Power* **2**(1), 7 (1967).

GINZTON, E. L., *Microwave Measurements*. McGraw-Hill, New York (1958).

HARVEY, A. F., *Microwave Engineering*, p. 14 Academic Press, New York (1963).

HEENAN, N. I., 'Travelling wave dryers', in *Microwave Power Engineering* (Edited by E. C. Okress), pp 126−144. Academic Press, New York (1968).

MONTGOMERY, C. G., DICKE, R. H., and PURCELL, E. M., *Principles of Microwave Circuits*. Dover Publications, New York (1965).

MORENO, T., *Microwave Transmission Design Data*. McGraw-Hill, New York (1948).

PÜSCHNER, H., *Heating with Microwaves*. Philips Technical Library (1966).

RAMO, S., WHINNERY, J. R., and VAN DUZER, T., *Fields in Communication Electronics*. John Wiley & Sons, New York (1965).

SUCHER, M., and FOX, J., *Handbook of Microwave Measurements*, Vols. I, II, III. Polytechnic Press, Brooklyn (1963).

WILLIAMS, N. H., and WARNER, H. C., 'Processing of dielectric sheets', in *Microwave Power Engineering* (Edited by E. C. Okress), pp. 175−188. Academic Press, New York (1968).

Chapter 6

Multimode oven applicators

6.1 Introduction

By far the most widely used microwave applicator is of the multimode type, used universally in domestic ovens, and for a large number of low power industrial units and many high power installations. Mechanically simple, it is versatile in being able to accept a wide range of heating loads, although heating uniformity is frequently a problem.

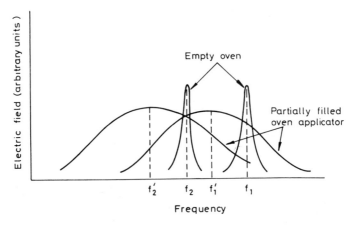

Fig. 6.1 *Frequency shift and damping of mode patterns due to the loading effect of the dielectric in the multimode oven*

In principle the multimode applicator is a closed metal box with some means of coupling in power from a generator, the dimensions of the box should be several wavelengths long in at least two dimensions. Such a box will support a large number of resonant modes in a given frequency range. For an empty applicator each of these modes is characterised by a sharp resonance response at a given frequency, as shown in Fig. 6.1. It is important to arrange for as many of these modes as possible to lie near the operating frequency of the magnetron or klystron source which feeds microwave energy to the oven applicator. However, when such an oven is partly filled with an absorbing workload, the Q-factor of each mode is reduced from that when it is empty and if the spectral density is high enough the resonance curves of the modes will overlap in frequency to give a continuous coupling into the load. As the dielectric constant of the workload will be greater than unity the mode spectral density will also be increased from the empty state which gives additional overlap to the modes. This loading effect is also shown qualitatively in Fig. 6.1 where, apart from mode damping giving rise to lower Q-factors, a frequency shift of the pattern is seen to occur as well.

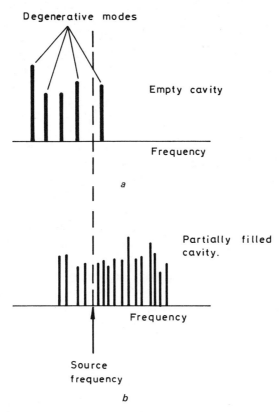

Fig. 6.2 *Effect of shift and split of a number of degenerative modes in an oven applicator*

It should be noted that the modes are not normally evenly spaced in frequency. In an oven applicator of regular geometrical shape (e.g., rectangle, cube) having planes of symmetry there are obvious sets of modes in which pairs or groups share the same resonant frequency referred to as degeneracy, as shown in Fig. 6.2(a). It is therefore desirable to avoid dimensions which bestow symmetry so as to separate these groups in frequency. A matched coupling structure also introduces a certain mode splitting effect of degenerate modes which is a desirable effect since it enhances the uniformity of heating within the applicator. However, the ideal situation would be to increase the mode spectral density of the applicator when it is partially loaded with the material to be heated, as shown in Fig. 6.2(b).

Provided the resonance curves overlap sufficiently to give a good impedance match (VSWR lower than 3:1) in the operating bandwidth of the generator, satisfactory power transfer to the load occurs. For example, in a cavity with a spectral density of five roughly equi-spaced modes in 10 MHz centred on 2450 MHz, provided the average Q-factor of the modes is not more than 1000, a satisfactory impedance match is obtained. As discussed in more detail later in this chapter, such a Q-factor would be obtained from a workload half filling the cavity with $\epsilon' = 3$ and $\tan \delta_{eff}$ of 2×10^{-3} which is low for most microwave heating applications. Obviously, as the $\tan \delta_{eff}$ of the load rises, the impedance match improves and the limiting performance arises when the penetration depth becomes comparable with the dimensions of the workload. In the case of foodstuffs this may not arise until $\tan \delta_{eff}$ is of order unity, so that satisfactory heating performance is possible with materials which have $\tan \delta_{eff}$ values varying over some three orders of magnitude which accounts for the great versatility of the multimode applicator.

6.2 Theoretical aspects of multimode oven applicators

Before we discuss qualitatively the electric field distribution and heating uniformity within a multimode oven applicator let us briefly examine some aspects of multimode theory. Despite the complexity involved in the theoretical treatment of microwave multimode applicators the following gives a fair understanding of their function and performance.

Oven-type applicators are capable of supporting a large number of resonant modes, E_{lmn} and H_{lmn}, within a narrow frequency spectrum, and in the case of a rectangular box with dimensions a, b and d each mode must satisfy the equation (Harvey, 1963)

$$\left(\frac{l\pi}{a}\right)^2 + \left(\frac{m\pi}{b}\right)^2 + \left(\frac{n\pi}{d}\right)^2 = \left(\frac{\omega_{lmn}}{c}\right)^2 \qquad (6.1)$$

where l, m and n are integers corresponding to the number of half-wavelengths of quasi-sinusoidal variation of field along the principal coordinate axes, ω_{lmn} is the angular resonant frequency of the l, m and n mode and c is the velocity of light. Successful operation of the oven depends on the separation of the resonant frequencies, ω_{lmn} being small compared with the bandwidth (i.e., ω_{lmn}/Q_L) of each mode. Maa (1939) and Bolt (1939) have solved eqn. (6.1), a form of which is shown qualitatively in Fig. 6.3. The discontinuous nature of this curve, where there are vertical rises, is due to mode degeneracies resulting in a number of modes at the same frequency. For a cubic cavity operating at exactly 2·45 GHz the maximum value of δn, the step height in Fig. 6.3, corresponds to 18 modes. Since most magnetron sources for ovens operate over a certain bandwidth, which should be maintained within allocated limits to avoid radio interference with communication systems, the total number of modes could therefore be much larger, more like fifty to sixty (James *et al.*, 1968).

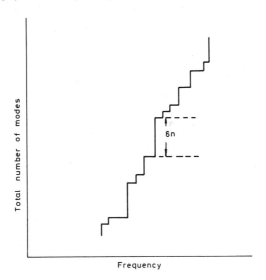

Fig. 6.3   *A form of the exact solution of eqn. (6.1)*
(After Bolt (1939) and Maa (1939), by permission of the Journal of Acoustical Society of America.)

An appreciation of the number of modes within a given cavity can be obtained through eqn. (6.1) which for a cubic cavity attains the form

$$l^2 + m^2 + n^2 = 4(d/\lambda)^2 \tag{6.2}$$

where $\omega/c = 2\pi/\lambda$. It is possible to obtain a mode integer chart, through the various permutations of (l, m, n) which satisfy eqn. (6.2). The various direct and indirect permutations of each set (l, m, n) give rise to first- or second-order degeneracies respectively. By careful choice of the oven dimensions one can obtain, for a given source spectrum, an appreciable mode density which should lead towards a fairly uniform energy distribution.

What is equally important, however, is the study of the frequency spectra, Q-factor (Shen Zhi-yuan, 1980) and mode pattern (Mihran, 1978) with a given lossy dielectric in the oven applicator. The inclusion of such a workload will undoubtedly perturb these parameters and therefore will alter the energy distribution within the oven applicator. Only the simplest of cases has been studied in an oven applicator and compared with measured data. For example, the mode pattern can be investigated using the network analyser as shown in Fig. 6.4. Alternatively, a less versatile but less expensive set-up can be used (Mihran, 1978,

1980). A signal is injected into the oven applicator and the transmitted signal is picked up via a probe and detected in a crystal diode. The transmitted power takes the form shown in Fig. 6.5 in the frequency range 900–1100 MHz for different volumes of water layer placed at the bottom of the oven. For simplicity most of the metallic objects such as mode stirrer (see Section 6.3.2), bake elements, thermostat, etc., were removed from within the oven. The measured data for two modes are shown in Fig. 6.6(*a*) as a function of the ratio of water to oven height, *h/d*. Two

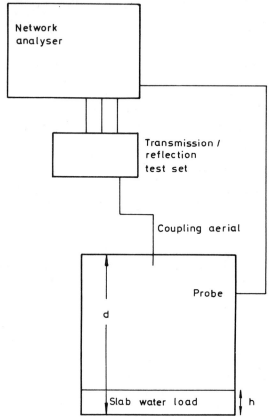

Fig. 6.4   *Analysis of oven applicator modes using a network analyser*

dominant modes both propagating in the *z*-direction are observed each showing different frequency tuning characteristics as the quantity of slab water is increased. The lower mode changes in frequency by as much as 50 MHz for a ratio *h/d* of about 0·24. The measured data have been compared with computed data following solutions of the Maxwell equations in oven applicators containing lossy dielectric slabs and shown in Fig. 6.6(*b*). The modes considered were the TM_{112}, TM_{311} and TM_{130}. Fairly good agreement is obtained particularly for large loads. However, for smaller loads the measured frequency deviations are considerably smaller than calculated.

To study the effects of the mode stirrer on the field distribution Ishii *et al.* (1979) have used the principle of geometrical optics which, apart from identifying design faults in the construction of the mode stirrer, the analysis pointed to ways of correcting.

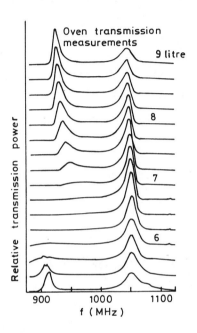

Fig. 6.5 *Transmission power response as a function of the frequency for various slab water insertions in litres*
(After Mihran (1978), Copyright 1978, The Institute of Electrical and Electronics Engineers Inc.)

Even though mode integer charts are useful for appreciating the possible permutations of *l*, *m* and *n* giving rise to a large number of different modes (Püschner, 1966; James *et al.*, 1968), it represents far from the ideal in theoretical desgin due to the complex nature of the combined effects of load perturbation, mode stirrers, methods of coupling the energy, etc. Ideally one requires a provision for altering the oven dimensions. However, although such an arrangement has been used in specific scientific applications (Spiegel *et al.*, 1980), it is too cumbersome and costly to apply to industrial oven applicators. Therefore in the first instance the dimensions of an oven applicator are chosen by a combination of the theoretical aspects discussed above coupled with the nature and size of material to be processed and other practical considerations.

6.3 Field distribution and heating uniformity

In a closed rectangular box the field distribution is given by the sum of all the modes excited at a given frequency, each mode giving a basic sinusoidal power variation in space along the principal coordinate axes, and satisfying the well-known field equations. There is, therefore, fundamentally a spatial non-uniform distribution of heating within a multimode oven. Although the presence of an absorbing

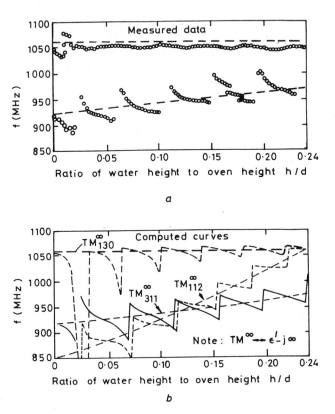

Fig. 6.6 *Comparison of experimental (a) and computed (b) data for a multimode oven*
(After Mihran (1978), Copyright 1978, The Institute of Electrical and Electronics Engineers Inc.)

workload reduces the VSWR of the resultant standing wave pattern, the non-uniformity remains a major feature, and much consideration has been given to minimise its effect. A further aspect is that a fair number of materials to be heated exhibit a rising loss-factor with temperature which causes the effect of non-uniformity to be amplified.

Several methods are available for improving the uniformity of heating.

6.3.1 Movement of the workload

This is the most obvious and effective technique since the position of the standing waves within the oven is determined mainly by the cavity walls and so to move the load inside the cavity so that it travels through the antinodes is clearly advantageous.

For straight line motion along one of the principal axes of a rectangular applicator only the field variations along that axis are compensated by this technique and in principle the vertical and traverse variations remain. One method of compensating these two components is to design the oven and conveyor so that the axis of travel is inclined to the principal axes of the oven.

In most small ovens movement of the workload is accomplished by placing it on a rotating turntable. This again provides an improvement in heating uniformity but it suffers from similar geometrical inexactitudes as a linear motion.

In practice the effect of moving the workload in the oven is more complicated than indicated above, because the characteristic impedance of the workload is substantially different from free space, giving rise to a shortening of wavelength within the material, and secondary reflections from the surfaces of the workload. There is therefore a perturbation of the field within the oven which will alter as the workload moves unless of course the load is geometrically continuous in the direction of travel (e.g., a continuous sheet for linear motion, or a concentric circular disc or ring for a turntable). In these circumstances the standing waves themselves are moved as the workload moves in the oven. Usually this is not advantageous because the standing waves tend to be located with respect to the boundaries of the load, an effect which becomes more severe as the dielectric constant increases, and gives rise to an effect described as 'field trapping'.

6.3.2 Mode stirrers

A mode stirrer is a moving device introduced into the oven to perturb the field distribution continuously. Usually it consists of a metal multi-blade fan rotated inside the oven applicator. Although a crude device it contributes significantly to the achievement of uniform heating, but its action is too complicated to warrant analysis. However, there are two ways in which it operates depending on the dielectric losses of the workload.

When the workload is highly absorptive the blades, usually bent to an angle of about $45°$, deflect the energy incident upon them from the generator feed, and since they are rotating the path of the reflected energy is constantly changing. In many ovens the blades pass close to the generator feed so as to maximise this action. For a low loss high workload the blades are tuned to resonance so that they behave as rotating parasitic antennae, exciting different modes to varying extent as they rotate.

There are many empirical designs of mode stirrer which generally aim to combine the two above-mentioned actions. Choice of rotation speed for a mode stirrer is a compromise of mechanical design (bearings and life) and process requirements. The product $1/N_0 \alpha_r$ (where N_0 is number of blades and α_r is number of revolutions per second) should be small compared with process time, the thermal time constant of a hot spot in the workload, and the thermal time constant of the magnetron cathode. Typically a four-blade mode stirrer rotates at 1–10 rev/sec.

6.3.3 Other methods of energy feed

A rotating antennae of narrow beamwidth set in the top of an oven provides the same effect as a rotating turntable and has particular value in small ovens where the dielectric insertions give low values of Q-factor. Alternatively, instead of a single slot or aerial feeding the microwave oven, a marked improvement in uniformity can be achieved by multi-slotted feeds beaming the energy through a rotating stirrer. This combination has a similar effect to having a distributed source system. A four slot waveguide feed has recently been described, the slots spaced λ_g apart and designed for equal impedance and equal free space radiation. A complementary stirrer was used in conjunction with such a feed which effectively blocked the radiation from two slots for a fraction of the heating period and the remaining two slots for another fraction of the heating period. The shape and position of the stirrer played a critical role and had to be optimised (Kashyap and Wyslouzil, 1977). Meredith (1979) achieved good heating uniformity with a plane polarised horn feeding a transverse E-field to a multimode oven heating a continuous flow web material.

6.4 Q-factor with a dielectric insertion

It will be evident from the above that an analysis of the field distribution and heating uniformity within a multimode oven is extremely complicated, and as a result, apart from the simplest of cases, empirical trials are universally necessary as part of the design of an oven. Nonetheless, simple calculations are possible to establish approximate values for the Q-factors of the modes, the electric and magnetic field intensities inside the oven and the amplitude of the wall currents.

An estimate of the Q-factor with the dielectric inserted into the oven can be made by expression (7.1) which is developed more fully in the next chapter:

$$Q = 2\pi \frac{\text{energy stored}}{\text{energy lost per cycle}} \tag{7.1}$$

If the cavity is completely filled with a dielectric the stored energy is given by

$$\int_{V_c} \tfrac{1}{2}\,\epsilon_0\,\epsilon'\,\hat{E}^2 dV \tag{6.3}$$

where $\hat{E}$ is the peak electric field intensity at a point in the cavity, and V_c is the volume of the oven cavity. However the energy lost per cycle in the dielectric derived using eqn. (4.10) is given by

$$\frac{2\pi}{\omega} \int_{V_L} \tfrac{1}{2}\,\omega\epsilon_0\,\epsilon'\,\tan\delta_{eff}\,\hat{E}^2 dV \tag{6.4}$$

Substitution of eqns. (6.3) and (6.4) into eqn. (7.1) yields

$$Q \approx \frac{1}{\tan \delta_{eff}} \qquad (6.5)$$

which is the same expression as that derived in Section 3.7. The wall losses have been ignored since, for medium lossy dielectrics, these are much smaller than losses in the workload. Relation (6.5) is independent of the dielectric constant because both the stored energy and dissipated energy are linearly proportional to it. However, the presence of the dielectric increases the effective volume of the cavity by shortening the effective wavelength, giving a corresponding increase in mode spectral density.

Where the cavity is partly filled the analysis is more complicated. Let v be the filling factor of the workload, V_L its volume and V_c the cavity volume so that $v = V_L/V_c$ and $\hat{E}_i$ and $\hat{E}_{ext}$ are the respective fields within the workload and external to it. Once again using eqn. (7.1) and ignoring wall losses we obtain

$$Q = 2\pi \frac{\int_{V_L} \frac{1}{2} \epsilon' \epsilon_0 \hat{E}_i^2 \, dV + \int_{V_c'} \frac{1}{2} \epsilon_0 \hat{E}_{ext}^2 \, dV}{\int_{V_L} \frac{1}{2} \omega \epsilon_0 \epsilon' \hat{E}_i^2 \tan \delta_{eff} \, dV} \cdot \frac{\omega}{2\pi}$$

$$= \frac{1}{\tan \delta_{eff}} + \frac{1}{\epsilon' \tan \delta_{eff}} \cdot \frac{\int_{V_c'} \hat{E}_{ext}^2 \, dV}{\int_{V_L} \hat{E}_i^2 \, dV} \qquad (6.6)$$

where V_c' is the volume integral relating to the clearance volume in the cavity $(V_c - V_L)$. For a fully filled cavity the integral in the numerator is zero giving the initial simple relation (6.5) and for an empty cavity the integral in the denominator is zero, yielding an infinite Q, as would be expected, since the wall losses have been ignored.

In practice the ratio of the two integrals depends on the dielectric constant of the load, its shape, size and its orientation and position within the oven, as all these factors affect $\hat{E}_{ext}$ and $\hat{E}_i$. We assume that the load has dimensions at least equal to $\lambda_0/\sqrt{\epsilon'}$ along the principal axes. In order to calculate the effect on the Q-factor due to the partial filling of the oven applicator let us assume that the fields $\hat{E}_{ext}$ and $\hat{E}_i$ are constant in space and that for a spherical load, as shown in Fig. 6.7, these two fields are related according to eqn. (4.42). Substitution of eqn. (4.42) in eqn. (6.6) yields the unloaded Q-factor*

$$Q_0 = \frac{1}{\tan \delta_{eff}} \left(1 + \frac{(\epsilon' + 2)^2}{9 \epsilon'} \frac{(1 - v)}{v} \right) = \frac{f''}{\tan \delta_{eff}} \qquad (6.7)$$

---

* This Q-factor has in the past been referred to as the loaded Q, Q_L of the cavity. However we have reserved the term 'loaded' Q to refer to a cavity physically connected to the external circuit supplying the power to the cavity. Since the above calculations neglect the coupling to the external circuit but do take into consideration a dielectric workload we call this the unloaded Q-factor Q_0 with the material inserted in the cavity. The symbol Q_0' is reserved for the Q-factor of an empty cavity (see Section 7.2).

where f'' is a geometrical factor depending upon the internal cavity and dielectric dimensions. Figures 6.8, 6.9 and 6.10 show some computed data for Q for a range of values in ϵ', $\tan \delta_{eff}$ and v using eqn. (6.7). As the filling factor and $\tan \delta_{eff}$ increase the Q factor decreases. Figure 6.11 shows that as the dielectric constant increases for a given $\tan \delta_{eff}$ and filling factor, v, the Q once again decreases.

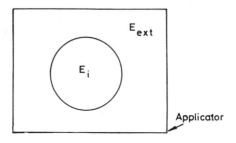

Fig. 6.7   *Spherical load in a multimode oven applicator*

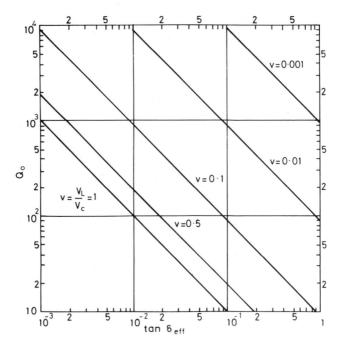

Fig. 6.8   *Q-factor of a partially filled oven applicator as a function of $\tan \delta_{eff}$ for various v and $\epsilon' = 1 \cdot 5$ calculated using eqn. (6.7)*

To a first approximation eqn. 6.7 provides a guide to the likely value of Q in partially filled multimode cavity, knowing the dielectric properties and all the dimensions. For example consider a block of natural rubber ($\epsilon' \approx 3$ and $\tan \delta_{eff} =$

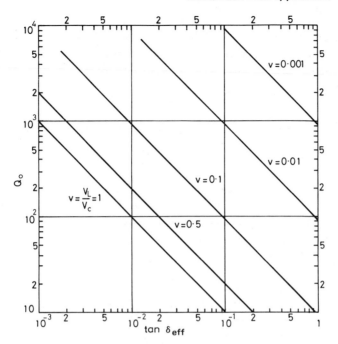

Fig. 6.9 *Q-factor of a partially filled oven applicator as a function of tan δ_eff for various v and ε' = 3 calculated using eqn. (6.7)*

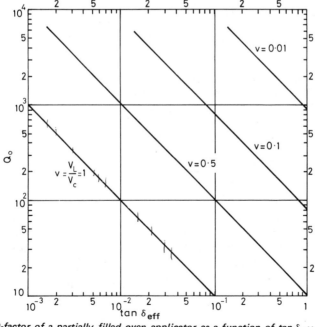

Fig. 6.10 *Q-factor of a partially filled oven applicator as a function of tan δ_eff for various v and ε' = 80 calculated using eqn. (6.7)*

3×10^{-3}) 1 m × 1 m × 0·7 m in size inserted in an industrial oven of dimensions 1·2 m × 1·2 m × 1 m. Figure 6.8 with $v \approx 0·5$ gives $Q_0 \simeq 650$. At 2450 MHz this gives a half power bandwidth of about 3·8 MHz which is more than the average mode spectral density in a cavity of the above dimensions half filled with dielectric material of $\epsilon' = 3$.

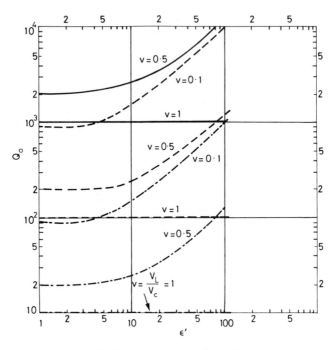

Fig. 6.11   *Q-factor of a partially filled oven applicator as a function of the dielectric constant ϵ' for various v and three tan δ_{eff} values calculated using eqn. (6.7)*
———— tan $\delta_{eff} = 10^{-3}$
———— tan $\delta_{eff} = 10^{-2}$
—·—·— tan $\delta_{eff} = 10^{-1}$

Similarly, consider a partial filling of a wet spherical dielectric of say 270 mm in diameter in a multimode oven of 0·45 m × 0·45 m × 0·5 m in dimension. Assuming the dielectric properties at 2450 MHz to be $\epsilon' = 10$ and tan $\delta_{eff} = 0·1$, Fig. 6.11 gives with $v = 0·1$ a Q-factor of about 155.

Equation (6.7) is based on the assumption that the microwave field fully penetrates the dielectric workload, indeed there being a high transmission of energy through it. For many workloads with high tan δ_{eff} the penetration depth is limited, as discussed in Section 4.3. Where such a dielectric insertion has dimensions comparable with or exceeding the penetration depth, and at the same time has a value of ϵ^* which is not so extreme as to give near total reflection of energy at the surface (e.g., a metal), then the resonant build-up of energy in the cavity does not occur to a significant extent. For example, in the case of a quantity of water of $275 \, cm^3$ used in a standard test on a domestic microwave oven, the oven can be considered aperiodic. In practice it is not important to define the conditions at which eqn. (6.7) is inapplicable because under aperiodic or very low Q conditions the E and H field intensities are readily calculable and at values which normally are easily handled. It is under high Q conditions where eqn. (6.7) applies and has great value to the oven design engineers.

It must also be emphasised that eqn. (6.7) is not valid for small filling factors, v, because as the wall losses have been ignored, eqn. (6.7) will give unrealistically high values of Q. To account for the wall losses eqn. (6.7) modifies to (Cook, 1973)

$$\left(\frac{1}{Q_0} - \frac{1}{Q_0'}\right)^{-1} = \frac{f''}{\tan \delta_{eff}} \tag{6.7a}$$

where Q_0' is the Q-factor of the empty cavity when the only form of power dissipation occurs on the metallic walls in the interior of the applicator.

6.5 Field intensity and wall currents

Calculation of the field intensity and wall currents is based on consideration of stored and dissipated energy as before, and provides data for assessing $I^2 R'$ heating effects in the walls, door chokes and contact gaskets, and electric field stress.

The stored energy U in terms of the electric and magnetic fields is

$$U = \int_{V_c} \tfrac{1}{2} \mu_0 \mu' H^2 dV = \int_{V_c} \tfrac{1}{2} \epsilon_0 \epsilon' E^2 dV \tag{6.8}$$

where E and H are the maximum time values of the electric and magnetic fields respectively and are variable in space.

In a resonant cavity the resultant fields are the vector sums of equal forward and reverse waves giving rise to a three-dimensional standing wave-pattern. Considering a one-dimensional variation, the vector diagram of Fig. 6.12 shows the resultant field H_{res} and the forward and reverse waves H_+ and H_-. The angle $\theta = 4\pi x/\lambda$ changes with position x along the axis. From Fig. 6.12 we obtain

$$H_{res}^2 = 2H_+^2 \left(1 + \cos\frac{4\pi x}{\lambda}\right) \tag{6.9}$$

By superposition H in eqn. (6.8) is the resultant of H_{res} in the three coordinates so that the total stored energy is therefore

$$U = \tfrac{1}{2} \mu_0 \mu' \int_0^a \int_0^b \int_0^d 2H_+^2 \left(1 + \cos\frac{4\pi l x}{\lambda}\right) \left(1 + \cos\frac{4\pi m y}{\lambda}\right)$$
$$\left(1 + \cos\frac{4\pi n z}{\lambda}\right) dx \, dy \, dz \tag{6.10}$$

where a, b and d are the oven dimensions and $V_c = abd$. Since the field pattern consists of integral numbers of half-lengths in each coordinate direction within the cavity the cosine integrals are all zero. Hence

$$U = \mu_0 \mu' H_+^2 \, abd \tag{6.11}$$

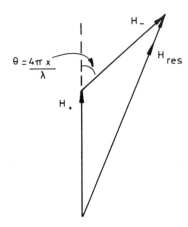

Fig. 6.12 *Superposition of the forward H_+ and the reflected H_- magnetic fields*

In terms of the maximum value in space, $H_{max} = 2H_+$, eqn. (6.11) yields

$$U = \tfrac{1}{4} \mu_0 \mu' H_{max}^2 \, abd \tag{6.12}$$

Now eqn. (7.1) can be expressed as

$$Q_0 = 2\pi \frac{\text{energy stored}}{\text{energy lost per cycle}} = 2\pi \left(\frac{U}{P}\right)\left(\frac{\omega}{2\pi}\right) = \frac{\omega U}{P} \tag{7.1a}$$

where P is the dissipated power.

Substitution of eqn. (6.12) into eqn. (7.1a) yields

$$H_{max} = 2\left(\frac{PQ_0}{\omega\mu_0\mu' abd}\right)^{1/2} = 2\left(\frac{PQ_0}{\omega\mu_0\mu' V_c}\right)^{1/2} \tag{6.13}$$

Since H_{max} is the maximum space value which occurs at the surfaces of the cavity, it is also the wall-current density in the appropriate units. Hence

$$J = 2\left(\frac{(P/V_c)Q_0}{\omega\mu_0\mu'}\right)^{1/2} \text{ A/m} \tag{6.14}$$

which shows that the current density J at a given ω is proportional to the square roots of the power density in the oven and the Q-factor.

As an example, consider the previous case of the multimode oven (1 m × 1 m × 0·7 m) with the rubber dielectric inserted within it. Previous calculation gave $Q = 650$ with $v = 0.5$. For a dissipation into the rubber of 1 kW at 2450 MHz and putting $\mu' \mu_0 = 4 \times 10^{-7}$ H/m substitution in eqn. (6.14) gives a wall current at density of 9.6 A/m or about 0·1 A/cm. For an increased dissipation of 10 kW the wall current, as given by eqn. (6.14), increases to about 30 A/m. Such currents are modest and present little difficulty in the design of the oven body, door chokes and contact gaskets.

6.6 Power density

An important parameter in the design of a multimode oven is the power volume density (units, kW/m^3) at which the oven is to operate, and there is the related question of how much power can be fed to a given oven before its performance becomes unsatisfactory. In practice there are two principal and separate limits to power density.

(a) Dielectric breakdown of the air or gas—vapour mixture in the oven.
(b) Destructive damage to the workload from excessive internal mechanical stress due, for example, to internally generated steam.

The second of these clearly depends solely on the material of the workload, its tensile strength, porosity, etc., and experimental evaluation is the only satisfactory method of setting the limit on power density in these cases.

An average value for the electric field stress in the volume outside the dielectric is readily obtained from eqn. (6.13) which after substituting for H_{max} through the well-known relation for plane waves

$$E_{max}/H_{max} = (\mu_0/\epsilon_0)^{1/2} \tag{6.15}$$

yields

$$E_{max} = 2 \left(\frac{(P/V_c)Q_0}{\omega \epsilon_0 \times 10^6} \right)^{1/2} \text{kV/m} \tag{6.16}$$

In the previous multimode cavity of $0·1 \text{ m}^3$ in volume operating at 2450 MHz with a spherical dielectric insertion of 0·27 m in diameter having $\epsilon' \simeq 10$ and $\tan \delta_{eff} = 10^{-1}$ the data of Fig. 6.10 gave a $Q_0 \simeq 155$. Substitution in eqn. (6.16) yields an electric field of 21·2 kV/m or 212 V/cm with a 10 kW power dissipation. At 30 kW and with otherwise similar conditions the electric field stress will be about 36·67 kV/m or 367 V/cm.

This expression must once again only be used to obtain a guideline figure because the electric field locally may be higher due to the conventional electro-static stress concentrations on corners, points and edges of the oven and of its workload. Particulate materials of high dielectric constant are specially prone to dielectric breakdown due to stress concentration at the points of contact between the particles; such local stress concentration may also cause excessive heating within the particles at the points of contact resulting in local heating of the air (by conventional heat transfer) with resultant reduction in its dielectric strength.

From the above it will be seen that the maximum permissible power density for satisfactory processing is much dependent on the conditions, but it should be stressed that in most cases it is not necessary to operate a plant close to the dielectric breakdown limit. In practice larger electric field strengths will be obtained in multimode ovens operating at the 900 MHz band on account of the higher power densities and smaller operating frequency compared to the 2450 MHz. As an example, consider 30 kW being dissipated in a material placed in a $\frac{1}{2}$ m³ multimode oven giving $(P/V_c) = 60$ kW/m³ and operating at 900 MHz. Substitution in eqn. (6.16) yields $E = 2 \cdot 2 \times 10^3 \ (Q_0)^{1/2}$. Assuming that the highest Q likely to be met in a particular application is about 1000, the maximum electric field stress calculated using eqn. (6.16) is $E \simeq 70$ kV/m or 700 V/cm. The uniform-field breakdown stress of air at normal ambient temperature and pressure is 3000 kV/m or 30 kV/cm giving an indication of the factor of safety (see Section 9.3.6).

6.7 Choice of wall material

The ideal requirements for the structure of a multimode oven applicator are that it should form an electrically closed volume with a material of infinite conductivity at all points. Current flows in the walls of the oven in a sheet and must not be impeded at any point if there is not to be a risk of local overheating, destructive arcing and a lowering of overall efficiency. It follows that all joints and seams in a multimode oven must be of the highest integrity and should be minimised in extent by careful design.

The choice of materials of construction is affected by limits to the overall efficiency in relation to the Q-factor of the product to be heated, and by other considerations such as hygiene and durability. For loads in which the loaded Q-factor is less than 100, non-magnetic stainless steel is an excellent material, being hard and substantially free of corrosion problems. However, for loaded Q-factors of over 200 the wall currents are such that stainless steel heats significantly resulting in a possibly unacceptable reduction in efficiency or an excessive temperature rise. In this case, as was explained in Section 4.8, materials of lower resistivity such as aluminium or copper must be used.

Aluminium can be used either as solid sheet or as a sprayed coating on steel. The solid sheet is much preferable electrically but in industrial plant the sheet must be thick for mechanical reasons. A sprayed metallic coating of aluminium has also been used but surface roughness impairs the effective resistivity slightly and is unsatisfactory hygienically. Adhesion of the coating and wear can also be a problem with sprayed aluminium. Copper is only recommended where the highest efficiency is required and cost is of secondary importance. Plated copper is not satisfactory due to oxidation, corrosion and wear and is only used for experimental plant.

6.8 Doors and door seals

In industrial equipment the door to the oven applicator may be single entry, or double entry having a door at each end of a tunnel. Many mechanical arrangements exist, doors being either hinged or sliding and the precise arrangement is determined by the mechanical convenience of the application. Hinged doors are generally inconvenient except on small ovens manually loaded, and for large industrial ovens vertical sliding doors are generally preferred since they do not impede the loading and unloading areas.

The principal requirement of the doors of an industrial microwave oven applicator are:

(*a*) To provide a closure from which ideally no microwave energy can escape in the interests of safety, efficiency, and in some cases radio interference. The integrity of the seal should remain throughout its life even when slightly damaged or contaminated with dirt.

(*b*) To provide internally a conducting surface identical to the body of the oven.

(*c*) To create a seal to the body of the oven with an electrical admittance as high as possible.

(*d*) To have a durable seal capable of carrying the wall currents imposed on it without risk of overheating or arcing, and able to withstand a large number (10^5 or 10^6) of openings and closings without wear.

(*e*) To provide a fail-safe interlock system which positively prevents microwave power being supplied to the oven unless all doors are closed and fully clamped.

By careful design these exacting requirements can be met. Obviously the most critical part of the door is the RF seal, usually based on arrays of quarter-wavelength choke ditches arranged to provide a virtual short circuit at the point on the door—oven interface where the main wall currents of the oven flow (Bucksbaum, 1977; Püschner, 1966). A novel choke structure has been reported (see Section 10.5.1) which utilises a periodic square wave shaped metallic sheet inserted into a groove of a choke and limits radiation to extremely low levels (Ohkawa *et al.*, 1978). Some earlier designs without chokes employed a plain metal-to-metal contact using spring fingers of beryllium copper to provide a continuous contact at the interface, but this arrangement is not entirely satisfactory because the gasket carries the main wall current and is rapidly damaged by arcing if it fails to make good contact even locally.

A typical choked door seal with its equivalent circuit is shown in Fig. 6.13. This seal makes use of the impedance transformation characteristics of a quarter-wavelength transmission line having two sections in cascade of which the first is terminated at a short circuit (at the bottom of the ditch) giving a virtual short circuit at the remote end of the second section by impedance transformation. Note that at the electrical mid-point of the choke, i.e., the junction between the two quarterwave sections, the impedance is very high giving a very small current.

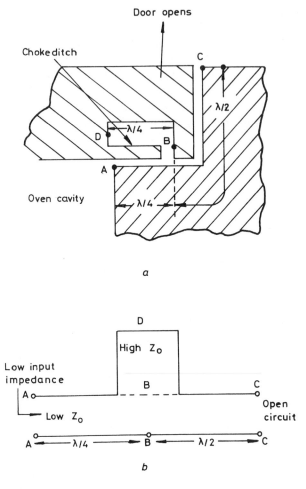

a

b

Fig. 6.13 *Typical multimode applicator door choke*
  a Plan cross-section
  b Equivalent circuit

An imperfect contact at this point is therefore acceptable; in some designs contact is specifically eliminated by allowing current to flow across a capacitive gap or by providing a low impedance path by a secondary choking system. Although the contactless choke is excellent in its relative freedom from the adverse effects of dirt, it is not a complete seal and it does allow some small, safe leakage which may make it unable to meet the most demanding interference-based leakage specification. When a total seal is required a metal-to-metal contact is essential, which is achieved by a contact gasket at the high impedance point where the current flow is small, imposing a light duty on the seal. In most cases such a seal can be of woven stainless steel wire.

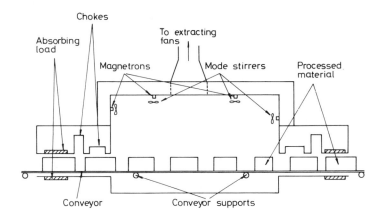

Fig. 6.14 *On-line multimode oven applicator with safety measures at the input and the output terminals*

In the case of on-line multimode oven applicators, there are no doors since openings are required at both ends for entry and exit of the continuous load. The methods used for safety, some of which are shown in Fig. 6.14 are similar to those employed in other on-line microwave applicators such as meanders and the details of the chokes used can be found in Section 10.5.2. Figure 6.14 shows a typical multimode tunnel applicator for processing bulk objects employing arrays of multiple generator feeds, as is described in the following section.

6.9 Multiple generator feeds

Industrial applications frequently require more power fed to the oven than is available from a single generator so that several sources must be used. An arrangement with four magnetrons feeding an applicator oven is shown in Fig. 6.14. In these circumstances it is necessary to evaluate the cross-coupling occurring between generators and its effect on generator life and uniformity of heating.

Each generator is subjected to incident (i.e., reflected) energy due to the imperfect match 'seen' by its own generated energy, plus a proportion of the energy of all the other generators. The total of these energy components must not exceed the permissible reflected energy the generator will accept in accordance with its own specification. An estimate of the cross-coupling effect can be made by calculating the wall current in the oven cavity by eqn. (6.14), and comparing it with the wall current at the feed aperture of a generator associated due solely to its own output power, so as to obtain a value for the effective reflection coefficient 'seen' by that generator. There is no fundamental limit to the number of generators which can feed a multimode oven, the number being controlled by the power density and Q_L of the cavity. Ovens of 75 kW have been built at 896/915 MHz using three magnetron generators, and at 2450 MHz ovens of 54 kW have been built using thirty-six magnetrons of 1·5 kW output.

With multiple generators the opportunity exists to distribute the power so as to give a better excitation of modes and better uniformity than can be achieved with a single feed, by distributing the feed points around the walls of the oven and by feeding at different polarisations.

6.10 References

BOLT, R. H., 'Frequency distribution of eigentones in a three-dimensional continuum', *Accoust. Soc. Am.* **20**, 228 (1939).

BUCKSBAUM, A. M., 'A durable microwave oven door', *J. Microwave Power* **12**(4), 293 (1977).

COOK, R. J., *Microwave Cavity Methods in High Frequency Dielectric Measurements* (Edited by J. Chamberlain and G. W. Chantry), pp 12–27. IPC Science and Technology Press Ltd, New York (1973).

HARVEY, A. F., *Microwave Engineering*, p 193. Academic Press, New York (1963).

ISHII, T. K., YEN, Y. H., and KIPP, R. J., 'Improvement of microwave power distribution by the use of the first order principle of geometrical optics for scientific microwave oven cavity', *J. Microwave Power* **14**(3), 201 (1979).

JAMES, C. R., TINGA, W., and VOSS, W. A. G., *Energy Conversion in Closed Microwave Cavities, Microwave Power Engineering* (Edited by E. C. Okress), Vol. 2, pp. 28–37. Academic Press, New York (1968).

KASHYAP, S. C., and WYSLOUZIL, W., 'Methods for improving heating uniformity of microwave ovens', *J. Microwave Power* **12**(3), 223 (1977).

MAA, D. Y., 'Distribution of eigentones in a rectangular chamber at low frequency range', *J. Accoust. Soc. Am.* **10**, 235 (1939).

MEREDITH, R. J., Magnetronics Ltd., Unit A, St Mary's Mills, Evelyn Drive, Leicester (1979).

MIHRAN, T. G., 'Microwave oven mode tuning by slab dielectric loads', *IEEE Transactions of Microwave Theory and Techniques* **MIT-26**, No. 6, 381 (1978).

MIHRAN, T. G., 'Measured TE and TM mode characteristics in slab loaded microwave oven cavities', *Proc. Microwave Power Symposium*, pp. 97–97c. Iowa City, Iowa (1980).

OHKAWA, S., WATANABE, M., and KANETO, K., 'High performance door seal for microwave oven', *Proc. IMPI Conference*, pp 2–4. Ottawa (1978).

PÜSCHNER, H. A., *Heating with Microwaves*. Philips Technical Library (1966).

SHEN ZHI-YUAN, 'An analysis of the loading effects of the box type microwave applicator', *J. Microwave Power* **15**(3), 155 (1980).

SPIEGEL, R. J., OAKEY, W. E., BRONAUGH, J. E. L., and KERNS, B. R., 'A dual mode variable volume cavity for irradiating biological objects', *Proc. Microwave Power Symposium*, pp. 48–51, Iowa City, Iowa (1980).

Chapter 7

Single mode resonant cavities

7.1 Introduction

Fundamental and higher order single mode microwave resonant structures are widely used in many branches of microwave engineering at low power such as frequency counters, interferometers or filters. Essentially a resonant cavity, or heater in the present context, consists of a metallic enclosure into which a launched microwave signal of the correct electromagnetic field polarisation will suffer multiple reflections between preferred directions. The superposition of the incident and reflected waves gives rise to a standing wave pattern which for some simple structures is very well defined in space. The precise knowledge of electromagnetic field configurations enables the dielectric material under treatment, to be placed in the position of maximum electric field for optimum transfer of the electromagnetic energy to it. A magnetic field is also set up inside the cavity with its maximum located at a different position to the electric field maximum which would be of use to heating magnetic materials. These cavities represent volumes of large stored energy which is transformed into heat via displacement and conduction currents flowing through the dielectric material as soon as it is placed within the heating zone. High absorption of the incident energy is a feature of these cavities. However, operation must be within narrow frequency bands in order to maintain high coupling efficiencies.

In general for the same power applied, a single mode resonant heater will establish much higher electric field strengths than a travelling wave or multimode applicator and for this the former is most useful for the treatment of low loss dielectrics. These fundamental mode heaters are in general more compact with extremely high power densities (10^7 kW/m^3). In the early evolution of microwave heating such cavities made a very small impact on the industrial scene because they lacked the versatility inherent in multimode or non-resonant microwave applicators. Moreover the development of associated microwave circulators and electronic control systems for automatic tuning and matching to make them workable in an industrial plant lagged behind. However, there are a number of specialised industrial applications which will benefit greatly from the unique advantages offered by single mode microwave resonant applicators.

7.2 Cavity modes and Q-factors

A resonant heating structure consists of a dielectric medium surrounded by conducting walls. The Maxwell equations contain all the parameters needed to define
the standing wave pattern that can be set up in a particular case. Solutions of the
Maxwell equations result in electromagnetic fields in time and space that satisfy
the existing boundary conditions. These solutions are called the normal modes
of the cavity and their number can be very large. Perhaps one would have thought
that no fields will be excited in a cavity, via an appropriate aperture, unless the
excitation frequency exactly matches the resonant frequency. In fact this is not
so, and a delta singularity does not occur around the resonant frequency, rather
there is a band of frequencies within which appreciable excitation occurs. The
origin of this spreading of the excitation frequency is the dissipation of energy
in the walls and in the dielectric within the cavity. A measure of the sharpness
of response of a cavity to external excitation is the quality factor Q which is
defined as

$$Q = 2\pi \frac{\text{total energy stored}}{\text{energy dissipated/cycle}} = \frac{\omega U}{P} \tag{7.1}$$

where P is the power dissipated in the cavity.* The energy stored in the cavity,
U, is independent of time and can be defined in terms of either the electric or
the magnetic fields since the energy is equally divided between the two, averaged
over a cycle. The energy dissipated includes wall losses as well as in the dielectric
within the cavity. Therefore treating the cavity in isolation from the rest of the
external circuit, the unloaded Q can be defined as

$$Q_0 = 2\pi \frac{\text{total energy stored}}{\text{energy dissipated in the walls and the dielectric/cycle}}$$
$$= \frac{\omega U}{P_s + P_w} \tag{7.2}$$

where P_s and P_w are the powers dissipated in the walls and dielectric insertion
respectively. A Q-factor can be assigned to an empty cavity, this being defined as

$$Q_0' = \frac{\omega' U}{P_s} \tag{7.3}$$

where ω' is the angular frequency of the same cavity without any dielectric inserted
within it. Assuming that $\omega \approx \omega'$, eqns. (7.2) and (7.3) yield, after inverting and
re-arranging,

$$\frac{1}{Q_0} = \frac{1}{Q_d} + \frac{1}{Q_0'} \tag{3.20}$$

* This is the same value of Q which is used for lumped element circuits where the definition
is more commonly $f_0/\Delta f$, where f_0 is the resonant frequency and Δf is the frequency separation at the half power points (see also eqn. (7.20)).

which is eqn. (3.20) with $Q_d \ (= \omega U/P_w)$ being the Q-factor due to the dielectric loading.

The characteristics of the cavity must be considered in terms of the circuit parameters external to the cavity for which two extra Q-factors are normally used, defined as

$$Q_{ex} = 2\pi \ \frac{\text{total energy stored}}{\text{energy dissipated in the external circuit/cycle}}$$

$$= \frac{\omega U}{P_{ex}}$$

(7.4)

and

$$Q_L = 2\pi \ \frac{\text{total energy stored}}{\text{energy dissipated in both the cavity and the external circuit/cycle}}$$

(7.5)

$$= \frac{\omega U}{P_{ex} + P_s + P_w}$$

where Q_{ex} and Q_L are the external and loaded Q-factors respectively and P_{ex} is the power dissipated in the external circuit.

It must be emphasized that the term 'loaded cavity' refers to the cavity physically connected to an external circuit whereas unloaded cavity refers to the cavity which is not connected to an external circuit but takes into consideration the insertion of the dielectric. It follows from the above that

$$\frac{1}{Q_L} = \frac{1}{Q_0} + \frac{1}{Q_{ex}}$$

(7.6)

7.3 Impedance matching

The behaviour of resonant cavities as elements in transmission lines can be analysed by means of equivalent shunt or series circuits, each representing the impedance of the cavity at specific planes on the connecting transmission line or waveguide. At a voltage node (standing wave minimum) a cavity detuned far away from resonance, where its admittance is very high, should be represented by a shunt equivalent circuit and this node position on the waveguide is referred to as the detuned short position. At a quarter of a wavelength away from this position the circuit should have a high impedance when the cavity is detuned sufficiently away from resonance and the appropriate representation at this position (voltage anti-node) is a series circuit. In the analysis that follows we will make use of the shunt equivalent circuit shown in Fig. 7.1(a) and the circuit referred to the primary of the transformer (coupling aperture) at the detuned short position, as shown in Fig. 7.1(c). The shunt resistance R_c' consists of losses in the dielectric as well as in the walls of the cavity. If the cavity is connected to a waveguide of characteristic impedance, Z_0, a coupling parameter β is defined as

$$\beta = R'_c/Z_0 \tag{7.7}$$

that is, the ratio of the cavity resistance to the waveguide characteristic impedance and controls the energy transfer from the power source to the heating cavity. For the condition

$$R'_c = Z_0 \tag{7.8}$$

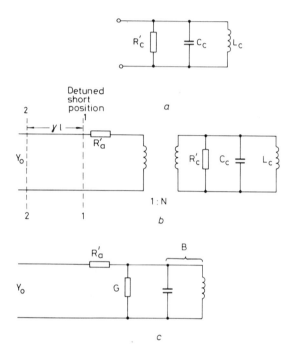

Fig. 7.1   (a) *Equivalent circuit of 'unloaded cavity', inclusive of the dielectric.* (b) *Equivalent circuit at the detuned short position of a resonant cavity loaded to a waveguide via a transformer.* (c) *Equivalent circuit at the detuned short position referred to the primary of the coupling transformer*

the cavity is said to be perfectly matched to the input waveguide and therefore to the rest of the circuit and all the available energy from the power unit is transferred to the cavity and dissipated within it.

   In general the cavity impedance (or resistance at resonance) is not equal to the characteristic waveguide impedance and a matching network must be incorporated between the cavity and waveguide impedance. This is accomplished by an aperture which offers a variable impedance depending upon its dimensions. Basically this means that the cavity impedance is transformed until it matches the characteristic impedance of the connecting waveguide and a convenient way of illustrating this is to assume that the coupling network acts as a transformer of turns ratio N, as shown in Fig. 7.1(b) (see Section 7.8). A series resistance, R_a', can be included to account for any additional losses of the coupling network. Neglecting R_a', the impedance referred to the primary at resonance, i.e., the coupled impedance, is now R_c'/N^2 yielding an effective coupling factor β':

$$\beta' = R_c'/N^2 Z_0 \tag{7.9}$$

or in terms of the characteristic admittance of the guide, Y_0,

$$\beta' = R_c' Y_0/N^2 \tag{7.10}$$

The condition for perfect coupling is obtained when $\beta' = 1$ yielding

$$N^2 = Y_0 R_c' \tag{7.11}$$

7.4 Measurements by reflected power

7.4.1 Input impedance of resonant cavity

Coupling of the electromagnetic energy into a resonant heating cavity can be experimentally investigated by means of the various techniques shown in Fig. 7.2. In Fig. 7.2(a) microwave power is transmitted to the cavity via a slotted line which incorporates a moveable E-field probe to sample the field within the slotted wave-guide. The high frequency signal is rectified and fed into a tuned amplifier which is usually calibrated in VSWR readings. Part of the output signal from the source is directed towards a frequency counter to monitor the frequency accurately. Alternatively, swept frequency techniques can be employed using a network analyser, as shown in Fig. 7.2(b) and (c). As regards measurements with a slotted line data involving the VSWR, S, or the phase angle can be used to characterise the resonant heater and to compute the various Q factors (Ginzton, 1957; Sucher and Fox, 1963; Montgomery *et al.*, 1965). In the following analysis we shall briefly describe the electrical characterisation of resonant cavities and the evaluation of its parameters through VSWR data using slotted lines or data using a network analyser.

The input admittance of the circuit shown in Fig. 7.1(b), connected at the end of the slotted line, and referred to the primary winding of the coupling network at the detuned short position at the plane 11 is given by

$$Y_{in} = \frac{N^2}{R_c'} + j\,\omega C_c N^2 \left(\frac{\omega}{\omega_0} - \frac{\omega_0}{\omega} \right) = G + jB \tag{7.12}$$

where G and B are the conductance and susceptance of the parallel circuit respectively, neglecting any losses of the coupling network. Equation (7.12) reduces to

$$Y_{in} = \frac{N^2}{R_c'} + jN^2 C_c \omega \frac{2\Delta\omega}{\omega_0} = G + jB \tag{7.13}$$

where $\Delta\omega = \omega_0 - \omega$ and $\omega_0 + \omega \simeq 2\omega$. Introducing the coupling factor β', eqn. (7.13) yields

$$Y_{in}/Y_0 = \frac{1}{\beta'}\left(1 + jQ_0\,\frac{2\,\Delta\omega}{\omega_0}\right) = (G + jB)/Y_0 \qquad (7.14)$$

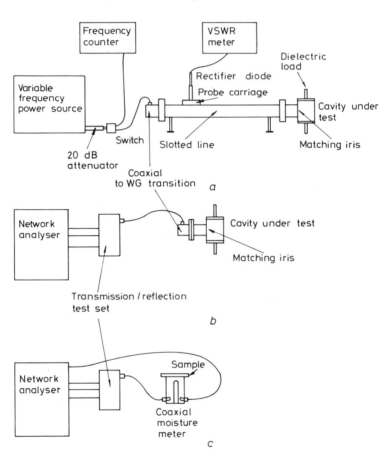

Fig. 7.2   *Measuring equipment for matching and tuning of resonant heating cavities. (a) Static measurement of VSWR as a function of frequency through a slotted line. (b) and (c) Dynamic measurements of impedance using swept frequency techniques*

where $Q_0 = \omega C_c R_c'$ is the unloaded Q-factor of the resonant cavity. Inverting eqn. (7.14) gives*

$$Y_0/Y_{in} = \frac{\beta'}{1 + j2Q_0(\Delta\omega/\omega_0)} = \frac{\beta'}{1 + jB/G} \qquad (7.15)$$

7.4.2 Determination of the Q_L and Q_0 factors

For a match $\beta' = 1$ and $\Delta\omega = 0$. However if such a cavity is not matched to the input waveguide, part of the energy will be reflected and the superposition of the incident and reflected waves will produce a standing wave pattern which could readily be examined using the slotted line. If the resultant pattern has a maximum voltage signal V_{max} and a minimum signal V_{min} (see Fig. 5.5), the standing wave ratio, S, is defined as

$$S = V_{max}/V_{min} \tag{7.16}$$

The standing wave ratio expressed in terms of the incident voltage $V_+ e^{j\omega t - \gamma z}$ and the reflected voltage $V_- e^{j\omega t + \gamma z}$ gives

$$S = \frac{|V_+| + |V_-|}{|V_+| - |V_-|} = \frac{1 + (|V_-|/|V_+|)}{1 - (|V_-|/|V_+|)} = \frac{1 + |\rho|}{1 - |\rho|} \tag{7.17}$$

where ρ is the reflection coefficient. The reflection coefficient is also defined in terms of the characteristic admittance Y_0 of a waveguide terminated by a load admittance Y_{in}, where

$$\rho = \frac{Y_0 - Y_{in}}{Y_0 + Y_{in}} \tag{7.18}$$

Substituting eqn. (7.12) into eqn. (7.18) yields

$$\rho = \frac{Y_0 - G - jB}{Y_0 + G + jB} \tag{7.19}$$

Typical data for the VSWR as a function of the frequency for resonant heating cavities are shown in Fig. 7.3. Such responses are used to calculate the loaded (Q_L) and unloaded (Q_0), Q-factors by determining the corresponding half power width Δf and using the following general expression

$$Q = \frac{f_0}{\Delta f} \tag{7.20}$$

Inspection of eqn. (7.15) shows that the admittance ratio when $\pm B = G$ is $\beta'/(1 \pm j)$ and at resonance this ratio is equal to β'. Thus the admittance at the two frequencies at which $|B| = |G|$ is $|1/(1 \pm j)|$ $(= 1/\sqrt{2})$ of the admittance

---

*It is more rigorous to consider the susceptance of the coupling aperture and refer the impedance to the detuned short position by cancelling this reactance by an appropriate length of waveguide. This leads to a modified eqn. (7.15) which includes an additional term in the denominator due to the normalised aperture susceptance b_i:

$$Y_0/Y_{in} = \cfrac{\beta'}{1 + j2Q_0 \left[\dfrac{\Delta\omega}{\omega_0} - \dfrac{b_i \beta' N^2}{2Q_0} \right]} \tag{7.15a}$$

at resonance so the field in the cavity will be reduced $(1/\sqrt{2})$ of its resonant value. Since the power is proportional to the square of the field intensity, the conditions $|\pm B| = G$ correspond to powers in the cavity equal to half the power at resonance. Making this substitution the reflection coefficient at the half power points for the unloaded cavity is

$$\rho_{\frac{1}{2}0} = \frac{Y_0 - G + jG}{Y_0 + G + jG} = \frac{(\beta' - 1) + j}{(\beta' + 1) + j} \tag{7.21}$$

The absolute value of $\rho_{\frac{1}{2}0}$ is

$$|\rho_{\frac{1}{2}0}| = [(\beta' - 1)^2 + 1]^{1/2} / [(\beta' + 1)^2 + 1]^{1/2} \tag{7.22}$$

The VSWR $S_{\frac{1}{2}0}$ at the 3 dB points is given by substituting eqn. (7.22) into eqn. (7.17) which yields

$$S_{\frac{1}{2}0} = \frac{1 + |\rho_{\frac{1}{2}0}|}{1 - |\rho_{\frac{1}{2}0}|} = \frac{2 + \beta'^2 + [4 + \beta'^2]^{1/2}}{2\beta'} \tag{7.23}$$

Similarly the VSWR at the half power points for the cavity connected to the external waveguide can be derived by assuming that the susceptance is equal to the total conductance present, i.e.,

$$|B| = |G + Y_0| \tag{7.24}$$

and substituting in eqn. (7.19) to determine $\rho_{\frac{1}{2}L}$. The VSWR follows by substitution of $\rho_{\frac{1}{2}L}$ in eqn. (7.17) to give

$$S_{\frac{1}{2}L} = \frac{1 + |\rho_{\frac{1}{2}L}|}{1 - |\rho_{\frac{1}{2}L}|} = \frac{1 + \beta' + [1 + \beta'^2]^{1/2}}{1 + \beta' - [1 + \beta'^2]^{1/2}} \tag{7.25}$$

The minimum VSWR's, S_0, shown in Fig. 7.3 are inherently related to the coupling factor β' as follows (Ginzton, 1957):

overcoupling case $\beta' = S_0$ $\qquad\qquad$ (7.26a)

undercoupling case $\beta' = 1/S_0$ $\qquad\qquad$ (7.26b)

After having established whether a resonant cavity is overcoupled or undercoupled the procedure for determing the Q's starts by calculating $S_{1/2}$ using the appropriate value for β' from eqn. (7.26) and the experimental data for S_0. The value of $S_{1/2}$ is then used in Fig. 7.3 to determine Δf at the 3 dB points, the latter of which is inserted in eqn. (7.20) to calculate the appropriate Q. Simple inspection of eqn. (7.25) shows that $S_{\frac{1}{2}L}(\beta') = S_{\frac{1}{2}L}(1/\beta')$ which shows that the determination of Q_L through the frequency width at the 3 dB points is independent of the degree of coupling. However, no similar equality exists for $S_{\frac{1}{2}0}$ and the degree of coupling must be known in order to calculate the unloaded Q-factor, Q_0.

The reflection coefficient at resonance is given by

$$\rho_0 = \frac{Y_0 - G}{Y_0 + G} = \frac{1 - (1/\beta')}{1 + (1/\beta')} \qquad (7.27)$$

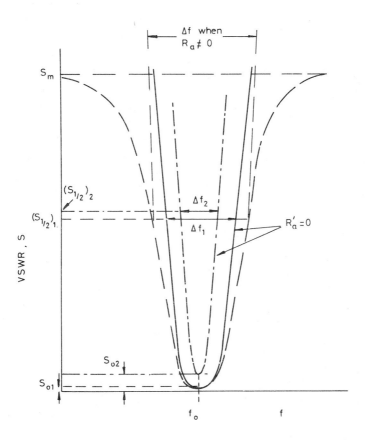

Fig. 7.3 *Typical voltage standing wave ratios (VSWR), S, vs. frequency response for a single mode resonant cavity with a dielectric inserted within it*

At critical coupling $\beta' = 1$ and $\rho_0 = 0$ whereas if the cavity is grossly overcoupled $\beta' \gg 0$ the reflection coefficient tend towards $+1$ and if grossly undercoupled -1. Table 7.1 shows the various parameters involved at resonance.

7.4.3 Coupling network losses

The calculation for Q_L follows the same procedure as in Section 7.4.2 except that eqn. (7.24) is modified to include losses in the coupling network at the 3 dB points.

Table 7.1

Degree of coupling	Coupling parameter	Reflection coefficeint at resonance	Minimum VSWR	Position of voltage minimum
Undercoupling	$\beta' < 1$	$-1 < \rho_0 < 0$	$\infty < S_0 < 1$	$\lambda/4$ from coupling plane
Critical coupling	$\beta' = 1$	0	1	Not applicable
Overcoupling	$\beta' > 1$	$0 < \rho_0 < +1$	$1 < S_0 < \infty$	At coupling aperture plane

The modified equation becomes

$$|B| = \left| G + \frac{1}{R'_a + 1/Y_0} \right| = \left| Y_0 \left(1/\beta' + \frac{S_m}{1 + S_m} \right) \right| \qquad (7.24a)$$

where $S_m = 1/R'_a Y_0$. The VSWR at the 3 dB points becomes

$$S_{\frac{1}{2}L} = \frac{(S_m + 1)(\beta' + 1) + [(S_m - \beta')^2 + (S_m \beta' - 1)^2]^{1/2}}{(S_m + 1)(\beta' + 1) - [(S_m - \beta')^2 + (S_m \beta' - 1)^2]^{1/2}} \qquad (7.28)$$

for $R'_a = 0$, $S_m \rightarrow \infty$ and eqn. (7.28) becomes identical to eqn. (7.25). The dotted line in Fig. 7.3 depicts the case where $R'_a \neq 0$ and coupling network losses cannot be safely ignored.

7.4.4 Graphical representation of cavity impedance

Problems involving complex impedances in waveguides or transmission lines can be readily solved using the Smith Chart which consists of loci of constant resistance and reactance plotted in polar form and relate to magnitude and phase of the reflection coefficient along any point in the waveguide. The transformation of impedance or admittance into reflection coefficient and vice versa is afforded through eqn. (7.18) where the impedance is normalised by dividing by the characteristic waveguide impedance. Apart from impedance and reflection coefficients other parameters such as VSWR, line travel (wavelengths), admittance, etc., can be also readily displayed.

Equation (7.15) represents a circle whose resonant impedance is equal to $\beta' Z_0 = R'_c/N^2$. Typical impedance loci of the shunt cavity are shown in Fig. 7.4 plotted on a simplified version of the Smith Chart. The frequency, increasing clockwise, is represented by the dots on the loci. For perfect coupling the circle crosses the centre of the Smith Chart and the cavity shunt resistance transformed to the primary, R'_c/N^2, is equal to the characteristic impedance of the input waveguide. If R'_c/N^2 is smaller than Z_0, then $\beta' < 1$, the cavity is undercoupled and its locus intersects the resistance axis to the left of the Smith Chart centre. On the other hand if R'_c/N^2 is made larger than Z_0 then $\beta' > 1$, the cavity is overcoupled and its locus intersects the resistance axis to the right of the centre.

The quality factors of the cavity Q_L and Q_0 can be determined by construction from such impedance locus by incorporating on the Smith Chart the impedance plots at the half power points corresponding to the equalities $|B| = |G|$ and $|B| = |G + Y_0|$. The intersection of impedance locus with these plots gives the corresponding frequencies for Q determination using eqn. (7.20) (Ginzton, 1957).

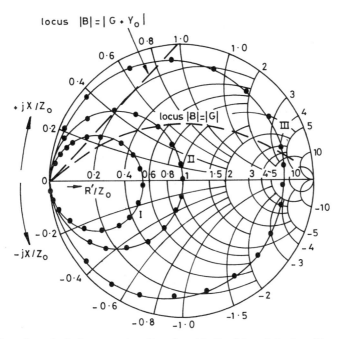

Fig. 7.4   *Impedance loci of a resonant cavity referred to the detuned short position*
   I Undercoupled cavity
   II Critically coupled cavity
   III Overcoupled cavity

The measurements described above refer mainly to techniques using a slotted line. Constructing the impedance locus of a resonant heater on the Smith Chart from data derived through slotted line techniques is a slow process. A faster technique is to use a network analyser as shown in Fig. 7.2(b) which sweeps the frequency along the required range and displays the complex reflection coefficient in polar form on a CRT screen. But since the input impedance of a cavity heater is inherently related to the reflection coefficient through eqn. (7.18) the network analyser can supply data of impedance by superimposing a simplified version of the Smith Chart on to the polar reflection coefficient display on the CRT.

It is important to remember that before displaying complex impedance data on the Smith Chart either using a network analyser or through slotted line techniques it is imperative to correctly establish the reference plane by following the relevant calibration procedures, otherwise the displayed impedances could be meaningless (Thomas, 1978). For example, because of the remoteness of a particular resonant cavity under test, it may be necessary to connect the network analyser to the waveguide and cavity assembly via an additional length of cable as shown in Fig. 7.2(b). The impedance at any point on a transmission line or waveguide Z_{22} relative to a terminating impedance Z_{in}, say, representing the cavity impedance in Fig. 7.2(b) at the detuned short position ($Z_{11} = Z_{in}$) is given by

$$Z_{22} = \frac{Z_0 \left[Z_{in} + \tanh\left(\gamma l\right) \right]}{Z_0 + Z_{in} \tanh\left(\gamma l\right)} \tag{3.6}$$

where Z_{22} represents the impedance looking towards the cavity at the terminals of the network analyser. To eliminate the influence of the interconnecting waveguide and cable we must ensure that $\tanh\left(\gamma l\right) = 0$ by adjustments in the attenuation and phase so that $Z_{22} = Z_{in}$.

Typical impedance loci referred to the detuned short position for some resonant heating cavities and obtained through a network analyser are shown in Figs. 7.5, 7.6 and 7.7. The latter two figures also include the return loss in dB as displayed on the network analyser through which the minimum VSWR, S_0, can be computed and correlated against data from the intersection of the impedance locus and the (R'/Z_0) axis.

7.4.5 Interpretation of Q_L and Q_0

An interesting dilemma occurs in the interpretation of eqn. (7.6). We can relate Q_L and Q_0 through the classical treatment of low power cavities (Ginzton, 1957; Sucher and Fox, 1963; Montgomery *et al.*, 1965) as

$$Q_L = \frac{\omega_0 N^2 C_c}{G + Y_0} = \frac{\omega_0 N^2 C_c / G}{1 + Y_0 / G} = \frac{Q_0}{1 + \beta'} \tag{7.29}$$

Expression (7.29) was used in eqn. (7.15) to substitute for Q_0 in terms of Q_L and determine the loci for Q_L on the Smith Chart. At critical coupling ($\beta' = 1$) eqn. (7.29) gives the well-known relation, $Q_L = \frac{1}{2} Q_0$. This implies through eqn. (7.6) that with a cavity matched at resonance $Q_0 = Q_{exc}$, which means that equal powers are dissipated in the dielectric inserted within the cavity and the external circuit.

Care must be taken in interpreting this result which applies only to the condition in which a generator is matched to a transmission line, i.e., $R'_G = Z_0$, and the cavity at resonance is also matched to the line $R'_L = Z_0$ (see Fig. 5.5 assuming resistive elements). On tune equal powers are then dissipated in R'_G and R'_L and the efficiency of the system is only 50%. This condition usually only exists when using low power test gear where the signal generator has an attenuator pad giving a source impedance $R'_G = Z_0$.

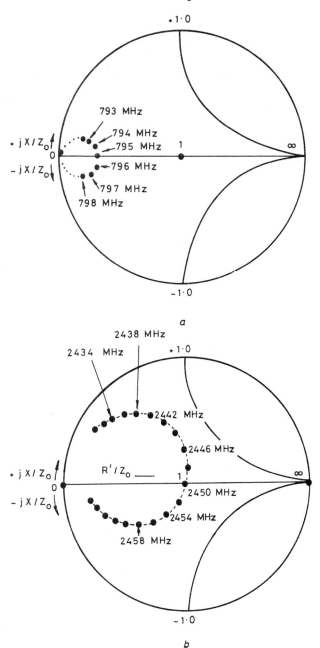

Fig. 7.5   *Impedance loci of resonant cavities*
a Coaxial cavity, $f_0 \approx 795$ MHz, coupled via a loop.
b TE_{10n} cavity, $f_0 \approx 2450$ MHz, coupling aperture: 37×22 mm, cavity length: 0.33 m, $S_0 \simeq 1.07$. Tufnol rod inserted in it, 6 mm in diameter

In the case of a magnetron or klystron generator the system is designed for maximum efficiency and in this case $R'_G \ll Z_0$ (typically $R'_G = 0.2 Z_0$). Because the load cavity is matched to the line, the length of the line is immaterial and can be omitted without affecting the fact that the cavity is fed from a generator in which

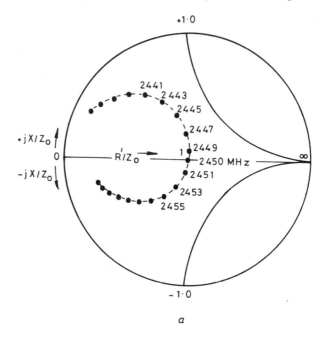

a

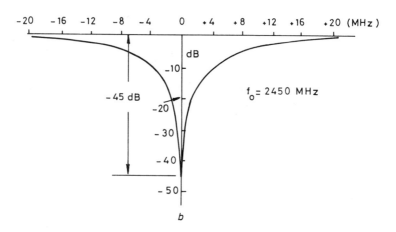

b

Fig. 7.6   *Characterisation of a TE_{10n} resonant cavity with a 2 mm diameter water pipe inserted in it. $f_0 = 2450$ MHz, coupling aperture: 50 × 20 mm, cavity length 0·4 m.*
a Impedance locus
b Return loss

$R'_G \ll R_L$, so that $\beta' \ll 1.0$ and the Q-factor, Q_L, now attains a value in the range

$$\tfrac{1}{2}Q_0 < Q_L < Q_0 \tag{7.30}$$

The value of Q_L then given by eqn. (7.6) is a real value which gives the frequency separation between the half power points of dissipation within the cavity as

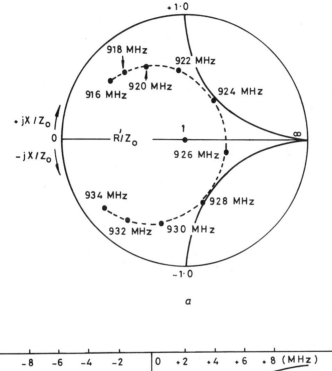

a

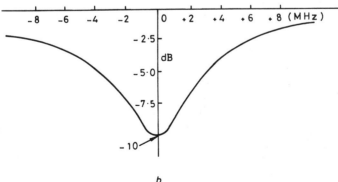

b

Fig. 7.7 *Characterisation of a TM_{010} resonant cavity with a 7 mm diameter water pipe inserted in its axis. Aperture: 120 × 30 mm, cavity diameter: 200 mm, $S_0 = 1.8$, $f_0 \approx 926$ MHz*
a Impedance locus
b Return loss

ω_0/Q_L. The power dissipated in the critically-coupled, resonant cavity on tune is the same as it would have been in a broadband matched load, and is not associated with any extra dissipated power. Although the transmission line is ascribed a characteristic impedance Z_0 (resistive), the resistance is really its cavity termination load at one end 'seen' from the other. There is no fundamental mechanism for power dissipation in the line other than small $I^2 R'$ losses in the waveguide wall.

If the circuit incorporates a microwave circulator the situation becomes more complicated because, looking back towards the generator from the cavity terminals, the impedance is Z_0. On tune the cavity is reflectionless and so no power is fed to the circulator load. The efficiency is then the same as it would have been without the circulator. However, off-tune power is reflected proportional to the (amount of detuning)^2. The system $Q_{L\,eff}$ now invokes the generator resistance R'_G, the cavity resistance $R'_L\,(1/G)$ and the circulator load which is equal to Z_0, and is a complicated relation in which

$$Q_{L\,eff} \neq Q_L = \frac{Q_0}{2} \qquad (7.31)$$

Evaluation of the utilization efficiency of a resonant heater requires to establish the value of the term $(1 - Q/Q'_0)$ (see eqn. (7.125)). With medium to high loss dielectric insertions both the Q_L and Q_0 are very much smaller than Q'_0 (the Q-factor of the cavity without any dielectric in it). Therefore substitutions of either Q_L or Q_0, for Q in the numerator of the efficiency term $(1 - Q/Q'_0)$, will produce a relatively small difference in utilisation efficiency even when $Q_L = \frac{1}{2}Q_0$.

7.5 Measurements using transmitted power

Most microwave resonant heating applicators possess only one port through which energy is coupled into them. Measurements to characterise such structures involving the use of transmitted power requiring an input as well as an output port are therefore normally not possible. However, such a limitation can often be overcome by inserting a small probe to sample the transmitted signal near the entry or exit ports through which the dielectric enters or leaves the applicator. The output signal through the probe as a function of the frequency yields the usual resonant response which peaks at the resonant frequency of the microwave applicator, as shown in Fig. 7.8. Although such a response can be obtained step by step through a variable frequency source and a meter to record the rectified transmitted signal, once again the complete curve in a given frequency range can be obtained instantaneously through the use of the network analyser in an arrangement as shown in Fig. 7.2(c). The bandwidth at the 3 dB points can be readily measured, yielding the Q_L of the cavity heater through eqn. (7.20). In addition the phase change of the signal through the heater can be displayed on a CRT and accurately determined.

Apart from the determination of Q-factor, the biggest advantage of measurements involving transmitted power using a network analyser is the visual display of cavity tuning. The resonant frequency can be seen to shift in accordance with the insertion of the dielectric and an applicator can be readily tuned to resonance at an industrial allocated frequency by adjustments of the relative sizes of the cavity and dielectric as well as by the use of tuning stubs.

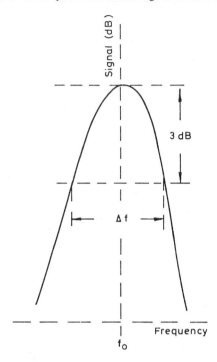

Fig. 7.8 *Measurement of the 3 dB bandwidth, Δf, of a resonant cavity heater using the transmitted signal*

Before proceeding to discuss some aspects of coupling apertures, the power transfer into the cavity and the overall efficiency, let us examine in detail some of the most common resonant cavities used in microwave heating and formulate simple design procedures.

7.6 TE_{10n} rectangular cavity

7.6.1 Introduction
The distribution of the electric and magnetic fields in the TE_{lmn} rectangular cavity can be derived from the solution in time and space of the Maxwell equations. The treatise which starts by separating the variables and assuming an $e^{j\omega t}$ variation, leads to the following field equations after considering the appropriate boundary

conditions for the fundamental TE_{10n} mode (Ramo *et al.*, 1965; Boudouris, 1971; Mooijweer, 1971):

$$E_y = E_{y0} \sin \frac{\pi x}{a} \sin \beta z \qquad (7.32a)$$

$$H_x = H_{x0} \sin \frac{\pi x}{a} \cos \beta z \qquad (7.32b)$$

$$H_z = H_{z0} \cos \frac{\pi x}{a} \sin \beta z \qquad (7.32c)$$

where a and b are the broad and narrow sides of the waveguide cross-section and d_c is along the z direction and represents the length of the cavity, as shown in Fig. 7.9. Before plotting the distribution of the field within the cavity let us consider the superposition of the forward and reflected waves at their maximum values, i.e., at the centre of the broader dimension where $x = a/2$. For that we assume a stationary forward wave E_+ as in Fig. 7.10(a), vectorially added to the reflected wave E_- rotating at an angle, θ, to the vertical axis.

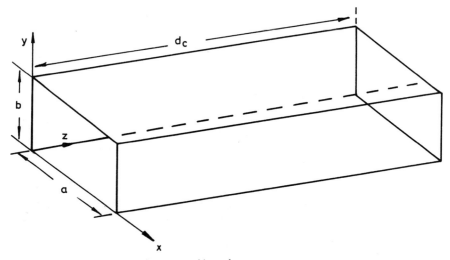

Fig. 7.9   *Coordinate system in a waveguide cavity*

The resultant field vector E_{res} is given by

$$E_{res}^2 = (E_+ + E_- \cos \theta)^2 + E_-^2 \sin^2 \theta \qquad (7.33)$$

Expanding eqn. (7.33) and bearing in mind that $E_+ = -E_-$ gives

$$E_{res} = \sqrt{2} E_+ (1 - \cos \theta)^{1/2} \qquad (7.34)$$

Substitution of (see Appendix III)

$$\sin \beta z = \left(\frac{1 - \cos 2\beta z}{2}\right)^{1/2} \qquad (7.35)$$

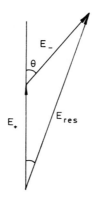

a

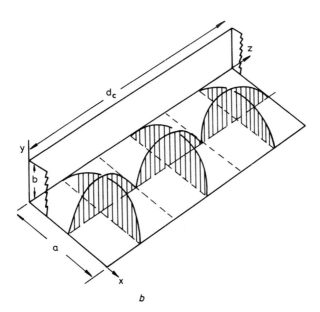

b

Fig. 7.10 *Development of standing waves*
    a Phasor diagram of forward and reflected waves
    b TE_{103} resonant cavity electric field pattern

in eqn. (7.32a), yields

$$E_{y(x=a/2)} = \frac{\sqrt{2}}{2} E_{yo}(1-\cos 2\beta z)^{1/2} \tag{7.36}$$

Comparison of eqns. (7.34) and (7.36) show that $\theta = 2\beta z$. The distribution of the field, for example, in the TE_{103} resonant cavity, taking also into account x-variations, is shown in Fig. 7.10(b). It should be noted that squaring eqn. (7.36) or eqn. (7.32a) gives E_y^2 which is proportional to the power dissipation in the workload. Equation (7.32a) shows clearly that this power dissipation is sinusoidally distributed along the z-axis.

7.6.2 Uniform field distribution in the twin TE_{10n} cavity heater

Most heating applications require a uniform field across the width of the product. However, careful examination of the field distribution along the z-direction, where the material is normally placed, reveals a cosine distribution which would result in highly non-uniform energy absorption into the material. To obtain a uniform field across the width, two such cavities are used side by side, as shown in Fig. 7.11, powered by a single source, splitting the power equally between the two cavities.

The resulting electric fields in each cavity shown in Fig. 7.11(b) are given by eqn. (7.34) where

$$E_{res\,I} = \sqrt{2}\, E_+ (1-\cos\theta_I)^{1/2} \tag{7.37}$$

and

$$E_{res\,II} = \sqrt{2}\, E_+ (1-\cos\theta_{II})^{1/2} \tag{7.38}$$

The heating effect is proportional to

$$\text{heating effect} \propto [E_{res\,I}^2 + E_{res\,II}^2] \tag{7.39}$$

Therefore

$$\text{heating effect} \propto 2E_+^2(2-\cos\theta_I-\cos\theta_{II}) \tag{7.40}$$

However ensuring a quarterwave displacement between the electric fields in the two cavities,

$$\theta_{II} = \theta_I + \pi \tag{7.41}$$

gives

$$\cos\theta_{II} = -\cos\theta_I \tag{7.42}$$

Equation (7.40) therefore reduces to

$$\text{heating effect} \propto 4E_+^2 \tag{7.43}$$

which shows that a constant field is obtained along the cavity length, having ignored any second order effects such as due to the insertion of the material. Such a twin cavity operation can be used to heat materials in sheet form where the thickness is a small fraction of the broader dimension a, otherwise the material will experience large field variations along its thickness (Holme and Metaxas, 1979).

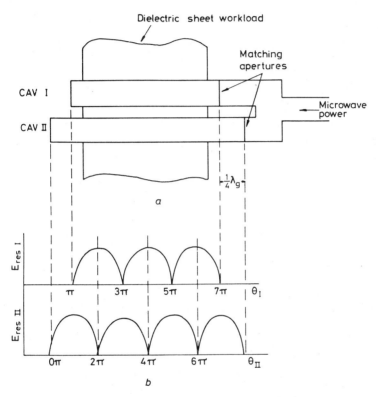

Fig. 7.11 *Development of a uniform electric field* TE_{10n} *twin cavity heater*
  a Twin TE_{10n} resonant cavity heater (Plan view)
  b Electric field distribution in the plane of the dielectric sheet in each cavity

7.6.3 Unloaded Q-factor Q_0 of the TE_{10n} cavity heater

The basic form of a rectangular TE_{10n} resonant cavity, shown in Fig. 7.12, consists of a short-circuited length of waveguide with a coupling aperture to afford impedance matching it to the rest of the microwave circuit. Leaving aside for the moment the matching considerations, let us concentrate on the resonant properties of the cavity and in particular the effect of the dielectric insertion to the Q-factor, Q_0, of the resonant structure.

The input impedance of a short-circuited waveguide of length d_c, characteristic impedance Z_0 and total attenuation constant α is given by

$$Z_{in} = Z_0 \tanh(\alpha + j\beta)d_c \tag{7.44}$$

Expansion yields

$$Z_{in} = Z_0 \frac{\tanh \alpha d_c + \tanh j\beta d_c}{1 - \tanh \alpha d_c \tanh j\beta d_c} \tag{7.45}$$

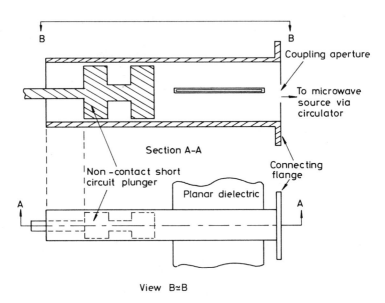

View B≈B

Fig. 7.12 *Fundamental mode TE_{10n} rectangular cavity partially filled with planar dielectric*

Assuming that the losses in the cavity are relatively small, that is, $\alpha d_c \ll 1$, then $\tanh \alpha d_c \approx \alpha d_c$ and substitution in eqn. (7.45) yields

$$Z_{in} \approx Z_0 \frac{\alpha d_c + j \tan \beta d_c}{1 - j\alpha d_c \tan \beta d_c} \tag{7.46}$$

With proper matching of the coupling aperture, resonance occurs when $\tan \beta d_c \to 0$ or $\beta d_c = n\pi$. However since $\beta = 2\pi/\lambda_g$ at resonance the length of the cavity becomes

$$d_c = \tfrac{1}{2}n\lambda_g \tag{7.47}$$

Close to resonance λ_g will change by a small amount $\Delta\lambda_g$:

$$d_c \approx n\left(\frac{\lambda_g + \Delta\lambda_g}{2}\right) \tag{7.48}$$

or

$$\beta d_c \approx \frac{2\pi n}{\lambda_g}\left(\frac{\lambda_g + \Delta\lambda_g}{2}\right) = \pi n\left(1 + \frac{\Delta\lambda_g}{\lambda_g}\right) \tag{7.49}$$

and

$$\tan \beta d_c = \tan\left(n\pi + n\pi\frac{\Delta\lambda_g}{\lambda_g}\right) \approx n\pi\frac{\Delta\lambda_g}{\lambda_g} \tag{7.50}$$

Substitution for $\tan \beta d_c$ from eqn. (7.50) in eqn. (7.46) yields

$$Z_{in} \approx Z_0 \frac{\alpha d_c + jn\pi(\Delta\lambda_g/\lambda_g)}{1 - j\alpha d_c n\pi(\Delta\lambda_g/\lambda_g)} \tag{7.51}$$

However, normally the product $\alpha\,\Delta\lambda_g$ in the denominator is very small, therefore

$$Z_{in} \approx Z_0\left(\alpha d_c + jn\pi\frac{\Delta\lambda_g}{\lambda_g}\right) \tag{7.52}$$

The cutoff and waveguide wavelengths are related by the equation

$$\lambda_g^2\left[1 - (\lambda/\lambda_c)^2\right] = \lambda^2 \tag{7.53}$$

Differentiation with respect to λ gives

$$(2\lambda_g)\frac{d\lambda_g}{d\lambda}\left[1 - (\lambda/\lambda_c)^2\right] + \lambda_g^2\left(-\frac{2\lambda}{\lambda_c^2}\right) = 2\lambda$$

yielding

$$\frac{d\lambda_g}{d\lambda} = (\lambda_g/\lambda)^3 \tag{7.54}$$

However, since $\lambda = 2\pi c/\omega_0$, differentiation yields

$$\frac{d\lambda}{d\omega} = -\frac{2\pi c}{\omega_0^2} = -\frac{\lambda}{\omega_0} \tag{7.55}$$

Combining eqns. (7.54) and (7.55) yields

$$\frac{d\lambda_g}{\lambda_g} = \frac{\Delta\lambda_g}{\lambda_g} = -\left(\frac{\lambda_g}{\lambda}\right)^2\frac{d\omega}{\omega_0} \tag{7.56}$$

Substitution in eqn. (7.52) the input impedance becomes

$$Z_{in} = Z_0 \left[\alpha d_c - jn \left(\frac{\lambda_g}{\lambda} \right)^2 \pi \frac{d\omega}{\omega_0} \right] \tag{7.57}$$

This is equivalent to a series resonant circuit of resistance R' and reactance $(\omega L - 1/\omega C)$, where

$$Z_{in} = R' + j \left(\omega L - \frac{1}{\omega C} \right) \tag{7.58}$$

Consider the reactance and assuming $\omega \simeq \omega_0$ and $\omega + \omega_0 \simeq 2\omega$, we obtain

$$\omega L - \frac{1}{\omega C} = \left(\frac{\omega^2}{\omega_0^2} - 1 \right) \frac{1}{\omega C} = -2Ld\omega \tag{7.59}$$

Equation (7.58) becomes

$$Z_{in} = R' - j\,2Ld\omega \tag{7.60}$$

Comparisons of eqns. (7.57) and (7.60) give

$$R' = Z_0 \alpha d_c \tag{7.61}$$

and

$$L = \tfrac{1}{2} \frac{n\pi Z_0}{\omega_0} \left(\frac{\lambda_g}{\lambda} \right)^2 \tag{7.62}$$

Therefore by definition

$$Q_0 = \omega_0 L/R' = \tfrac{1}{2} \frac{n\pi}{\alpha d_c} \left(\frac{\lambda_g}{\lambda} \right)^2 \tag{7.63}$$

However, since $d_c = \tfrac{1}{2} n \lambda_g$, eqn. (7.63) reduces to

$$Q_0 = \frac{\pi}{\alpha} \frac{\lambda_g}{\lambda^2} \tag{7.64}$$

The attenuation constant contains all the information about the material properties and is strongly related to the loss factor ϵ''_{eff}. Altman (1964) has derived an expression for a partially filled waveguide with a sheet material of thickness, t, which is given by eqn. (5.11). Ignoring wall losses and assuming $x_1 = a/2$, eqn. (5.11) yields

$$\alpha = 2\pi \epsilon''_{eff} \frac{t}{a} \frac{\lambda_g}{\lambda^2}$$ (7.65)

Substitution of eqn. (7.65) into eqn. (7.64) yields

$$Q_0 = \tfrac{1}{2} \frac{a}{t} \frac{1}{\epsilon''_{eff}}$$ (7.66)

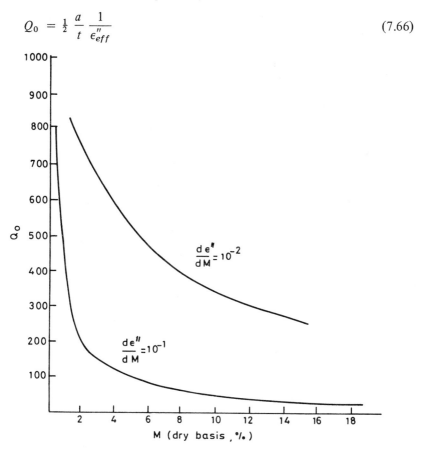

Fig. 7.13 *Cavity Q-factor Q_0 as a function of the moisture content of a dielectric material inserted in it. The moisture content M in the slope $(d\,\epsilon''/dM)$ is expressed in %*

Equation (7.66) shows a strong dependence of the Q-factor of the partially filled cavity with the ϵ''_{eff} of the inserted material. In drying applications this becomes very significant because it results in large variations of the Q-factor with moisture content, M, since ϵ''_{eff} is a function of moisture (see Section 3.4.1). As an example, Fig. 7.13 shows the dependence of Q_0 on moisture for a material of thickness 2·5 mm inserted into a WG4 cavity of $a = 0·25$ mm with $\epsilon''_0 \approx 0·05$ and for two values of $d\epsilon''_{eff}/dM$, 10^{-1} and 10^{-2}, where M is expressed in %, dry basis (see Appendix I and Appendix VII). The ϵ''_{eff} is assumed to vary linearly with moisture in the range $2 < M < 18\%$.

7.6.4 Impedance matching

Energy is coupled to a resonant structure by means of an aperture which can be inductive or capacitive and therefore can also be resonant depending upon the shape. The aperture affords proper impedance matching in order to use the available power most effectively. We shall consider an inductive aperture as a means of coupling the TE_{10n} structure to the rest of the heating circuit as shown in Fig. 7.14(a). The admittance of the aperture is represented by $-jb_iY_0$, where b_i is its normalised susceptance and therefore the total input admittance is given by

$$Y_{in} = -jb_iY_0 + \frac{Y_0}{\tanh(\alpha + j\beta)z} \tag{7.67}$$

where z is the cavity length and is given by

$$z = d_c + z_0 \tag{7.68}$$

where d_c is the resonant length with a dielectric inserted in the cavity and z_0 is a distance above or below that which gives resonance. Expanding eqn. (7.67) yields

$$Y_{in}/Y_0 = -jb_i + \frac{(1 - j\tanh\alpha z\tan\beta z)(\tanh\alpha z - j\tan\beta z)}{(\tanh\alpha z)^2 + (\tan\beta z)^2} \tag{7.69}$$

For resonance the imaginary part of the admittance is zero:

$$\text{Imag.}\left(\frac{Y_{in}}{Y_0}\right) = 0 \tag{7.70}$$

and $z = d_c$. Therefore the condition for resonance can be obtained from eqn. (7.69) by equating its imaginary part to zero:

$$\tan^2\beta d_c + \frac{1}{b_i}\tan\beta d_c\,[1 + (\tanh\alpha d_c)^2] + (\tanh\alpha d_c)^2 = 0 \tag{7.71}$$

Solution of eqn. (7.71) yields

$$2\tan\beta d_c = -\frac{1}{b_i}(1 + \phi_0^2)\left[1 \pm \left[1 - \left(\frac{2\phi_0 b_i}{1 + \phi_0^2}\right)^2\right]^{1/2}\right] \tag{7.72}$$

where

$$\phi_0 = \tanh\alpha d_c \tag{7.73}$$

Assuming that $\phi_0^2 \ll 1$, eqn. (7.72) yields

$$2\tan\beta d_c \approx -\frac{1}{b_i}[1 \pm [1 - (2\phi_0 b_i)^2]^{1/2}] \tag{7.74}$$

Furthermore, assuming that $(2\phi_0 b_i)^2 \ll 1$, then eqn. (7.74) reduces to

$$\tan \beta d_c \approx -\frac{1}{b_i} \quad \text{or} \quad \tan \beta d_c = 0 \tag{7.75}$$

However, these two solutions are equivalent because for larger apertures $1/b_i$ tends to be a very small quantity.

Substituting eqn. (7.75) into the real part of eqn. (7.69), assuming once more that ϕ_0^2 and $(\alpha d_c)^2$ are very small compared to unity, yields

$$Y_{in}/Y_0 \approx \frac{\alpha d_c}{(\tanh \beta d_c)^2} = \alpha d_c b_i^2 \tag{7.76}$$

For a match, $Y_{in}/Y_0 = 1$, and therefore the aperture matches the partially filled resonant structure when its susceptance is such as to satisfy the following resonance condition:

$$\alpha d_c b_i^2 = 1 \tag{7.77}$$

It is evident that the reflection coefficient at resonance is zero and therefore the voltage standing wave ratio defined by eqn. (7.17) is unity.

Let us now examine the situation when the variable plunger in Fig. 7.14(a) is moved appreciably away from resonance. Assuming that the susceptive part of the admittance dominates over the conductive part. Equation (7.69) reduces to

$$Y_{in}/Y_0 = -jb_i - j\frac{1}{\tan \beta (d_c + z_0)} \tag{7.78}$$

having assumed again that ϕ_0^2 is very small and $(\alpha z)^2 \ll (\tan \beta z)^2$.

Expansion of eqn. (7.78) yields

$$Y_{in}/Y_0 = -j\left[b_i + \frac{(1/\tan \beta z_0) - \tan \beta d_c}{1 + (\tan \beta d_c/\tan \beta z_0)} \right] \tag{7.79}$$

Let us consider $z_0 \to \lambda_g/4$, therefore $\tan \beta z_0 \to \infty$ and eqn. (7.79) yields

$$Y_{in}/Y_0 = -j (b_i - \tan \beta d_c) \tag{7.80}$$

However $\tan \beta d_c \to 0$, therefore

$$Y_{in}/Y_0 = -jb_i \tag{7.81}$$

This leads to high reflection coefficients which give rise to very high values of S. Figure 7.14(b) depicts some typical VSWR, S, responses as the plunger moves to and fro from resonance. It also shows the case when the cavity is not matched where $\alpha b_i^2 d_c \neq 1$. If the plunger moves far enough another minimum point will be reached and the distance between the two minima will approximate to $\frac{1}{2}\lambda_g$.

7.6.5 *Shift of the VSWR vs. cavity length responses*

As the properties of the material inserted within the TE_{10n} resonant cavity change, so should the cavity length, d_c, in order for the resonance condition to be maintained. Equation (7.74) gives the value of $\tan \beta d_c$ at resonance from which we can deduce the angle βd_c. Therefore

$$\beta d_c = n\pi + \tan^{-1}\left[-\frac{1}{2b_i}(1 \pm [1-(2\phi_0 b_i)^2]^{1/2})\right] \qquad (7.82)$$

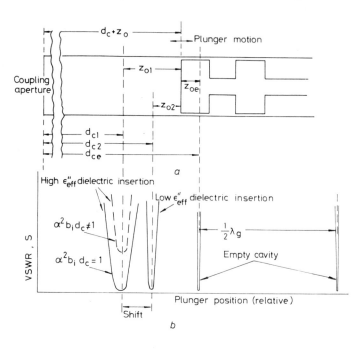

Fig. 7.14 *Matching a TE_{10n} rectangular cavity and dielectric to the external circuit*
   d_{ce} cavity resonant length with no dielectric insertion in it (high Q)
   d_{c2} cavity resonant length with a low loss dielectric inserted in it (medium Q)
   d_{c1} cavity resonant length with a medium to high loss dielectric inserted in it (low Q)

where n is an integer. Substitution for $\beta \, (= 2\pi/\lambda_g)$ in the above expression yields

$$d_c = \frac{n\lambda_g}{2} - \frac{\lambda_g}{2\pi}\tan^{-1}\left[\tfrac{1}{2}\frac{1}{b_i}(1 \pm [1-(2\phi_0 b_i)^2]^{1/2})\right] \qquad (7.83)$$

The cavity length at resonance is therefore made up of two terms, the first being applicable when no dielectric is inserted in the cavity and gives a fixed number of half guide wavelengths, while the second term is totally due to the insertion of the thin dielectric which has the effect of changing the effective waveguide wavelength and shifts the VSWR response by a small amount away from its response when the cavity is operated without any dielectric. The second term is controlled by the dielectric properties of the material (since $\phi_0 = \tanh \alpha d_c$ and α is related to ϵ''_{eff} via expression (7.65)) and by the susceptance of the aperture which in turn is controlled by the material properties. For dry low loss materials α is very small and b_i is relatively large which combine to effectively reduce the shift to zero and simplify eqn. (7.83) to eqn. (7.47). Figure. 7.15 shows a typical set of VSWR

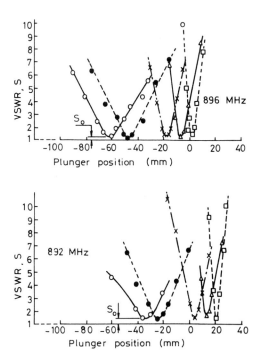

Fig. 7.15 *VSWR, S, as a function of the plunger position for various moistures of tufted carpet loads and the corresponding aperture widths for effective matching, at two frequencies*

	Moisture content	Aperture width
ooo	17·5%	130 mm
•••	15%	117 mm
xxx	9%	105 mm
▵▵▵	5%	9·5 mm
▫▫▫	1%	9·2 mm

(After Metaxas and Meredith, 1978 Copyright *The Journal of Microwave Power*, 1978.)

responses as a function of the plunger positions for a wet foam backed carpet inserted at the centre of a TE_{10n} resonant cavity (Metaxas and Meredith, 1978).

Let us calculate the shift between two such responses, using eqn. (7.83), and compare with the data shown in Fig. 7.15. The aperture susceptance has been calculated for various dimensions using eqn. (7.107) (see Section 7.8.3) and its variation is shown in Fig. 7.16. For a moisture content of 4%, $d_i \approx 93$ mm, yielding a value of b_i of about 4. Given that $\lambda_g \approx 0.455$ m (900 MHz band in WG4), the second part of eqn. (7.83), once again assuming that $2\phi_0 b_i^2 < 1$, is about 17 mm. Similarly, at a moisture content of 9%, the second term of eqn. (7.83) is about 26 mm. Therefore the shift between responses at 9% and 5% is $(26 - 17) = 9$ mm. Figure 7.15 shows a shift of about 10 mm. For larger moisture contents the assumption $2\phi_0 b_i^2 < 1$ is not valid.

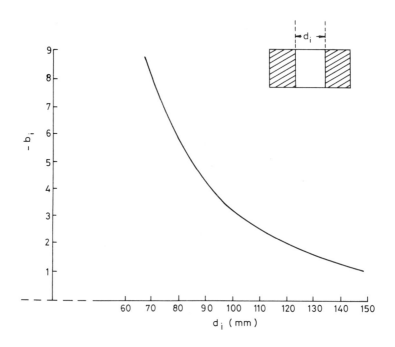

Fig. 7.16 *Normalised aperture susceptance b_i of an inductive aperture as a function of its width, d_i in WG4 (248 × 124 mm)*

7.6.6 Q-factors from VSWR vs. frequency response

Once the correct plunger position is found for a given set of conditions the corresponding VSWR, S, vs. frequency response can be measured. Such data using slotted line techniques are shown in Figs. 7.17(a) and (b), for various carpet structures and moisture contents. The Q-factor can be determined through the half power width as was explained in Section 7.4.2. In general, as the moisture content increases, the responses become broader and the Q-factors decrease. Typical bandwidths for the heaviest carpets are between 2–15 MHz yielding Q-values in the range $60 < Q_L < 450$. The VSWR vs. frequency responses for the lighter carpets are much sharper giving smaller bandwidths and corresponding higher Q-values.

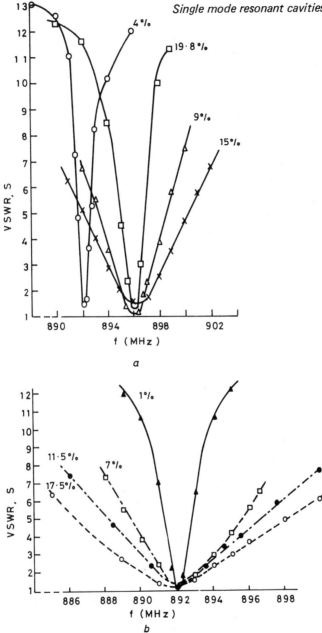

Fig. 7.17 *Input VSWR, S, as a function of the frequency of a TE$_{106}$ resonant cavity partially filled with various carpet structures at different moisture contents (dry basis)*
a 4% and 19·8%, primary backed carpets (0·52 kg/m^2)
*   9% and 15%, foam backed carpets (1·9 kg/m^2)*
b Foam backed (2·2 kg/m^2).
(After Metaxas and Meredith (1978), Copyright *The Journal of Microwave Power*, 1978.)

Variations of the Q with moisture content for four carpet structures, using data similar to Figs. 7.17(a) and (b), are shown in Fig. 3.24.

7.6.7 Electric field strength in a TE_{10n} cavity heater

An expression for the electric field strengths developed in the unloaded TE_{10n} resonant cavity can be obtained by considering the stored and dissipated energies and using eqn. (7.1). The energy within a resonant cavity oscillates between the magnetic and electric field. By considering the instant when the magnetic energy is zero, its counterpart is at its maximum and assuming also that the energy is stored mainly in the electric field outside the dielectric insertion because of its small filling factor, the total energy stored in the volume V of the cavity becomes

$$U_{max.E} = \int_V \tfrac{1}{2}\epsilon_0 |E_y|^2 \, dV \tag{7.84}$$

where E_y is the maximum value of the electric field. Using eqn. (7.32a), eqn. (7.84) yields the following triple integral:

$$U_{max.E} = \int_0^a \int_0^b \int^{\frac{1}{2}n\lambda g} \epsilon_0 E_{y0}^2 \sin^2\left(\frac{\pi x}{a}\right) \sin^2 \beta z \, dx \, dy \, dz \tag{7.85}$$

Integration yields

$$U_{max.E} = \frac{1}{16}\epsilon_0 E_{y0}^2 \, abn \, \lambda_g \tag{7.86}$$

By definition

$$Q_0 = \frac{\omega U_{max.E}}{P} \tag{7.1}$$

Substituting for the stored energy in eqn. (7.1) and using eqn. (7.64) with $\lambda = \lambda_0$, the maximum value of the electric field becomes, with $\omega = 2\pi f_0 = (2\pi/\lambda_0)$ $(1/\sqrt{\epsilon_0\mu_0})$,

$$E_{y0}^2 = \frac{8P}{\alpha \lambda_0 \, abn}\left(\frac{\mu_0}{\epsilon_0}\right)^{1/2} \tag{7.87}$$

It is evident from eqn. (7.87) that for a given power dissipation the electric field depends strongly on the attenuation constant which, of course, is a function of the dielectric properties and particularly the loss factor ϵ''_{eff} of the material. For example, in a particular case where $P = 20\,\text{kW}$, $n = 6$, $\sqrt{\mu_0/\epsilon_0} = 377\,\Omega$ in a WG4 at 900 MHz, the field developed for an attenuation of 0·01 Np/m is about 375 kV/m or 3·75 kV/cm.[*] As the loss factor and hence the attenuation

---

[*] If one considers the partially filled cavity to the connecting waveguide and assumes that at critical coupling, $Q_L = \tfrac{1}{2}Q_0$, then the resulting field is 2·65 kV/cm.

of the dielectric reduces the electric field increases and tends to reach its highest value when the cavity is processing a very low loss material.

7.7 Cylindrical cavities

7.7.1 Introduction

Similarly to the TE_{10n} cavity described in the previous section the field equations developed in a cylindrical cavity can be obtained theoretically from the solutions of Maxwell equations (Mooijweer, 1971; Boudouris, 1971). Cavity operation in the TM modes should be very useful since the electric field is along the direction of propagation and therefore in most cases along the dielectric material placed along its axis. Equally the TE_{11} mode has useful characteristics when circularly polarised. By letting l be the number of tangential full periods, m the radial half period and n the axial half period variations of the electromagnetic field, the mode designation becomes TM_{lmn} and the resonant wavelengths of this mode are given by (Harvey, 1963)

$$\lambda = 2\left[\left(\frac{x_{l,m}}{\pi R}\right)^2 + \left(\frac{n}{h}\right)^2\right]^{-1/2} \tag{7.88}$$

where $x_{l,m}$ is the mth root of the $J_l(x) = 0$, R is the cavity internal radius and h the cavity height. Values of these roots for some of the lower order modes are given in Table 7.2.

Table 7.2

	Roots of $J_l(x)$		
l	m	n	x_{lm}
0	1	n	2·4048
1	1	n	3·832
2	1	n	5·1356
0	2	n	5·5201
3	1	n	6·3802
1	2	n	7·0156
4	1	n	7·588
2	2	n	8·4172
0	3	n	8·6537

7.7.2 TM_{010} cavity heater

Of all possible TM modes that can develop in a cylindrical waveguide most attention has been paid to the cylindrical cavity operating in the TM_{01n} mode. This obvious preference is best explained upon considering the distribution of the electric and magnetic field within such a cavity, the most important aspect being the existence of an axial electric field. This renders such a cavity very suitable for treating filamentary materials such as textiles, fibres, yarns, tow, etc., as well as liquids placed along its axis. Furthermore, coupling the microwave energy into such a cavity is facilitated by the existence of a circumferential magnetic field, thus enabling to transfer the power from the generator to the cavity via a rectangular waveguide operating in its dominant TE_{10} mode. Such H–H coupling can be obtained by connecting the waveguide and TM_{010} cavity, as shown in Fig. 7.18, via an appropriate matching aperture.

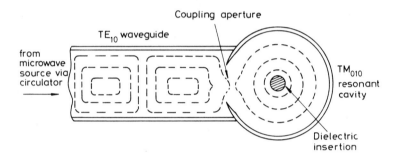

Fig. 7.18 *H–H coupling of a cylindrical cavity to the connecting waveguide*
– – – –, magnetic field lines

The solution of Maxwell's equation leads to a set of six field equations which for the TM_{010} mode reduce to the following two fields. An axial electric field which varies radially according to a Bessel function of zero order and is given by (Boudouris, 1971)

$$E_z(r) = E_{max} J_0(Kr) \tag{7.89}$$

where E_{max} is the maximum value of the electric field at the axis, r is the radius and K is a constant. The circumferential magnetic field attains the form

$$H_\phi(r) = jE_{max} \sqrt{\frac{\mu_a}{\epsilon_a}} J_1(Kr) \tag{7.90}$$

Figure 7.19 shows the distribution of the two fields in TM_{010} cavity. The factor K is given by (see Appendix IV)

$$K = \frac{\omega}{c} (\epsilon^* \mu^*)^{1/2} = 2.405/R_d \tag{7.91}$$

where R_d is the cavity radius. For a cavity without a dielectric inserted in it, the value of the factor K follows from eqn. (7.91), which yields

$$K = \omega/c = 2.405/R_0 \tag{7.92}$$

where R_0 is the radius of an empty cavity operated at the same frequency. Therefore, eliminating the angular frequency from eqns. (7.91) and (7.92) yields

$$R_d = \frac{R_0}{\sqrt{\epsilon^*}} \qquad (7.93)$$

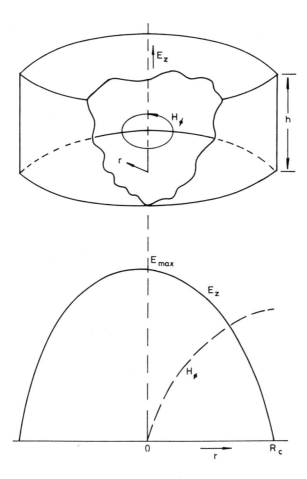

Fig. 7.19 *Electric and magnetic field strength distribution in an empty* TM_{010} *resonant cavity*

Figures 7.20(*a*) and (*b*) show the two cases just cited. In designing a TM_{010} cavity for a heating application, the material being processed usually fills a small volume around the axis, as shown in Fig. 7.20(*c*), the dielectric radius being a small fraction of the cavity radius to ensure that the dielectric does not experience wide variations of the electric field across its width. We therefore seek to design a cavity of radius R_c such that $R_0 > R_c > R_d$, for treating filamentary type of dielectrics of radii R_w, where R_w is a small fraction of R_c usually satisfying the relation $R_c \geqslant 10 R_w$.

7.7.3 *TM$_{010}$ cavity design*

Consider the cavity and two coaxial dielectrics as shown in Fig. 7.21. The electromagnetic fields in region I, i.e., within the inner dielectric, where $r < R_w$, are given

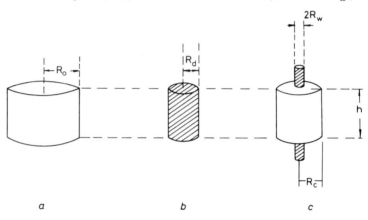

Fig. 7.20 *Development of a cylindrical TM$_{010}$ resonant cavity heater upon insertion of an axial dielectric, where $R_0 > R_c > R_d > R_w$*
a Empty cavity
b Completely filled cavity
c Partially filled cavity

by eqns. (7.89) and (7.90). However, in region II, the relevant field equations assume the form (Metaxas, 1974a)

$$E_z(r) = A_2 J_0(K_2 r) + B_2 Y_1(K_2 r) \tag{7.94}$$

and

$$H_\phi(r) = j \left(\frac{\epsilon_{a2}}{\mu_{a2}}\right)^{1/2} [A_2 J_1(K_2 r) + B_2 Y_1(K_2 r)] \tag{7.95}$$

where J and Y are Bessel's functions of the first and second kind. By applying the continuity of the tangential E and H fields at the air/dielectric interface and observing that $E_z = 0$ at $r = R_c$ we obtain the following conditional equation which controls resonance (Teasdale and Crawford, 1952):

$$f_1(p_r K_0) = f_2(K_0, R_c K_0 / R_w) \tag{7.96}$$

where f_1, and f_2 are known Bessel functions of the first and second kind, p_r^2 is the dielectric constant ratio of the inner to that of the outer dielectric. Also, K_0 is given by

$$K_0 = \omega(\epsilon_2^*)^{1/2} R_w/c \qquad (7.97)$$

Since in the majority of cases, $|\epsilon'| \gg |\epsilon''_{eff}|$, the error introduced by substituting ϵ^* in the above equations by the real part of the dielectric constant ϵ' is very small, being only 3% when $\epsilon' = 80$ and $\epsilon''_{eff} = 20$, as is the case for foodstuffs.

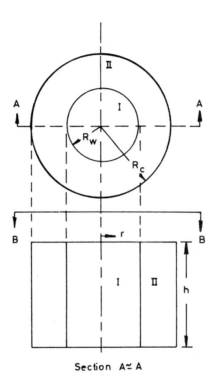

Section A≃A

Fig. 7.21   *Coaxial dielectric cylindrical cavity*

Therefore, the computations are simplified by making this assumption. Equation (7.96) is solved graphically by plotting the two functions f_1 and f_2 as a function of the parameter K_0 for a range of values of p_r^2 and R_c/R_w, as shown in Fig. 7.22. The intersections of the two family of curves give the values of K_0 which satisfy the above conditional equation.

The curves of Fig. 7.22 are general and apply for cavities with two coaxial dielectrics operating at any frequency. Specific data can be obtained for cavities operating with a single coaxial dielectric operating at the two industrial allocated frequencies in the microwave regime. For example eqn. (7.97) yields, with $\epsilon'_2 = 1$, $\omega = 2\pi f$ where f is 2.45×10^9 Hz and with R_w in mm

$$K_0 = 0.0514 R_w \qquad (7.98)$$

The intersection of an R_c/R_w curve with a $p_r^2 = \epsilon_1'$ curve gives the value of K_0 and therefore R_w is calculated using eqns. (7.98). The value of R_c can then be easily computed since the ratio of the radii is known. This is repeated for a range of R_c/R_w and a curve of R_c as a function of R_w is plotted for a given ϵ_1' (Metaxas, 1974b). Figures 7.23(a) and (b) show such diameter data for various dielectric constants in the range $1 < \epsilon_1' < 80$ at the 896/915 MHz band and 2·45 GHz respectively. As is expected, for a given dielectric constant, the larger the dielectric diameter the smaller is the cavity diameter. To give an example, for an axial dielectric workload of 12 mm in diameter and dielectric constant of 4, Fig. 7.23(a) shows that the cavity internal diameter should be about 245 mm for a TM_{010} cavity operating in the 896/915 MHz band. On the other hand, an axial water column of 7 mm in diameter ($\epsilon_1' \approx 80$), the internal cavity diameter reduces to 200 mm for operating in the same frequency band.

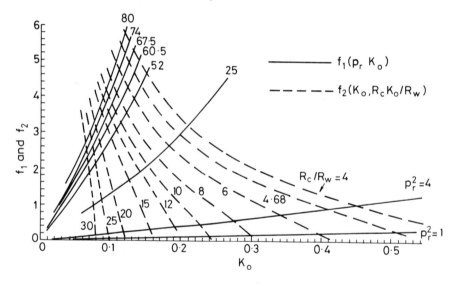

Fig. 7.22   *Curves for solving the conditional equation (7.96)*
(After Metaxas, 1974b, Copyright *The Journal of Microwave Power*, 1974. Data for $p_r^2 = 1, 4$ and 25 from Teasdale and Crawford, 1952.)

The length of a TM_{010} resonant cavity does not enter into the design because it is not a critical parameter. As a direct contrast to the previous cavity considered, namely the TE_{10n} where tuning was simply afforded by changing the plunger position and hence the cavity length, alternative tuning arrangements have to be employed with the TM_{010} cavity.

7.7.4 Impedance matching of TM_{010} cavity heater

The determination of the cavity internal diameter as a function of the dielectric diameter at a particular frequency, is a necessary step towards final design of a

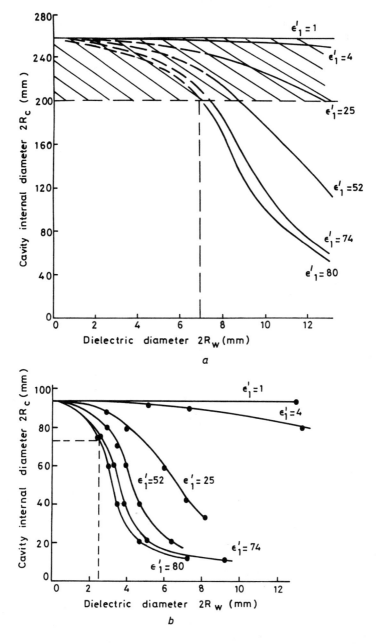

Fig. 7.23   *Internal diameter of a TM$_{010}$ cavity as a function of the dielectric diameter for various ϵ', at (a) 896/915 MHz band and at (b) 2450 MHz, assuming that $\epsilon' \gg \epsilon''_{eff}$ (After Metaxas (a) 1976b and (b) 1974b, Copyright The Journal of Microwave Power.)*

TM_{010} resonant cavity as a heating device. The design is completed by the optimisation of an aperture which effectively couples the microwave power into the cavity, where a large fraction of this power is dissipated in the dielectric workload. Figure 7.24 shows a family of VSWR vs. frequency curves for various aperture dimensions, leading to the optimum value of 44 x 23 mm and at 2·45 GHz for an axial quartz pipe filled with water of about 2·5 mm in internal diameter (Metaxas, 1974*b*). Similar data are shown in Fig. 7.25 for a TM_{010} cavity heater operating in the 900 MHz band (Metaxas, 1976*b*).

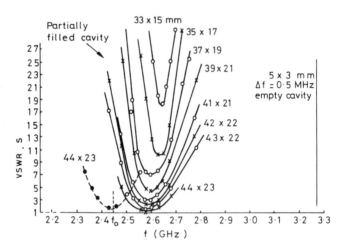

Fig. 7.24 *Input VSWR, S, as a function of the frequency for various aperture dimensions of a TM_{010} resonant heater*
(After Metaxas, 1974*b*, Copyright *The Journal of Microwave Power*, 1974.)

The loaded Q-factors, Q_L, at near critical coupling, $\beta' \simeq 1$, calculated from the data shown in Figs. 7.24 and 7.25, are about 12 and 25 respectively, giving a stable cavity operation and requiring no automatic tuning or matching control. On the other hand TM_{010} cavities treating low loss filamentary materials will have Q values of the order of a few thousand which makes their operation much more unstable and sensitive to frequency changes and such cases necessitate the use of automatic tuning and matching devices for ensuring good overall utilization efficiency (see Section 9.7). Very high electric field strengths can be established in the dielectric insertions if its ϵ''_{eff} is very low.

7.7.5 Field distribution in the dielectric region of a TM_{010} cavity heater

The radial variation of the axial electric field in a TM_{010} resonant cavity is given by eqn. (7.89) and its reduction from its maximum value of E_{max} is related to the dielectric properties of the axial dielectric through the parameter K (Metaxas, 1976*b*). Let us calculate an expression for the average axial electric field, E_{av}, which is defined from the heating formulae, eqns. (4.10) and (4.11):

$$P = \omega \epsilon_0 \epsilon''_{eff} E^2_{av} V \qquad (4.11)$$

and is shown in Fig. 7.26. Starting from eqn. (4.10) and bearing in mind that $E \cdot E^* = E^2$, we obtain

$$P = \omega \epsilon''_{eff} \epsilon_0 \int_V E^2 \, dV = \omega \epsilon_0 \epsilon''_{eff} \int_0^{R_w} E^2 \, 2\pi r h \, dr \qquad (7.99)$$

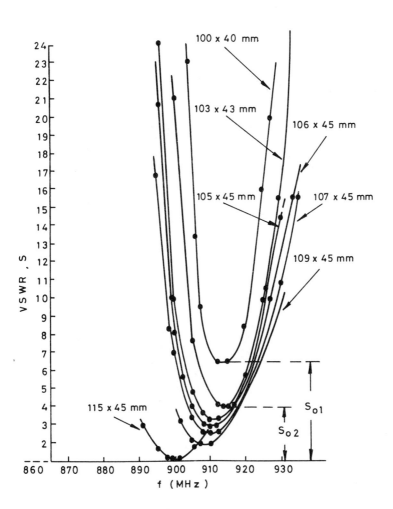

Fig. 7.25 *Input VSWR, S, as a function of the frequency for various aperture dimensions, for a TM$_{010}$ resonant cavity with a 7 mm water column inserted in its axis* (After Metaxas, 1976b, Copyright *The Journal of Microwave Power*, 1976.)

where h is the cavity's axial length. Substitution of the electric field from eqn. (7.89) and integrating yields

$$P = \omega\epsilon_0 \epsilon''_{eff} E^2_{max} \left[\frac{r^2}{2} \left(J_0^2(Kr) + J_0'^2(Kr) \right) \right]_0^{R_w} (2\pi h)$$

$$= \omega\epsilon_0 \epsilon''_{eff} \pi h R_w^2 E^2_{max} J_0^2(KR_w) \left[1 + \frac{J_0'^2(KR_w)}{J_0^2(KR_w)} \right]$$

(7.100)

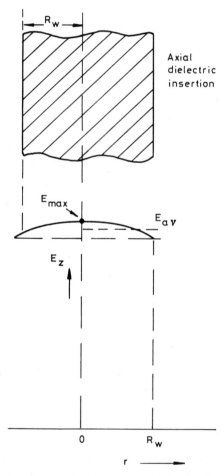

Fig. 7.26   *Variation of the axial electric field strength $E_z(r)$ across the dielectric cross-section*

Combining eqn. (7.100) with eqn. (4.11) yields, with $V = \pi R_w^2 h$,

$$E_{av} = E_{max} J_0(KR_w) \left[1 + \frac{J_0'^2(KR_w)}{J_0^2(KR_w)} \right]^{1/2} \qquad (7.101)$$

For example, consider the case of a 3·5 mm radius of water column, inserted axially into a TM_{010} cavity operating at the 896/915 MHz band. Assuming that $\epsilon' = 80, KR_w \approx 0·59$, giving $J_0 = 0·915$ and $J_0' = 0·28$ (see Appendix IV) and using eqn. (7.101), $E_{av} = 0·96 E_{max}$. The electric field therefore drops by 4% from its maximum value at the centre of the cavity.

7.7.6 Higher order mode cavities

The major drawback of the TM_{010} resonant cavity is its overall restricted area of uniform field capacity which makes it unsuitable for processing relatively large and lossy dielectric loads. One way to treat such large loads is to sacrifice some field uniformity across the sample's cross-section by employing a resonant cavity operating in a higher order mode and using other methods to equalize the uneven temperature distribution that might result due to the large electric field variations. For example, the heat treatment of large diameter pumpable fluids might be possible by utilizing their turbulence to produce a more uniform temperature distribution.

7.7.7 TM_{020} resonant cavity heater

Risman and Ohlsson (1975) have designed a resonant cylindrical cavity operating in the TM_{020} mode for processing pumpable liquids of large diameters. As a way of introducing this cavity, consider the limiting case of a completely filled TM_{010} resonant cavity with a lossy dielectric of $\epsilon' = 80$. The parameter K follows from eqn. (7.91) with $f = 915$ MHz giving $K = 0·17$/mm. For resonance the inner cavity diameter, R_c, which in this case is equal to the dielectric diameter, R_w is given by $KR_w = 2·405$ giving $2R_w = 28$ mm. However, if a dielectric of 28 mm in diameter and $\epsilon' = 80$ is now placed in a larger diameter cylinder, the latter will support the TM_{020} mode if the electric field is zero at the walls as shown in Fig. 7.27. Table 7.2 shows that the relevant root is equal to 5·5201. The cavity and dielectric load radii R_c and R_w, respectively, are related through the following expression which controls resonance:

$$J_0(KR_w) Y_0(KR_c) = J_0(KR_c) Y_0(KR_w) \qquad (7.102)$$

and using eqns. (7.102) and (7.89) their relationship for various $\tan \delta_{eff}$'s and at a frequency 2450 MHz is shown in Fig. 7.28. Multiplying the diameter scale by a factor of 2·68 enables the data to be used also at 915 MHz. The loaded Q-factor of the TM_{02n} cavity parially filled with foodstuffs was about 25, with VSWR of up to 2 yielding utilisation efficiencies of 95% with load diameters of up to 13·5 mm.

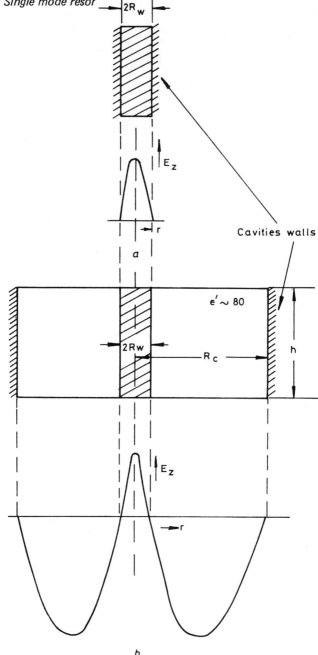

Fig. 7.27   *Development of a TM$_{020}$ resonant cavity*
   a TM$_{010}$ cavity completely filled with a dielectric
   b Partially filled TM$_{020}$ resonant cavity
   (Adapted from Risman and Ohlsson, 1975, Copyright *The Journal of Microwave Power*, 1975.)

Finally when the loss tangent of the workload is appreciable the parameter K in eqn. (7.89) becomes complex and is given by

$$K^* = \frac{\omega}{c} \sqrt{\epsilon^* \mu^*} = \frac{\omega}{c} \sqrt{\frac{\epsilon'}{2}}$$

$$[[(1 + \tan \delta_{eff}^2)^{1/2} + 1]^{1/2} - j [(1 + \tan \delta_{eff}^2)^{1/2} - 1]^{1/2}]$$

$$(7.103)$$

with $\mu^* = 1$.

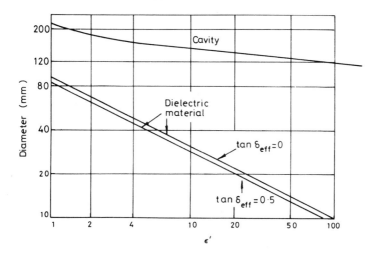

Fig. 7.28   *Dielectric material and cavity diameters as a function of ϵ' and at $f_0 = 2450$ MHz*
(After Risman and Ohlsson, 1975, Copyright *The Journal of Microwave Power,*
1975.)

7.7.8 TM_{11n} resonant cavity heater

Some improvement in the uniformity of the electric field across the cross-section of the dielectric, as well as heating larger workloads, has been achieved by employing a cylindrical cavity operating in the TM_{11n} mode. To sustain this mode the overall dimensions of the cavity are larger than those operating in the lower TM modes dealt with in the previous sections. The electric and magnetic field distributions across the cross-section are shown in Fig. 7.29. The axial electric field exhibits two maxima as the diameter of the cavity is traversed. Unlike the TM_{02n} cavity, the axial electric field variation across the cross-section of the dielectrics does not exhibit wide variations from zero to maximum value.

Bhartia *et al.* (1971) and Metaxas (1976*a*) have investigated experimentally the use of TM_{11n} mode circular cavities for heating of foodstuffs where dielectric inserts were placed at positions of E_z field maxima in the cavity. The TM_{11n}

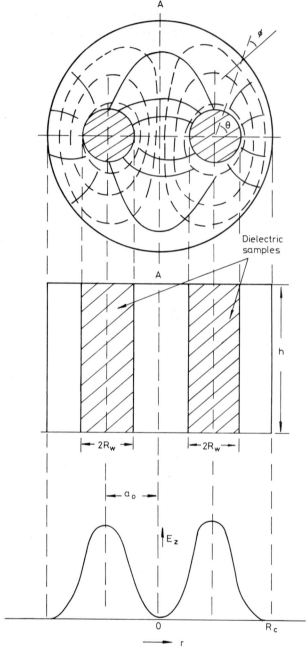

Fig. 7.29 *Electric and magnetic field distribution across the diameter of a TM_{11n} mode resonant cavity*

——— electric field lines

– – – – magnetic field lines

mode has the property that the tangential components of the electric field E_z and E_θ vanish at the cavity walls and that the tangential components of the electric field of the plane of anti-symmetry, through $A{-}A$ shown in Fig. 7.29, E_z and E_r also vanish, where the components refer to the polar co-ordinates (r, θ, z) with origin at the centre of the dielectric insert. Hodgkins (1976) has considered the theoretical design of such a cavity and derived the critical parameters for dielectric inserts with a wide range of dielectric constants and at 896/915 MHz and 2450 MHz. The inserts were centred on the nodes of the TM_{11n} mode of the empty cavity although this may not be the optimum position for large perturbations of the original mode.

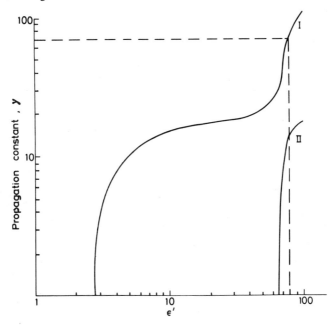

Fig. 7.30 *Design data for a TM_{11n} mode cavity*
(After Metaxas, 1976*a*.)

Figure 7.30 shows the propagation factor as a function of the relative dielectric constant for a cavity of 178 mm in internal radius while the diameter of the dielectric inserts was 50 mm. For a dielectric constant of foodstuffs of about 80 the dominant mode, curve I of Fig. 7.30, yields

$$\gamma = \frac{n\pi}{h} = 70 \qquad\qquad (7.104)$$

where h is the length of the cavity in metres. Therefore, $h = 0\cdot045\,n$ and with $n = 10$, $h = 0\cdot45\,m$, giving $TM_{11(10)}$ as the operating mode. Figure 7.30 also shows the design data for another mode, curve II, which, with the above parameter, yields the TM_{312} mode. This would be an interfering mode. So is the TE_{01n} mode since it is degenerate with the TM_{11n} mode having the same eigenvalue. Bhartia *et al.* (1971) have suggested various methods to eliminate these interferring modes which includes the insertion of metallic screens at appropriate positions inside the cavity.

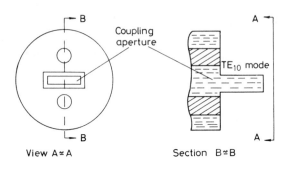

a

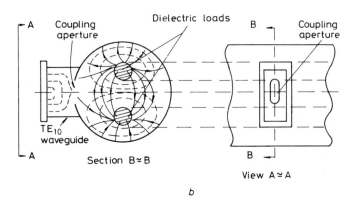

b

Fig. 7.31   *Aperture coupling of the energy into the TM_{11n} cavity*
——— electric field
– – – magnetic field
(Adapted from (*a*) Bhartia *et al.*, 1971, Copyright *The Journal of Microwave Power*, 1971 and (*b*) Metaxas, 1976*a*.)

Special mode transducers to couple the energy into the cavity are not necessary since the TM_{11n} mode can be readily set up through the magnetic field of the external connecting waveguide operating in the fundamental TE_{10} mode. Two such methods of coupling the power in the cavity via an aperture is shown in Fig. 7.31. The aperture can be experimentally optimised to obtain the correct matching.

The origin of hot spots in the treatment of high loss liquids in the TM_{11n} mode has been closely investigated (Metaxas, 1976a) by inserting a 5 mm tube within one of the 50 mm load diameter tubes, as shown in Fig. 7.32. A separate supply of water was passed through the smaller bore tube and its flow and temperature rise monitored. With a given power dissipation in the TM_{11n} cavity any changes in the power dissipated in the smaller tube as its position shifted within the large load tubes was indicative of hot spots. It was assumed that any hot spots regions were due to a load-electric field assymetries and by relocating the two load tubes the temperature differences, even at 20 kW total power dissipation within the cavity, were not higher than 7%.

7.8 Coupling apertures

7.8.1 Introduction
Before completing the treatise on coupled cavities let us briefly look more closely into the nature of apertures as coupling networks between the resonant heater and its connecting waveguide. As has been mentioned throughout this chapter the microwave energy can be transferred to the cavity from the output waveguide via an aperture. Coaxial cable feeds have also been used but these are limited to relatively low powers due to excessive cable losses. The apertures may take any shape, for example rectangular, circular, square, etc. In the neighbourhood of the aperture higher order mode fields are set up when a wave is incident upon it so that the total field satisfies the boundary conditions. A dominant mode wave is reflected from the aperture (when not matched) and some of the incident power is transmitted through the opening. The dimensions of the apertures in heating cavities are so chosen, more often than not by empirical methods, as to allow maximum transmission of the microwave energy from the source to the cavity where it is dissipated into the processed material.

7.8.2 Function of the aperture as a coupling network
The function of an aperture employed in a microwave cavity used for heating or drying purposes is to act as a coupling network for matching the impedance of the cavity with the dielectric inserted in it to the impedance of the connecting waveguide. To investigate this impedance transformation Slater (1950) has used circuit theory arguments to derive a relation between the two impedances to be matched in terms of three complex impedance coefficients of the coupling network. Alternatively, another way to characterise a lossless coupling network, such as an aperture, is to consider the transformation of the complex reflection coefficient rather than impedance when matching two waveguides of different impedances, such as in the case of a TE_{10n} resonant cavity connected at the end of a rectangular waveguide operating in the fundamental TE_{10} mode. It is possible to set up a bilinear transformation between the complex reflection coefficients in the two waveguides which has three independent parameters and is thus the

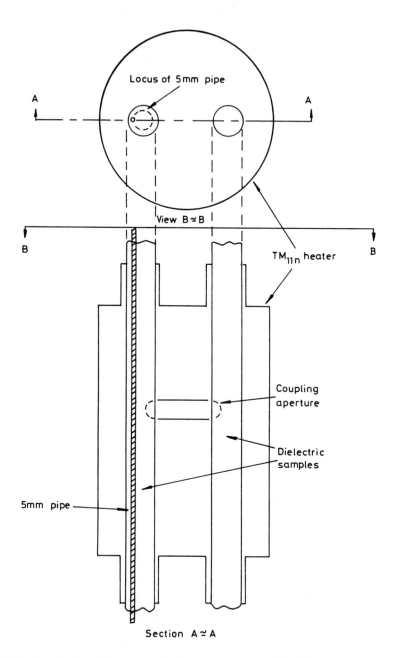

Fig. 7.32 *Investigation of hot spots during the treatment of liquid dielectrics in a TM_{11n} resonant cavity*

general expression for the impedance transformation. The three parameters can be determined experimentally by standing wave measurements.

Another way to express the transformation ratio of a coupling network is to write an expression involving the reflection coefficients and a fixed susceptance basically stating that there are corresponding points in the two waveguides such that the admittance across the left-hand point is the shunt combination of the admittance across the right-hand point and a fixed shunt susceptance jB_a. This gives justification to having more often than not described the equivalent circuit representation of an aperture by a susceptance. However, we have also represented an aperture by an ideal transformer of turns ratio N (see eqn. (7.12)). In fact both

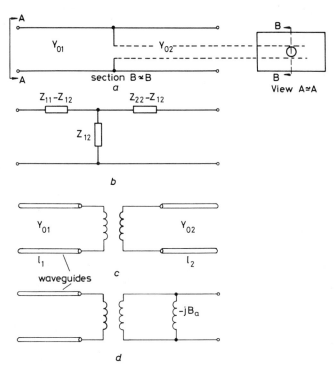

Fig. 7.33 *Aperture coupled waveguides*
   a Aperture
   b *T*-equivalent circuit for a two port
   c Equivalent circuit in terms of two waveguides and a transformer
   d Equivalent circuit in terms of a transformer, a length of a waveguide and a shunt
      susceptance

these representations are valid either individually or in combination depending upon the choice of the reference planes in the waveguides either side of the aperture, shown in Fig. 7.33 (Montgomery *et al.*, 1965; Ramo *et al.*, 1965). A *T*-equivalent circuit representation comprising the three complex impedances is shown in Fig. 7.33(*b*) while two other possible representations of the aperture involving an ideal transformer, sections of waveguides and a shunt element are shown in Figs. 7.33(*c*) and (*d*). The circuit parameters of the equivalent circuits, that is the line lengths, the turn ratio N and the susceptance are related to the three complex impedances Z_{11}, Z_{22} and Z_{12}. It can be shown, for example, that by shifting the arbitrary reference planes either side of the aperture, the latter can be represented only by an ideal transformer whose turns ratio is equivalent to $|B_a|$. Also the equivalent circuit representation does not necessarily lie in the same plane as the actual aperture. It is therefore important to realise that any desired impedance transformation can be achieved between a resonant heater and its connecting waveguide by an appropriate choice of a coupling network, such as an aperture, which transforms a matched load ($\rho_1 = 0$) in the cavity into a matched load ($\rho_2 = 0$) in the waveguide.

7.8.3 Aperture susceptance

The scattering of the microwave energy through very thin apertures, whose dimensions are small compared with the operating wavelength, connecting a waveguide to a cavity heater can be theoretically investigated using a general perturbation theory (Bethe, 1944). This entails the computation of the electric and magnetic dipole moments induced in the aperture by the exciting fields, followed by estimating the electric fields radiated by these dipole moments. The magnetic dipole moment set-up in the aperture by the exciting magnetic field will radiate power in a given mode in the cavity, only when the magnetic field of the mode to be excited within the cavity has a component parallel to the magnetic dipole moment and a similar argument follows for the electric dipole moment set-up in the aperture.

Field estimates are based on the fact that if the electric and magnetic fields vary very rapidly with the coordinates they are practically the same as can be derived from solutions of electrostatic or magnetostatic problems. Slater (1950) used such approximations to reduce the wave equations which govern the fields to their equivalent Laplacian form and sought solutions for the distribution of the electric and magnetic fields near the edges of the aperture opening corresponding to those of a static problem. These solutions can then be used to calculate the susceptance of the aperture through a relation which involves the tangential and normal components of the electric and magnetic fields in the plane of the aperture. The resulting susceptance becomes more and more accurate in the limit where the aperture opening becomes smaller and smaller compared to the wavelength. Cohn (1951) has used such theoretical arguments to derive the equivalent circuit representation of aperture coupled waveguides of the same cross-section. The normalised susceptance for a symmetrical small aperture, as shown in Fig. 7.33(*a*), is given by

$$b_i = B_a/Y_0 = \frac{ab\lambda_g}{4\pi M_p'} \tag{7.105}$$

where M_p' is the magnetic polarisability of the aperture. Apertures which contribute a positive imaginary part to the admittance are termed capacitive and those which contribute a negative part are termed inductive, as shown in Figs. 7.34(a) and (b)

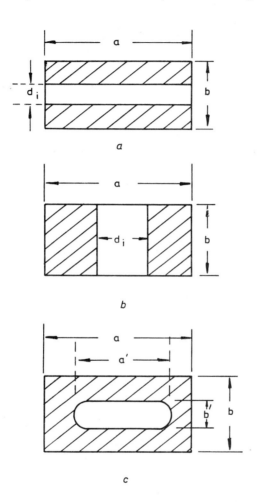

Fig. 7.34 *Typical coupling apertures in rectangular waveguides*
a Capacitive aperture
b Inductive aperture
c Resonant aperture

respectively. The normalised susceptance of the symmetrical capacitive aperture shown in Fig. 7.34(a) can be approximated to (Montgomery *et al.*, 1965)

$$b_i = \frac{4b}{\lambda_g} \, \text{lncsc} \, \frac{\pi d_i}{2b} \tag{7.106}$$

Similarly for the inductive aperture shown in Fig. 7.34(*b*), the normalised susceptance is approximately equal to

$$b_i = -\frac{\lambda_g}{a} \cot^2 \frac{\pi d_i}{2a} \tag{7.107}$$

Since apertures can exhibit capacitive or inductive susceptances it is possible,
therefore, to construct an aperture which contains both inductive and capacitive
elements, as shown in Fig. 7.34(*c*), which exhibits resonant features. The normalised susceptance of such an iris has been calculated by Cohn (1952) through Fosters
theorem and is given by

$$b_i = -ab\,\lambda_g\,(1-(f/f_0)^2)/4\pi M_p' \tag{7.108}$$

where f_0 is the resonant frequency of the aperture. At $f \ll f_0$ the susceptance
becomes equal to that given by eqn. (7.105). For resonance the following empirical
relation has been suggested by Slater (1942):[*]

$$\frac{a}{b}\sqrt{1-(\lambda_0/2a)^2} = \frac{a'}{b'}\sqrt{1-(\lambda_0/2a')^2} \tag{7.109}$$

giving the dimensions of the aperture. Such apertures are necessary to match
cavities with dielectric insertions of high loss factors, ϵ_{eff}'', such as in microwave
drying applications. These apertures are experimentally optimised to obtain a
condition of as near critical coupling as possible. For narrow, rounded end apertures, where $b'/a' \ll 0.1$, the resonant length follows the empirical relation

$$a' = \frac{\lambda_0}{2} + 0.273b' \tag{7.110}$$

For very small (b'/a') ratios eqn. (7.109) reduces to $a' \simeq \lambda_0/2$.

7.9 Power transfer into a resonant cavity heater

Energy utilisation is of paramount importance in any application involving microwave power. Loss of available power could arise due to aperture mismatching. In
terms of the reflection coefficient ρ and its conjugate value ρ^*, the power input
to a resonant heater is given by

$$P_{in} = P_0(1-\rho\rho^*) \tag{7.111}$$

where P_0 is the incident power at the plane of the aperture. Using eqns. (7.15)
and (7.18) the two reflection coefficients become

$$\rho = \frac{(\beta'-1)-j2Q_0(\Delta\omega/\omega_0)}{(\beta'+1)+j2Q_0(\Delta\omega/\omega_0)} \tag{7.112}$$

[*] From *Microwave Transmission* by J. C. Slater. Copyright © 1942, McGraw-Hill. Used with the
permission of McGraw-Hill Book Co.

and

$$\rho^* = \frac{(\beta' - 1) + j\, 2\, Q_0(\Delta\omega/\omega_0)}{(\beta' + 1) - 2j Q_0(\Delta\omega/\omega_0)} \tag{7.113}$$

Combining eqns. (7.112) and (7.113) yields

$$\rho\rho^* = \frac{(\beta' - 1)^2 + (2 Q_0(\Delta\omega/\omega_0))^2}{(\beta' + 1)^2 + (2 Q_0(\Delta\omega/\omega_0))^2} \tag{7.114}$$

Substitution of eqn. (7.114) into eqn. (7.111) yields

$$P_{in} = \frac{P_0\, 4\beta'}{(\beta' + 1)^2 + (2(\Delta\omega/\omega_0)Q_0)^2} \tag{7.115}$$

At resonance

$$P_{in} = \frac{4 P_0 \beta'}{(1 + \beta')^2} \tag{7.116}$$

Since $P_{in}(\beta') \equiv P_{in}(1/\beta')$ at resonance the degree of coupling does not influence the power transfer into the heater and for both under and over coupling we have

$$P_{in} = \frac{4 S_0 P_0}{(S_0 + 1)^2} \tag{7.117}$$

Off resonance, however, the expressions for the power transfer become for under coupling

$$P_{in} = \frac{4 S_0 P_0}{(S_0 + 1)^2 + (2(\Delta\omega/\omega_0)Q_0 S_0)^2} \tag{7.118}$$

and for over coupling

$$P_{in} = \frac{4 S_0 P_0}{(S_0 + 1)^2 + (2(\Delta\omega/\omega_0)Q_0)^2} \tag{7.119}$$

7.10 Efficiency of resonant cavity heaters

Figure 7.35 shows a typical heating circuit using a microwave heater to dissipate some power, P_w, in the dielectric from an input power P_{in}. The mains power, P_m,

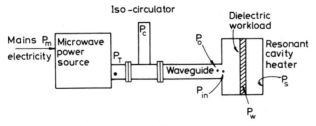

Fig. 7.35 *Power parameters for derivation of efficiency*

is transformed to useful microwave output power, P_T, through a magnetron source and assuming some external power dissipation, P_c, the remainder power, P_0, is incident on the cavity's matching aperture. The term, P_c, includes direct dissipation in an external circuit such as a ferrite circulator as well as reflected power, $(P_0 - P_{in})$, arising due to the aperture mismatch. The overall circuit efficiency is defined as

$$n_0 = 100 \frac{P_w}{P_m} = 100 \frac{P_T}{P_m} \times \frac{P_0}{P_T} \times \frac{P_{in}}{P_0} \times \frac{P_w}{P_{in}}$$

$$= 100 n_g \times n_t \times n_c \times n_a \tag{7.120}$$

where the subscripts g, t, c and a, stand for generation, transfer, coupling and applicator. The last three efficiencies can be grouped together to form the total utilization efficiency, n_u, or

$$n_u = n_t \times n_c \times n_a \tag{7.121}$$

The first two terms of eqn. (7.120) depend on the particular type of generator and ancillary equipment used, such as a circulator. An expression for the ratio

$$P_w/P_{in} = \frac{P_w}{P_s + P_w} \tag{7.122}$$

can be derived as follows. Eliminating U from eqns. (7.2) and (7.3) and bearing in mind that $P_s = \sqrt{\omega_0}\, F$, while $P_s' = \sqrt{\omega_0'}\, F$, through the use of eqn. (4.70), we obtain

$$\sqrt{\omega_0}\, F + P_w = \frac{\omega_0}{\sqrt{\omega_0'}} \frac{Q_0'}{Q_0} F \tag{7.123}$$

Re-arranging eqn. (7.123) yields

$$P_w = P_s \left(\frac{\sqrt{\omega_0}\, Q_0'}{\sqrt{\omega_0'}\, Q_0} - 1 \right) \tag{7.124}$$

Making use of eqn. (7.124), eqn. (7.122) yields

$$P_w/P_{in} = \left(1 - \sqrt{\frac{\omega_0'}{\omega_0} \frac{Q_0}{Q_0'}} \right) \tag{7.125}$$

Substituting for the ratio P_{in}/P_0 from eqn. (7.117) and for the ratio P_w/P_{in} from eqn. (7.125), eqn. (7.120) yields at resonance

$$n_0 = \frac{100 n_g n_t 4 S_0}{(S_0 + 1)^2} \left(1 - \sqrt{\frac{\omega_0'}{\omega_0} \frac{Q_0}{Q_0'}} \right) \tag{7.126}$$

Ideally $S_0 \to 1$, $\omega_0 \approx \omega_0'$ and $Q_0' \gg Q_0$, yielding $n_0 = 100\, n_g n_t$. However, in a typical 900 MHz industrial resonant system the following values could be applied: $n_g = 0.8$, $n_t = 0.95$, $S_0 = 1.5$, $\omega_0' = 1000$ MHz, $\omega_0 = 915$ MHz, $Q_0 = 50$ and $Q_0' = 10^3$. Substitution gives an overall efficiency of about 70%. With less efficient magnetrons a more typical value would be about 50%.

7.11 References

ALTMAN, J. L., *Microwave Circuits*. Princeton, New Jersey (1964).

BETHE, H. A., 'Theory of diffraction by small holes', *Phys. Rev.* 66, 163 (1944).

BHARTIA, P., KASHYAP, S. C., STUCHLY, S. S. and HAMID, M. A. K., 'Tuning, coupling and matching of microwave heating applicators at higher order modes', *J. Microwave Power* 6(3), 221 (1971).

BOUDOURIS, G., *Cavites Electromagnetiques*. Dunod, Paris (1971).

COHN, S. B., *Proc. IRE* 39, 1416 (1951).

COHN, S. B., 'Microwave coupling by large apertures', *Proc. IRE* 40, 696 (1952).

GINZTON, E. L., *Microwave Measurements*. McGraw-Hill, New York (1957).

HARVEY, A. F., *Microwave Engineering*, p. 194 Academic Press, New York (1963).

HODGKINS, W. R., 'Modes in cylindrical waveguides containing a circular cylindrical dielectric insert', Electricity Council Research Centre Internal Report No. MM26, March (1976).

HOLME, I., and METAXAS, A. C., 'Microwave drying of nylon tufted carpets. III. Field trials', *J. Microwave Power* 14(4), 367 (1979).

JACKSON, J. D., *Classical Electrodynamics*. John Wiley & Sons, New York (1962).

METAXAS, A. C., Design of a TM_{010} resonant cavity to process liquid foodstuffs at 896 MHz, Electricity Council Research Centre, Capenhurst, Chester, U.K., Internal Report No. M767, September (1974a).

METAXAS, A. C., 'Design of a TM_{010} resonant cavity as a heating device at 2.45 GHz', *J. Microwave Power* 9(2), 123 (1974b).

METAXAS, A. C., 'Design of equipment for microwave sterilisation of food particulates', Electricity Council Research Centre, Capenhurst, Chester, U.K., Internal Report No. M/921, July (1976a).

METAXAS, A. C., 'Rapid heating of liquid foodstuffs at 896 MHz', *J. Microwave Power* 11(2), 105 (1976b).

METAXAS, A. C., and MEREDITH, R., 'Microwave drying of nylon tufted carpets. II. Electrical characterisation of a modified TE_{10n} resonant cavity', *J. Microwave Power* 13(4), 315 (1978).

MONTGOMERY, C. G., DICKE, R. H., and PURCELL, E. M., *Principles of Microwave Circuits*, pp. 163–167. Dover Publications, New York (1965).

MOOIJWEER, H., *Microwave Techniques*. McMillan Press Ltd./Philips (1971).

RAMO, S., WHINNERY, J. R., and VAN DUZER, T., *Fields and Waves in Communication Electronics*. John Wiley & Sons, New York (1965).

RISMAN, P. O., and OHLSSON, D. T., 'Theory for and experiments with a TM_{02n} applicator', *J. Microwave Power* 10(3), 271 (1975).

SLATER, J., *Microwave Transmission*, pp. 185–187. McGraw-Hill, New York (1942).

SLATER, J., *Microwave Electronics*, Van Nostrand, Princeton, New Jersey (1950).

SUCHER, M., and FOX, J., *Handbook of Microwave Measurements*, Vol. 2, Third Edition. Polytechnic Institute of Brooklyn (1963).

TEASDALE, R. D., and CRAWFORD, G. N., 'Cut off frequency for circular waveguides containing two coaxial dielectrics', *Proc. National Electricity Conference* 8, 296 (1952).

THOMAS, R. L., *A Practical Introduction to Impedance Matching*. Artech House, Inc., Second Printing (1978).

Chapter 8

Special applicator structures

8.1 Introduction

In the last three Chapters 5, 6 and 7 we have described applicators which broadly adhere to recognised categories, namely travelling wave, multimode and single mode resonant cavities respectively. Although the majority of industrial microwave equipment does in general conform to one of these three categories, there are other forms of applicator which are special by the nature of their construction or are hybrids of the recognised categories discussed previously. The aim of this chapter is to give breadth to the subject of microwave applicators by selecting from numerous publications in the literature, a small number of atypical design configurations which can, under special circumstances, be used successfully in industry. The order in which these applicators are presented in the following paragraphs is not significant since we are dealing with special designs of individual applicators which may be treated in isolation. Because of the vast number of different design configurations available it would not be possible to present a comprehensive survey and those presented will not be treated rigorously but qualitative descriptions of design and performance will be given to highlight the important features.

8.2 A modified TE_{10n} twin-cavity resonant dryer

In Section 7.6.2 we have discussed a twin cavity applicator based on two TE_{10n} resonant cavities displaced by $\frac{1}{4}\lambda_g$ to superimpose the electric fields and produce a near-constant field distribution. Such a twin cavity arrangement, used singly or in multiple units is useful for processing planar materials, such as paper, board, wood planks, etc. However this system as it stands cannot be used for some textile fabrics or carpets since these are transported by the edges using sharp metallic pins, called stenter pins, which may promote arcing in the presence of high electric field strengths in the TE_{10n} cavities. Stentering, as it is called, also ensures the dimensional stability of the fabrics in the final stages of drying.

To overcome the arcing problem when drying tufted carpets, Metaxas and

Meredith (1978) have suitably modified the above twin cavity applicator to suit the special configurational requirement presented by stenter held fabrics, such as tufted carpets. The feed and remote ends of the single cavity have been replaced with *T*-junctions, as shown in Fig. 8.1 (Meredith and Metaxas, 1978). The

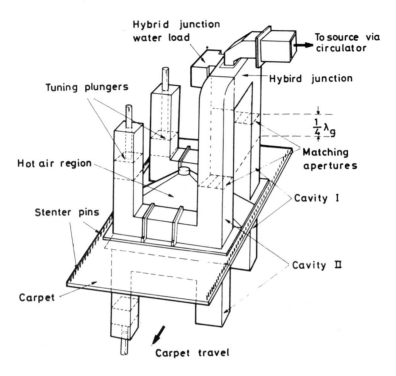

Fig. 8.1   *Modified TE$_{10n}$ twin cavity dryer*
(After Holme and Metaxas, 1979, Copyright *The Journal of Microwave Power*, 1979.)

impedances looking outwards from the *T*'s, along the two directions of the cross bar are arranged to be symmetrical by equi-distant short-circuit plates. At the feed end one short-circuit plate houses the matching aperture through which the microwave energy is fed into the cavity and its symmetrical twin is a plane short-circuit plate. At the remote end there are a pair of mechanically ganged non-contacting short-circuit plungers arranged to be symmetrical about the *T*-junction. The slotted *T*-junctions ensure that the stenter chains with their associated sharp pins reside outside each cavity, where the fringing fields are very much smaller than those set up in the cavity, thus reducing the hazard of arcing. Inevitably the part of the carpet near the pins is underdried. However, this part is finally trimmed away to produce the required width.

Typical data for drying foam-backed carpets of about 2 kg/m^2 in specific weight is shown in Fig. 8.2. The effectiveness of microwave energy in levelling the moisture distribution across and along the carpets has been analysed using such drying data. Figure 8.3 shows plots of percentage moisture difference and percentage mean before and after microwave drying. The construction of a typical line *AB* is as follows. *A* is a point representing the moisture content conditions of the

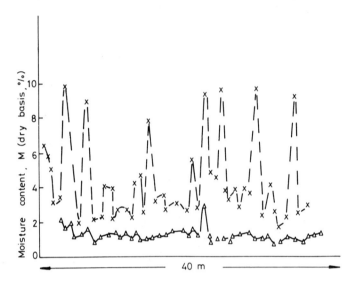

Fig. 8.2 *Final drying of foam-backed carpets at 1 m/min with microwaves for a 40 m long sample, 0·9 m in width*
xxx before microwave drying
△△△ after microwave drying
(After Holme and Metaxas, 1979, Copyright *The Journal of Microwave Power*, 1979.)

samples entering the microwave dryer, i.e., 20% mean moisture with a peak variation 18%, while *B* represents the moisture content conditions of the carpet after it has been treated in the microwave dryer where there was a 10% mean moisture and a 13% peak to peak variation. The steeper the slope of *AB* the better the degree of levelling, while a nearly horizontal line represents microwave drying with no levelling at all. The degree of levelling was found to be much better in the mean moisture range 1–10% (Holme and Metaxas, 1979).

8.3 A TE_{10n} applicator near cutoff

A way of overcoming the cosine field distribution in a TE_{10n} resonant cavity and to produce a constant field distribution along its entire width is to operate such a cavity near cutoff conditions. Figure 8.4 shows qualitatively the variation between the angular frequency and the phase constant, β. Point O represents a typical operating condition where the frequency is about one and a half times the cutoff value, at which the angle $\theta < \pi/2$ radians. Moreover the waveguide wavelength in terms of the cutoff wavelength takes the following form, shown also qualitatively in Fig. 8.5,

$$\lambda_g = \frac{\lambda_0'}{[1 - (\lambda_0'/\lambda_c)^2]^{1/2}} \qquad (8.1)$$

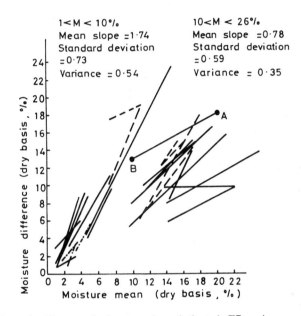

Fig. 8.3 *Moisture levelling on tufted carpets through the twin TE_{10n} dryer*
——— 1 m/min
— — — 3 m/min
—.—.—. 6 m/min
(After Holme and Metaxas, 1979, Copyright *The Journal of Microwave Power*, 1979.)

where λ_0' is the free space wavelength. In a rectangular waveguide the cutoff wavelength λ_c is equal to $2a$ where a is the larger cross-sectional dimension. Figures 8.4 and 8.5 show, therefore, that as the operating point is moved from O to A, $\beta \to 0$, $\theta \to \pi/2$ and $\omega \to \omega_c$ which results in a guide wavelength which is many times its value at normal operation. This has been mechanically accomplished (Curran, 1966) by careful choice of the larger dimensions $2a$, such as to give adequate broadening of the waveguide wavelength but at the same time to be far enough removed from the assymtotic value, $\lambda_0' = 2a$, to ensure stable operation. A schematic presentation of such an applicator is shown in Fig. 8.6, which includes $\frac{1}{4}\lambda_g$ terminations on either side in order to satisfy the required

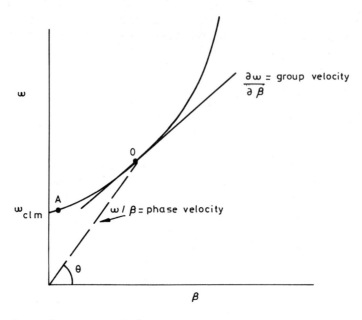

Fig. 8.4   *Angular frequency on a function of phase constant*

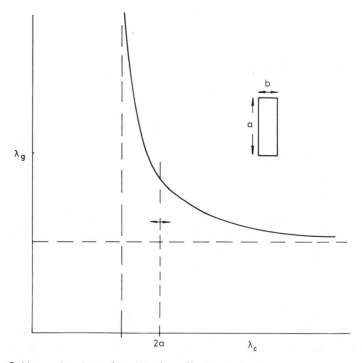

Fig. 8.5   *Guide wavelength as a function of cutoff wavelength*

boundary conditions. The input impedance Z_{in} prior to each short circuit termination is given by

$$Z_{in} = jZ_0 \tan \beta x = jZ_0 \tan \frac{2\pi x}{\lambda_g} \qquad (8.2)$$

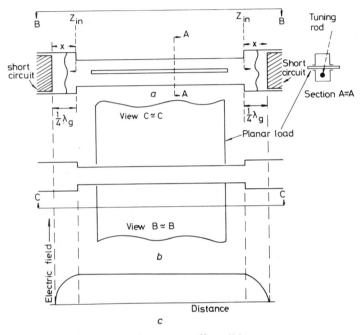

Fig. 8.6   *Waveguide applicator operating near cutoff conditions*
  a Side elevation of applicator
  b Plan of applicator
  c Electric field distribution across the applicator width
  (After Curran, Symposium of Electrical Power and Chemical Engineering, 1966.)

where x is the length of the termination, as shown in Fig. 8.6. For $x = \frac{1}{4}\lambda_g$ the input impedance is infinity, which permits the E field to be established at a high value along the length of the near cutoff waveguide.

Such a system was effectively used to process a thin web of plastic film, with fine tuning being afforded by an off-centre dielectric rod in the waveguide applicator.

8.4 Periodic structure applicator

The characteristics of transmission lines and waveguides are altered when these are periodically loaded with inductive or capacitive reactances. VanKoughnett *et al.* (1975) have described such a structure for drying low loss web materials. The propagation of an electromagnetic wave along such a structure differs greatly in the absence of the periodic loading. In a particular example of a parallel plate loaded with inductive posts all along its length, as shown in Fig. 8.7, the propagation constants γ' and γ in the rodded structure and in the absence of the rods take the form

$$\cosh(\gamma' x) = \cosh(\gamma x) + j\,\frac{B}{2\,Y_0}\,\sinh(\gamma x) \qquad (8.3)$$

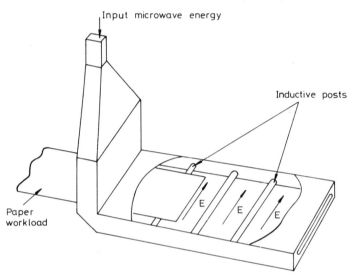

Fig. 8.7   *Periodic structure applicator*
(After VanKoughnett *et al.*, 1975, Copyright *The Journal of Microwave Power*, 1975.)

where B is the susceptance of the rods. Waves are propagating in different directions within the rods setting up a standing wave field, whose absolute value depends upon the susceptance. This electric field can be strongly coupled to a thin low loss paper load provided conditions are so chosen to ensure operation within a frequency band where strong wave attenuation occurs. The rod susceptance and the periodicity is chosen to match the structure to the input waveguide over a broad band of frequencies. The attenuation per unit length is much larger than in the equivalent rodless structure, this ratio being equal to $\operatorname{cosec}(\beta x)$ for a structure which is broadly matched to the input waveguide by a two-section quarter-wave transformer.

The length of the periodic structure is governed by the characteristics of the workload to be processed and the attenuation required. Extremely high utilization efficiencies have been attained with such a periodic structure applicator when operated in a resonant configuration to dry low loss craft paper. The efficiency drops considerably as the input moisture is reduced which necessitates tuning of the structure for every moisture content level.

8.5 Rectangular TEM applicator

Most applicator structures discussed so far were operated either in the transverse electric (TE) or the transverse magnetic (TM) mode. However waveguides operated in the TEM mode can also be effectively used in microwave heating or drying applications. In such a mode both the electric and magnetic fields are transverse to the propagation, as shown in Fig. 8.8(a), and the wall currents flow longitudinally (in Fig. 8.8(a) into the paper), that is parallel to the wave propagation. A

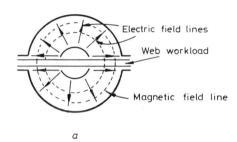

a

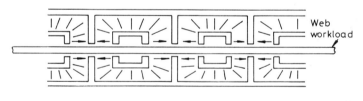

.b

Fig. 8.8   (a) *Coaxial circular waveguide supporting the TEM mode.* (b) *Development of a coaxial rectangular meander applicator supporting the TEM mode*
(After Johnston and Voss, Proceedings IMPI Symposium, Monterey, 1971.)

non-radiating longitudinal slot can be used to insert planar materials into the TEM structure for processing. Such a structure can be made an effective dryer by constructing it in a meander form described in Chapter 5. Moreover the centre conductor can be used as the air knife in combination drying resulting in improved efficiency.

Johnston and Voss (1971) described a combination dryer based on the above principle, using rectangular coaxial waveguides supporting the TEM mode as shown in Fig. 8.8(b). Altman's (1964) perturbation theory is used to compute the degree of attenuation and is found to be critically dependent on the dimensions of the inner rectangular conductor relative to the surrounding waveguide. Higher or lower attenuation per unit length, α, can be achieved compared to the equivalent waveguide operating in the TE mode by careful choice of the relative dimensions. It is advisable to round the corners of the inner conductor to prevent arcing.

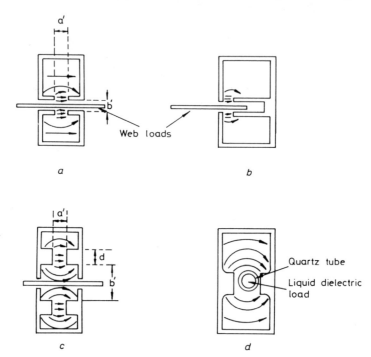

Fig. 8.9   *Ridged applicators*
(After Bleackley *et al.*, 1972, Copyright *The Journal of Microwave Power*, 1972.)

8.6   Ridged applicators

The electric field distribution in waveguides can be modified by the insertion of metallic ridges. When meander type applicators were discussed in Chapter 5 it was pointed out that a serious limitation in their use is often encountered when processing materials which are either too lossy or alternatively have a very low loss indeed. This is because in the lossy case the power dissipated per pass is too great for uniform distribution of energy across the web width while in a low loss case the number of passes become unacceptably large for a reasonable absorption of the available power. Although the resonant structures discussed in Chapter 7 and in Sections 8.2 and 8.3 can offer alternative solutions, the introduction of ridges of given dimensions into meander applicators widens their range of application.

Bleackley *et al.* (1972) have presented various configurations of ridged structures, as shown in Fig. 8.9. In (*a*) the electric field around the ridge is increased which enhances the coupling to the low loss web material and results in a more compact applicator than the equivalent structure without ridges. The width a' must be properly chosen. If it is too large the coupling does not increase appreciably, whereas a too small width increases the likelihood of breakdown. Effective coupling is produced if $1.5 b' < a' < 2 b'$.

Figure 8.9(b) shows a similar field enhancement, due to ridge insertion, which gives good coupling for drying narrow strips of glue, etc. Reduction of the effective field can be achieved by insertion of two double ridges away from the processed load, as shown in Fig. 8.9(c). In this way the effective coupling is reduced rendering meander type applicators suitable for treating very lossy planar webs. The coupling can be reduced by any amount by proper optimisation of the parameters a', b' and d. For bulk materials or liquids flowing through low loss tubes the ridge arrangement in Fig. 8.9(d) can be used.

Assymmetrical ridges can also be used in waveguides in order to design applicators with specific characteristics. One such system designed to dry ink lines on paper backed by a metallic drum (VanKoughnett and Wyslouzil, 1972) is shown in Fig. 8.10. A single ridge pointed out at the tip extends the side dimension of the waveguide and concentrates the field lines in to the paper. Chokes are used at either side to limit the radiation escaping due to the establishment of a TEM mode between the metallic substrate and the waveguide walls.

8.7 Disc loaded applicator

The difficulty with processing filamentary materials with microwaves lies in the fact that one has to concentrate the power within narrow geometrical limits in order to attain extremely high coupling efficiency. Geometrically the most obvious applicator to use is a circular cavity operated in the TM_{010} mode. However the Q-factor of such a cavity when processing filamentary materials is usually quite high ($> 10^3$) which necessitates automatic tuning devices to maintain resonance.

VanKoughnett *et al.* (1974) have described a disc loaded TM_{01} cavity which largely overcomes the above problem, exhibits high efficiency and is free from mode control problems. It can also be operated in both resonant and non-resonant mode. Such a cavity, shown schematically in Fig. 8.11, possesses the following frequency dispersion relation, assuming a small bandwidth,

$$f \simeq \tfrac{1}{2} (f_{c2} + f_{c1}) - \tfrac{1}{2} (f_{c2} - f_{c1}) \cos (\beta x_d) \qquad (8.4)$$

where f_{c2} and f_{c1} are the upper and lower cutoff frequencies of the pass band, βx_d is the phase shift per period of the structure for the fundamental space harmonic at frequency f and x_d is the spacing of the discs. The aperture diameter $2a'$ determines the cavity's bandwidth while the diameter $2b'$ determines the lower cutoff frequency.

Such a cavity is free of other modes when TM_{01} mode pass band is centred at the operating frequency 2450 MHz and for this the phase shift per period is

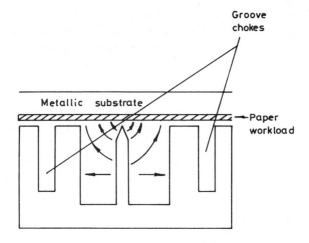

Fig. 8.10   *Ink line dryer applicator*
(After VanKoughnett and Wyslouzil, 1972, Copyright *The Journal of Microwave Power*, 1972.)

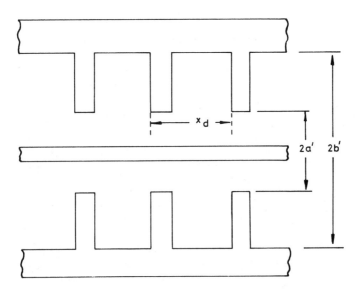

Fig. 8.11   *Disc loaded circular applicator*
(After VanKoughnett *et al.*, 1974, Copyright *The Journal of Microwave Power*, 1974.)

chosen such that $\beta x_d = \pi/2$ at the centre frequency. When the phase velocity in the disc loaded cavity is equal to the free space propagation velocity, c, the electric field across a diameter, $2a'$, is essentially constant and is given by

$$E = \lambda_0' (480P)^{1/2}/\pi(a')^2 \tag{8.5}$$

where P is the r.m.s. value of the power flow along the structure. Since the fringing fields at the edges exceed E and cause breakdown of the surrounding air, an adequate safety margin is incorporated in the design when choosing the dimension, $2a'$. The overall diameter $2b'$, and bandwidth were determined experimentally using a short-circuited disc loaded structure.

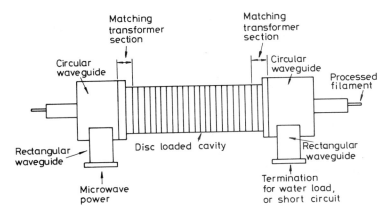

Fig. 8.12   *External view of disc load cavity with feed and termination sections* (After VanKoughnett *et al.*, 1974, Copyright *The Journal of Microwave Power*, 1974.)

The energy is fed into the cavity via a rectangular to circular transition while matching the circular waveguide to the circular TM_{01} disc loaded cavity was accomplished via a two-section quarter-wave transformer. The latter has to match a VSWR of about 50 which is presented to the circular waveguide by the disc loaded cavity terminated by its characteristic impedance. The complete applicator is shown in Fig. 8.12 while Fig. 8.13 shows the input VSWR when its output is terminated in a matched load.

The overall design of such an applicator enhances its versatility in that it can be used in a variety of ways depending upon the properties of the processed filaments. For example the output end may be short-circuited or terminated by a matched load or the applicator may be used in a resonant system. In the latter case an automatic tuning system should be incorporated between the exciting source and the applicator. The major disadvantage of such an applicator is its structural design complexity and consequent cost.

8.8 Dielectric loaded applicators

An alternative but far simpler applicator for heating filamentary materials has
been presented by VanKoughnett (1972) and is shown in Fig. 8.14. This utilizes
a slotted waveguide terminated at both ends by *E*-plane bends, for feeding the

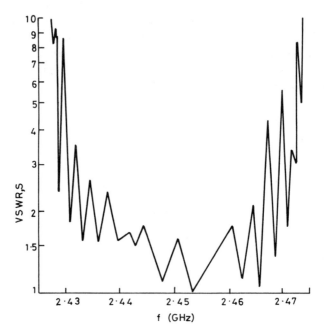

Fig. 8.13   *Input VSWR of the disc loaded cavity as a function of the frequency*
(After VanKoughnett *et al.*, 1974, Copyright *The Journal of Microwave Power*,
1974.)

power in and connecting appropriate terminations as described in Section 8.8. The
straight portion of the waveguide is partially filled with a low loss dielectric. The
filament to be processed rests in a small groove at the centre of the dielecttic
surface. An axial electric field is established at the filament/dielectric interface
which couples the energy to the material. For an exponential decay of the power
along the filament surface and assuming the losses in the rectangular waveguide
and low loss dielectric to be negligible, the fractional loss of power in a length l is
$P = 2\alpha l$ which is equal to the power dissipated in the filament, given by eqn.
(4.11), of volume $A_c l$, where A_c is its cross-section. The attenuation per free
space wavelength in dB units is given by $\Gamma A_c \epsilon_{eff}''$ where Γ is a coupling factor
given by

$$\Gamma = \frac{8 \cdot 686 \, (E_{rms})^2}{120 \, P} \tag{8.6}$$

where P is the total power carried in the waveguide and E_{rms} is the r.m.s. value of the axial electric field. The coupling factor, which must be maximised for optimum transfer of power to the filament, was computed theoretically. It was found that for maximum Γ the height of the dielectric was half way up the waveguide. To process low loss filaments the applicator must be operated in a resonant mode and with higher dielectric constant filling materials. In a particular example at 2·45 GHz the coupling factor increased by about five times by increasing the dielectric constant from 2 to 4.

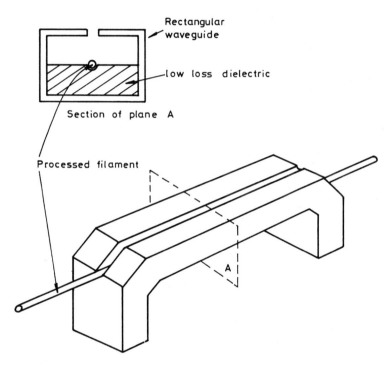

Fig. 8.14 *Dielectric loaded applicator*
(After VanKoughnett, 1972, Copyright *The Journal of Microwave Power*, 1972.)

Low loss dielectrics have also been used in circular waveguides to construct applicators with specific characteristics. Figure 8.15 shows two such examples. In Fig. 8.15(*a*) a hollow dielectric tube is inserted axially into the circular waveguide into which microwave power is launched via another small circular waveguide operating in the TE_{11} mode (the latter is not shown in the diagram). The particulates to be processed are fed from the top and are compacted around the dielectric tube due to gravity flow. The dielectric tube is transparent to the microwave energy which allows its passage through it in a manner of a leaky waveguide (Hamid *et al.*, 1975). Therefore the microwave power propagates down the outer layer forming the load in a hybrid EH_{11} mode and power is thus absorbed by the

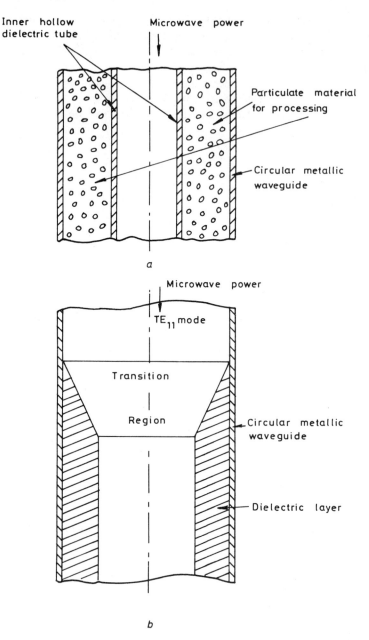

Fig. 8.15   *Circular waveguide applicators partially loaded with low loss dielectric*
        a (After Hamid *et al.*, 1975, Copyright *The Journal of Microwave power*, 1975.)
        b (After Gupta, 1976, Copyright *The Journal of Microwave Power*, 1976.)

particulate material. This hybrid mode contains coupled components of both the electric and magnetic fields in the axial direction. Such an applicator was used to process soya beans and other seeds for destroying the growth inhibitor enzyme antitripsin by roasting at above 105°C for two minutes.

Figure 8.15(*b*) shows another example of a partially loaded cylindrical waveguide. In this applicator a low loss dielectric layer is placed in contact with the walls of a circular waveguide to improve the power flow distribution when it is operated in the dominant TE_{11} mode. In the absence of the dielectric layer the power flow is maximum at the centre and falls monotonically with the radial distance, following a $[J_1'(Kr)]^2$ variation. Any dielectric load, of comparable size to the diameter of the circular waveguide, allowed to pass through it would thus undergo non-uniform heating due to this power flow variation. In an attempt to rectify this Gupta (1976) proposed the insertion of the dielectric layer which essentially modifies the boundary conditions at the surface and avoids the transverse decay of the fields by adjusting the thickness of the dielectric layer. The dielectric layer must be tapered at the top in order to modify the launched TE_{11} mode in the empty waveguide into the lowest order hybrid mode in the layered waveguide. This, however, results in non-uniform power flow distribution in radial, azimuthal as well as in the axial direction in the transition region. Alternative coupling schemes have been put forward to overcome this. Such a layered waveguide can also be used in circular waveguide applicators operating in the symmetric TM_{01} mode previously described in Chapter 7.

8.9 Travelling resonator applicator

A loop circuit wave recirculation to enhance the electric field distribution has been proposed for microwave processing of low loss web and thread type materials (Suzuki *et al.*, 1975). Essentially the system consists of a ring into which microwave energy is launched via a directional coupler, as shown in Fig. 8.16. The applicator, into which the material is inserted, forms part of the ring. The aim is to increase the power flow circulating in the ring many-fold relative to the incident power in the main waveguide by suitable choice of the circuit parameters without creating a standing wave condition. If a_1 to a_4 and b_1 to b_4 are the voltages of the input and the output waves from the directional coupler respectively, assuming no reflections in the ring circuit, the ratio of the circulating wave in the ring circuit to the incident wave in the main waveguide is given

$$(b_4/a_1) = jC_{14}/[1 - T_v(1 - C_{14}^2)^{1/2} e^{-j\beta x}] \tag{8.7}$$

where C_{14} is the voltage coupling coefficient, T_v is the voltage transmission coefficient defined as (a_3/b_1), β is the phase constant and x is the length of the loop circuit. This ratio and hence the power ratio $(|b_4|/|a_1|)^2$ attain maximum values when $\beta n = \pi x$. In the presence of reflections in the ring circuit the required power ratio is $|b_4 - a_4|^2/|a_1|^2$ and plotted against the phase constant exhibits the sharp twin response shown in Fig. 8.16(b), with a TE_{10} mode cavity as the applicator forming the upper part of the ring. In the absence of any reflections the power ratio would exhibit a single sharp response as shown dotted in the same figure. In this particular case the two peaks correspond to resonances at 2448 MHz and 2524 MHz and the power ratio was 14.

In order to achieve a power flow ratio as high as this the attenuation around the ring including the material to be processed, defined as $20 \log T_v$ dB, must not

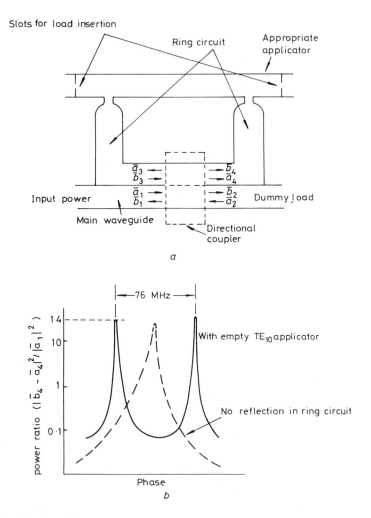

Fig. 8.16 *Principle of ring resonator*

    a Schematic diagram of applicator and ring circuit

    b Power ratio of circulating to incident wave including the effects of reflection in the ring circuit

    (After Suzuki *et al.*, 1975, Copyright *The Journal of Microwave Power*, 1975.)

exceed a certain limit depending upon the coupling coefficient C_{14}. For example, the attenuation must not exceed 0·45 dB and 0·3 dB for a coupling coefficient of 10 and 20 dB respectively. Larger losses are allowed only if smaller power ratios can be tolerated.

The travelling wave resonator has been used to heat synthetic rope much more efficiently than the equivalent non-resonant travelling wave resonator. Uneven heating due to standing waves was found to be negligible. With 1·1 kW incident power and a 10·5 dB directional coupler the power ratio was found equal to about 9, giving an attenuation loss of 0·5 dB in the ring circuit.

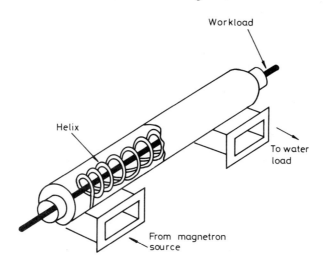

Fig. 8.17 *Helix applicator*
(After Shute, 1971, Copyright *The Journal of Microwave Power*, 1971.)

8.10 Helix applicator

Applicators based on the helix slow wave structure have been used for curing rubber extrusions (Shute, 1971) and proposed for heating yarn or wool sliver (Smith and Minaee, 1976). The essential features of such an applicator are shown in Fig. 8.17. A helix structure is incorporated in a circular waveguide which is fitted with rectangular waveguide at right angles at either end for feeding and dissipating any remaining power in a water load. As in the travelling wave tube the principle is to slow down the flow of the electromagnetic signal fed at the input waveguide by forcing it to travel along the helix structure, thus increasing its residence time between the input and output waveguide resulting in better coupling of the energy to the processing material which flows through the helical structure. The effective speed of the signal in the applicator is governed by the pitch of the helix. The launching structure at the input of the helix is experimentally optimised for minimum reflections of the incoming signal. However, its main difficulty is in access for internal cleaning.

8.11 Radiator applicators

8.11.1 Horn applicator

Sectoral horns are used extensively in attenuation type moisture meters for sending and receiving the electromagnetic signal through the wet dielectric sample. Such horns can also be used in applicators for processing with microwave energy. Figure 8.18 shows the cross-section of a single sectoral horn connected to a tunnel applicator carrying a load on a conveyor belt. Rigorous analysis of such a system is fairly complex and for this reason only the principle interactions will be outlined. Proper function of such an applicator requires that the load and horn assembly is properly matched to the output waveguide, say at $B–B$, which transfers the energy from the power unit and circulator ultimately to the load. If matching is not perfect a part of the incident power, P_0, will be reflected giving an input power into the load, P_{in}, as described by eqn. (7.111), i.e., $P_0(1 - |\rho|^2)$ where ρ is the reflection coefficient at the load/air interface. The impedance at $B–B$ looking towards the processed load is given by

$$Z_{B-B} = Z_{01} \frac{1 + \rho_e}{1 - \rho_e} \tag{8.8}$$

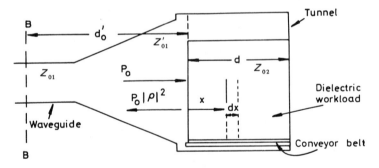

Fig. 8.18   *Horn load assembly*

where ρ_e is the effective reflection coefficient at $B–B$. Neglecting the mismatch due to the horn, ρ_e is given by $\rho e^{-j2\beta d_o'}$ where $\beta = 2\pi f/c$ and d_o' is the equivalent distance of the load/air interface from the reference plane $B–B$. The impedance at the load/air interface is given by an expression similar to eqn. (8.8) which can be equated to $Z_{02} \tanh(\gamma_2 x)$ assuming the load is short-circuited at the far end. Using eqn. (3.4) this gives the equation, assuming $Z_{01}' \simeq Z_{01}$,

$$(j\beta/\gamma_2) \tanh(\gamma_2 d) = \left(\frac{1 + \rho}{1 - \rho}\right) \tag{8.9}$$

from which ρ can be computed. As we can see, the dielectric properties of the processed load (which are in turn dependent on the moisture content, temperature, etc.) do play a vital role in the proper matching of the load/horn assembly since these govern the propagation constant, γ_2, in the load through expressions (4.21), (4.22) and (4.23).

The input power is attenuated according to

$$P_x = P_0 (1 - |\rho|^2) e^{-2\alpha_2 x} \tag{8.10}$$

where P_x is the remaining power at a distance x from the load/air interface. The loss of power in a small distance dx, $P_x 2\alpha_2 \, dx$, can be equated to that given by eqn. (4.10), to give an expression for the peak electric field strength at a distance x in the load, with $E \cdot E^* = E_x^2$,

$$E_x = \left[\frac{4 P_x \alpha_2}{\omega \epsilon_0 \epsilon_{eff}'' A} \right]^{1/2} \tag{8.11}$$

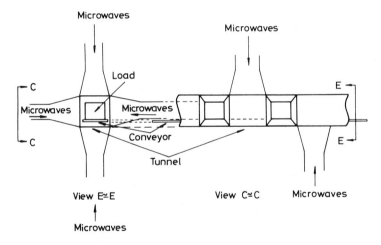

Fig. 8.19 *Multihorn applicator*

where A is the load cross-sectional area. When this non-resonant assembly is used as an applicator for processing thick loads, the power must necessarily be dissipated in the load after its first traverse to avoid large reflections from the short-circuit plate at the far end. This will give a highly non-uniform power distribution in the load. To compensate for this, an industrial processor of this kind has to supply the power to the load alternately from four sides, as shown schematically in Fig. 8.19. The division of the available power into four equal parts can be achieved via a set of 3 dB couplers. In addition standing waves are created by the short-circuit plates. Meredith (1978) demonstrated that the standing wave can be compensated by choice of short-circuit position.

8.11.2 Multislot waveguide applicator

Unless great care is taken in the design of the slot/stirrer structure in large multi-mode type of applicators powered by a single source, such systems may exhibit large non-uniformities in the distribution of the power within its enclosure. As was discussed earlier in Chapter 6, a multimagnetron type of approach can reduce such non-uniformities. However, Rueggeberg (1980) has put forward another method of launching the energy uniformly into an oven applicator. This is accomplished through the specific design of a number of field radiating slots cut in the broad dimensions of a waveguide. A schematic diagram of a slot pair is shown in Fig. 8.20 with their displacements from the centre line indicated as x_1' and x_2'. The rounded slots were about one half of a free space wavelength long and about 6 mm in width at 2450 MHz. These dimensions were chosen so as to avoid arcing across the centre of the slot. The distances x_1' and x_2' were determined experimentally through power reflection measurements to give equal power drop per slot pair, which required that each successive slot pair is located further away from the centre line. Moreover, to minimise cross-coupling between slots and avoid slot reflections, the distance between slot centres was found equal to $\frac{3}{4}\lambda_g$. These preliminary measurements also indicated that the slots of a matched pair ought to be located on opposite sides of the waveguide centre line. The waveguide was terminated by a low reflecting slot located at the outer edge and 60·2 mm from the short circuit plate at the end of the waveguide.

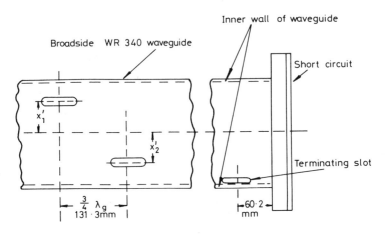

Fig. 8.20   *A typical slot pair and a terminating slot of the waveguide radiator*
(After Rueggeberg, 1980, Copyright The Institute of Electrical and Electronics Engineers, Inc., 1980.)

Figure 8.21 shows the radiated power characteristics of two identically constructed slotted waveguide radiators comprising seven slot pairs and a terminating slot. The input VSWR of a range of such multi-slotted applicators lies between 1·1 and 1·4 indicating that even in the worst case over 95% of the available input power to the waveguide is transmitted into the processing enclosure. Arrays of such radiating applicators have been used to process planar dielectrics up to 4 m in width with temperature variations within ± 5% of the required 100°C mean value.

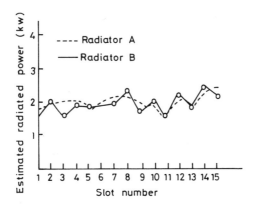

Fig. 8.21 *Radiated power of two identically constructed slotted-waveguide antenna systems* (After Rueggeberg, 1980, Copyright The Institute of Electrical and Electronics Engineers, Inc., 1980.)

8.11.3 Broadside array applicator

Multisections of open ended rectangular waveguides operating in the TE_{10} mode, arranged as shown in Fig. 8.22, has been used to process materials on conveyor belts (Püschner, 1967). The resulting E field is unidirectional due to a phase reversal in adjacent sections by turning the coupling loops by 180°. The steep rise of the electric field at the ends give rise to uniform heating of the edges as well. Microwave power is fed to the sections via a common short-circuited coaxial line.

The systems described in Section 8.11.2 and 8.11.3 are only effective where the load in the oven has a high absorption as the power division amongst the slots is a function of the loading impedance of the slots which is very variable under conditions of light loading in a closed cavity or oven.

8.12 Elliptical and spheroidal applicators

Elliptical and spheroidal type of resonant cavities have been considered in the past for specific microwave heating applications on account of the fact that such cavities do exhibit well defined electric field distributions and high power densities.

A major advantage of cylindrical cavities with elliptical cross-section is that fewer mode degeneracies are obtained compared with those of circular cross-section. For example, in the former cavities the TE_{10n} and TM_{11n} modes are not degenerate. Of all the available modes in waveguides with elliptical cross-section particular attention has been given to the $TM_{\binom{e}{o}11n}$ because of the well defined electric field distribution (Kretzschmar and Pietermaat, 1971). An example of the latter is shown schematically in Fig. 8.23 where two maxima of the longitudinal electric field E_z are obtained. The subscripts e and o refer to even and odd modes which have the E_z field symmetrical with respect to the major and minor axes respectively.

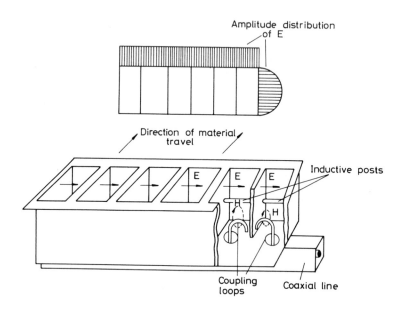

Fig. 8.22 *Broadside array applicator*
(After Püschner, 1967, Copyright *The Journal of Microwave Power*, 1967.)

The insertion of a dielectric load will undoubtedly perturb the field distribution and lower its absolute value. However, since such partially loaded cavities are exceedingly complex to treat theoretically, Kretzschmar (1972) has confined the analysis on empty cylindrical cavities with elliptical cross-section and computed the maximum longitudinal electrical field, E_{zm}, in terms of other parameters such as the dimension of the major axis, $2a$, the eccentricity e, the wall conductivity, σ, and the power coupled into the cavity, P. A typical example of such data is shown in Fig. 8.24 for the TM_{e115} mode through which a number of conclusions can be readily made. For example there is an optimum value of the major axis, $2a$, for a given eccentricity for maximum E_z/P value. Since this optimum value is near the cutoff condition care must be taken to allow for the effect of the inserted load by choosing an operating point to the right of the maximum curve when the cavity is empty. Figure 8.24 shows clearly that for a given mode and eccentricity the maximum field decreases with increasing major axis while for a constant major axis the maximum electric field increases with eccentricity. Similarly to cylindrical

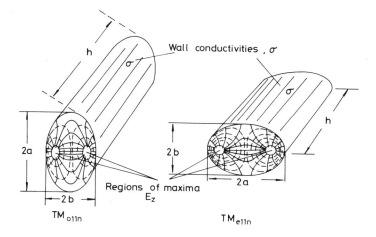

Fig. 8.23 *Cylindrical cavities with elliptical cross-section*
———— Electric field lines
– – – – Magnetic field lines
(After Kretzschmar, 1972, Copyright *The Journal of Microwave Power*, 1972.)

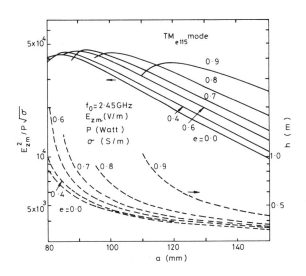

Fig. 8.24 *Variation of the maximum longitudinal electric field and the length of the cavity (broken line) of the TM_{e115} mode as a function of the major semi-axis a with the eccentricity e as a paramer*
(After Kretzschmar, 1972, Copyright *The Journal of Microwave Power*, 1972.)

cavities with circular cross-section, the maximum electric field in the $TM_{(\substack{e\\o})01n}$ modes occurs on the axis while the $TM_{(\substack{e\\o})11n}$ modes exhibit two positions of maxima electric field symmetrically away from the centre and situated on the major axis for even modes and on the minor axis for odd ones, respectively.

The quality factor, Q, for a number of modes in cylindrical cavities with elliptical cross-section has been determined theoretically by Rengarajan and Lewis (1980) by perturbation techniques through eqn. (7.2). Their data show an increasing Q factor with eccentricity for the TE_{o111} mode and large values of Q for the TE_{e01n} modes.

In situations requiring even greater focussing of the microwave energy, such as in the heating of small biological, geological or food samples of spherical geometry cavities of alternative geometry may be used. Hamid *et al.* (1971) have explored the feasibility of using a prolate spheroidal applicator operating in the TM_{e11n} mode which, as expected, exhibits two focal points where the microwave energy is primarily focussed and thus is suitable for heating two samples simultaneously. Using Jackson's (1962) approximate methods they calculate the Q factor of the cavity through the equation

$$Q = (\mu_a/\mu_c)(V_c/A_s\delta_s)F \tag{8.12}$$

where μ_c is the permeability of the cavity walls, V_c and A_s are the volume and surface area of the cavity, δ_s is the skin depth and F a geometrical constant. For the TM_{e11n} mode this equation yields a value of $0{\cdot}732(a'/\delta_s)$ where a' is the dimension of the semi-major axis. A prototype cavity was built at 915 MHz using a magnetic loop placed in the plane perpendicular to the magnetic field lines for coupling the energy into it, in preference to an electric probe which may excite other unwanted modes. To account for detuning effects of the load, the cavity was tuned by inserting a metal post along the major axis where the highest magnetic field exists. The input impedance of the prolate spheroidal cavity is shown in Fig. 8.25 which gives an unloaded Q of 860. This is considerably lower than the theoretical value calculated using eqn. (8.12) due to conductivity and other surface imperfections.

Through conventional perturbation techniques using a metal bead of 5 mm in diameter the electric field along the major axis was mapped identifying two field concentrations where the dielectric ought to be located for maximum coupling efficiency. A plastic sphere 25 mm in diameter filled with water was used for simulating conditions with a load yielding a VSWR shown dotted in Fig. 8.25.

8.13 Miscellaneous cavity and coupling slots configurations

In order to process large diameter ceramic or glass samples effectively, Thiebaut *et al.* (1979) have described a closed groove-guide type resonant cavity, as shown in Fig. 8.26. In this cavity the *E*-field is polarised axially with the workload, whilst its *H* field is concentrated within the groove giving an enhancement to the magnetic

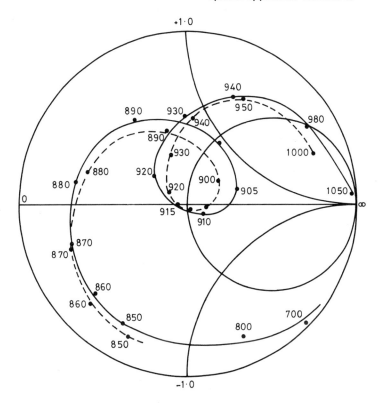

Fig. 8.25 *Input impedance of prolate spheroidal resonator as a function of the frequency
in MHz*
———— without dielectric insertion
– – – – with dielectric insertion
(After Hamid *et al.*, 1971, Copyright *The Journal of Microwave Power*, 1971.)

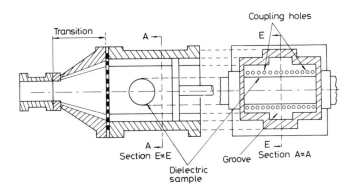

Fig. 8.26 *Groove cavity with multihole coupling*
(After Thiebaut *et al.*, 1979, Copyright *The Journal of Microwave Power*, 1979.)

stored energy. Since the electric field stored energy must equal the magnetic stored energy, the E-field is increased in value by the presence of the groove. Because the surface area within which wall currents flow is increased by a smaller proportion than the volume increase, the Q-factor rises and 10 000 can be achieved. Single hole coupling suffers from a low coupling coefficient in addition to the introduction of spurious modes into the groove cavity. To avoid this the microwave energy is transferred into the cavity via a transition which converts the TE_{10} mode into the groove-guide mode and the coupling coefficient is increased by a multiple aperture arrangement.

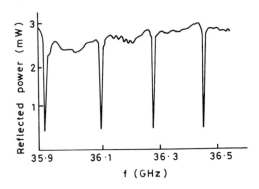

Fig. 8.27   *Reflected power as a function of the frequency for a groove cavity showing selective coupling of TE_{11n} modes*
(After Thiebaut *et al.*, 1979, Copyright *The Journal of Microwave Power*, 1979.)

The reflected power as a function of the frequency for a small prototype cavity operating at around 36 GHz is shown in Fig. 8.27. The response is free of any modes apart from the desired TE_{11n} injected into the cavity through the selective multihole coupling system. Moreover, the profile of the electric field can be easily adjusted by careful choice of the groove dimensions. A groove cavity at 2450 MHz processed ceramic or glass samples of 30 mm in diameter until melting in 3 min using 500 W of microwave power.

A novel coupling prototype arrangement at X-band has also been introduced by Vuorinen and Bruce (1976) in a TM_{010} cylindrical resonant cavity designed for liquid heating applications. Coupling is afforded via the longitudinal components of the magnetic field in a rectangular waveguide through two symmetrical curved slots, as shown in Fig. 8.28. This coupling system discriminates against the unwanted TE_{11} mode being set up in the TM_{010} cavity. Additional coupling is also obtained from the central feed hole through which the sample to be processed is inserted into the resonant cavity. The circular cavity is excited symmetrically using another coupler, while the two rectangular waveguides are fed from a short-slot, top wall 3 dB hybrid, this combination providing a more uniform heating than coupling through a single coupler. The cavity dimensions were chosen such

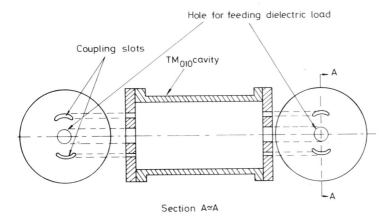

Fig. 8.28   *TM$_{010}$ resonant cavity with double slot coupling from either side*
(After Vuorinen and Bruce, 1976, Copyright *The Journal of Microwave Power*, 1976.)

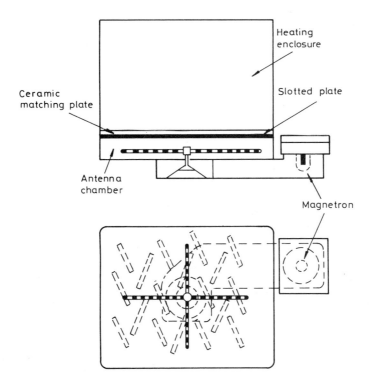

Fig. 8.29   *Microwave cavity using a near field applicator*
(After Risman, 1974, Copyright *The Journal of Microwave Power*, 1974.)

that the TM_{010} mode is cutoff with no load insertion into it. The cavity appropriately scaled up for 2450 MHz operation has been used to treat water–sulphur and other liquid mixtures.

A multi-slotted applicator of the near field type has been described by Risman (1974) which feeds the microwave power into a non-resonant oven enclosure for reheating small food samples wrapped in paper or plastic containers. The essential features of such an arrangement is shown in Fig. 8.29. Microwaves are transferred to an antenna chamber of small height from the magnetron source via a door knob type transition connected to an x-shaped antenna. The energy is in turn fed into the non-resonant enclosure via a number of $\frac{1}{2}\lambda$ resonant slots arranged in a pattern, as shown in Fig. 8.29. An interesting feature of this system is that each individual slot is badly matched to the antenna chamber under no load conditions resulting in low electric field strengths in the heating enclosure when it is operated empty. Furthermore, the field distribution in the cooking oven is symmetric and it only varies slightly with different food loads. However, this system was specifically designed for processing thin loads in paper or plastic containers. The one-sided penetration of the microwave energy and its near field configuration causes the oven to be unsuitable for heating thick loads.

Risman (1978) has also described a compact cylindrical applicator for drying small food samples. The mode of excitation changes during the drying period, with the TM_{011} mode dominant at the initial stage giving way to the TM_{010} mode for the last stages of drying.

This chapter has covered a broad spectrum of individual cavities in order to highlight the vast possibilities that exist. Earlier designs of microwave applicators can be found in Okress (1968), Püschner (1967) and Gerling (1973).

8.14 References

ALTMAN, J. L., *Microwave Circuits*, van Nostrand, Princeton, New Jersey (1964).

BLEACKLEY, W. J., VANKOUGHNETT, A. L., and WYSLOUZIL, W., 'Ridged waveguide microwave applicators', *J. Microwave Power* 7(1), 23 (1972).

CURRAN, J. E., *Symposium of Electrical Power and Chemical Engineering*, University of Birmingham, Organised by the Institution of Chemical Engineers, Midlands Branch (1966).

GERLING, J. E., 'Applicators and their design', *Trans. IMPI* 1, Paper No. 10, 144 (1973).

GUPTA, K. C., 'Cylindrical waveguide applicators with uniform power flow distribution', *J. Microwave Power* 11(4), 321 (1976).

HAMID, M. A. K., MOSTOWY, N. J., and BHARTIA, P., 'Microwave bean roasters', *J. Microwave Power* 10(1), 110 (1975).

HAMID, M. A. K., STUCHLY, S. S., and BHARTIA, P., 'Tuning and excitation of prolate spheroidal cavity resonator for microwave heating', *J. Microwave Power* 6(3), 213 (1971).

HOLME, I., and METAXAS, A. C., 'Microwave drying of nylon tufted carpets. III. Field trials', *J. Microwave Power* 14(4), 367 (1979).

JACKSON, J. D., *Classical Electrodynamics*. John Wiley & Sons, New York (1962).

JOHNSTON, D. A., and VOSS, W. A. G., *Proc. IMPI Symposium*, Monterey, Paper No. 8.4 (1971).

KRETZSCHMAR, J. G., 'Maximum longitudinal electric field strength in hollow cylindrical cavities with elliptical cross-section', *J. Microwave Power* 7(1), 35, (1972).

KRETZSCHMAR, J. G., and PIETERMAAT, F. P., 'Concentrated microwave heating in elliptical waveguides', *J. Microwave Power* 6(3), 207 (1971).

MEREDITH, R., Magnetronics Ltd., Unit A, St Mary's Mills, Evelyn Drive, Leicester, LE3 2BV (1978).

MEREDITH, R., and METAXAS, A. C., *Microwave Carpet Dryer*, British Patent No. 6531/77, Case No. 2111, Feb. (1978).

METAXAS, A. C., and MEREDITH, R., 'Microwave drying of nylon tufted carpets. II. Electrical characteristics of a modified TE_{10n} resonant cavity', *J. Microwave Power* 13(4), 315 (1978).

OKRESS, E. C., *Microwave Power Engineering*. Academic Press, New York (1968).

PÜSCHNER, H. A., 'Microwave heating techniques in Europe', *J. Microwave Power* 2(2), 31 (1967).

RENGARAJAN, S. R., and LEWIS, J. E., 'Quality factor of elliptical cylindrical resonant cavities', *J. Microwave Power* 15(1), 53 (1980).

RISMAN, P. O., 'A commercial microwave oven using a near field applicator', *J. Microwave Power* 9(2), 163 (1974).

RISMAN, P. O., 'A microwave applicator for drying food samples', *J. Microwave Power* 13(4), 97 (1978).

RUEGGEBERG, W., 'A multislotted waveguide antenna for high powered microwave heating systems', *IEEE Transactions of Industry Applications* 1A-16, No. 6, Nov/Dec, pp. 807–813 (1980).

SHUTE, R. A., 'Industrial microwave systems for the rubber industry', *J. Microwave Power* 6(3), 193 (1971).

SMITH, R. B., and MINAEE, B., 'Microwave heating of yarn', *Proc. IMPI Symposium, Leuven, J. Microwave Power* 11(2), 189 (1976).

SUZUKI, T., OSHIMA, K., NAGAO, T., OKAKURA, T., and HASEGAWA, K., 'A travelling wave resonator for industrial microwave heating', *J. Microwave Power* 10(2), 199 (1975).

THIEBAUT, J. M., BERTEAUD, A. J., and ROUSSY, G., 'A new microwave resonant applicator', *J. Microwave Power* 14(3), 218 (1979).

VANKOUGHNETT, A. L., 'A microwave applicator for filamentary materials', *J. Microwave Power* 7(1), 17 (1972).

VANKOUGHNETT, A. L., DUNN, J. G., and WOODS, L. W., 'A microwave applicator for heating filamentary materials', *J. Microwave Power* 9(3), 195 (1974).

VANKOUGHNETT, A. L., KASHYAP, S. C., and DUNN, J. G., 'Parallel plate microwave applicator', *J. Microwave Power* 10(4), 451 (1975).

VANKOUGHNETT, A. L., and WYSLOUZIL, W., 'A microwave dryer for ink lines', *J. Microwave Power* 7(4), 347 (1972).

VUORINEN, P. A., and BRUCE, P. N., 'X-band model of a microwave applicator for liquid heating applications', *J. Microwave Power* 11(2), 99 (1976).

Chapter 9

The microwave heating circuit, breakdown phenomena and vacuum processing

9.1 Introduction

The last four chapters described the various forms of applicators that are being used to process many different materials inserted into them. It now remains to describe the function of the other components which form part of the typical

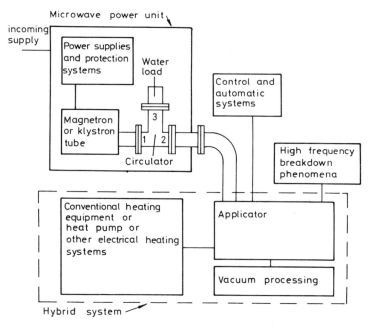

Fig. 9.1 *Microwave heating apparatus showing the topics discussed in this chapter*

microwave heating equipment shown diagramatically in Fig. 9.1. The heart of the equipment is in the generation of microwaves usually through a magnetron or a klystron. Only a brief insight into the principle of such tubes will be given as these and similar devices such as the amplitron, the triode, etc., have been fully described in the past (Okress, 1968). The analysis will be focussed on aspects which are relevant to their use in a microwave heating circuit such as the associated power supply and protection schemes.

Some industrial materials are heat sensitive and cannot be dried at atmospheric pressure. By taking advantage of the lower boiling temperature at reduced pressure such materials can be effectively processed in a partial vacuum atmosphere. Microwaves can considerably enhance the processing of materials under vacuum and the principles and practice of such an application are given special emphasis. However, to appreciate the constraints within which such a process must operate, the analysis on vacuum processing is preceeded by an insight into high frequency gas discharge phenomena with particular emphasis on the mechanism of breakdown at pressures where diffusion to the vessel walls is the dominant charge loss mechanism. The critical parameter in microwave vacuum processing is the electric field strength established in the applicator which must be well below that required for gas breakdown at a particular set of conditions, by allowing for an adequate safety margin at the initial design stage.

It has by now been established that successful processing with high frequency power in industrial applications requires a measure of hybridisation with other forms of conventional energy such as hot air, steam, etc., as well as with other electrical methods such as infra-red and heat pumping techniques. The principles behind such schemes will be outlined together with simple hybrid systems. Finally, the principles behind automatic control systems will be briefly outlined.

9.2 Microwave power sources

Many industrial heating plants using microwave energy have a large heat requirement, some installations being over 100 kW, which requires that the generator efficiency must be high to avoid energy wastage. Moreover, the generated power must be stable in frequency and free of harmonics and other spurious frequencies to meet international regulations on frequency allocation. To these stringent requirements must be added that of low capital cost. These factors lead naturally to the choice of the magnetron as the generator principally used because of its high power output, its efficiency, frequency stability and relatively low cost. Other microwave valves are inherently too expensive and complex for industrial process equipment except possibly the klystron which, as a driven amplifier, has excellent frequency stability. Klystrons would have to be used if the statutory allocated frequency bands were less than 2% in bandwidth, but the extra cost would eliminate most installations economically. Finally, solid-state generators are as yet of insufficient power to merit consideration for industrial heating plants, even when used in large arrays.

Since most industrial microwave processing plants do use the magnetron as the source of power, we will describe its construction and performance in some detail. However, for the sake of completeness, we shall begin with a very brief qualitative description of klystrons.

9.2.1 Klystron amplifier

Essentially the klystron is an amplifier of microwave signals using an electron beam as the amplification medium. Its main features are shown schematically in Fig. 9.2.

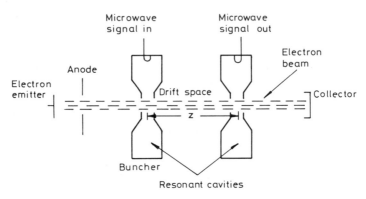

Fig. 9.2   *Klystron amplifier*

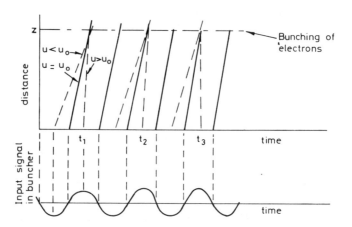

Fig. 9.3   *Principle of velocity modulation in a klystron*

A low power microwave signal is injected into a resonant cavity, called the buncher, through which a beam of high energy electrons of velocity u_0 passes. The buncher is so designed as to set up an axial alternating electric field which interacts with the incoming beam and accelerates ($u > u_0$) or decelerates ($u < u_0$) the individual electrons depending upon the phase of field, as shown qualitatively in Fig. 9.3. The individual electrons emerge through the buncher and travel at constant but different velocities across the drift space. The integral effect of such an interaction is to velocity modulate the electron beam and cause the electrons to bunch together at a given distance z away from the buncher at times t_1, t_2, t_3, etc. If a second cavity is placed at around the distance z, the modulated beam will now induce a microwave electric field across its grid structure whose polarity is such as to slow down the electron beam and power is thus extracted from the device. In simple terms, the d.c. energy of the electron beam has been converted to microwave energy at a frequency of the induced signal.

To optimise the power gain and efficiency klystrons are constructed with more than two cavities (Maloney and Fallion, 1974). Large high powers c.w. (in excess of 100 kW) have been designed at S-band (2450 MHz) and in X-band around 10 GHz (Pohl, 1968).

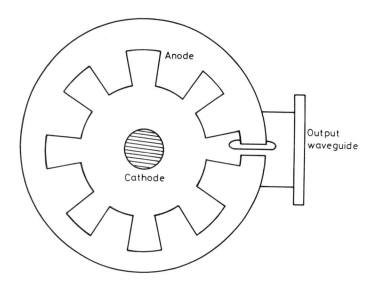

Fig. 9.4 *Eight vane magnetron structure*

9.2.2 Magnetron oscillator

Constructionally the magnetron is a high vacuum electronic valve consisting of a hollow copper anode incorporating a resonant microwave structure, at the centre of which is an electron emitting cathode, as shown in Fig. 9.4. Only a qualitative description of the magnetron operation and performance will be given here and more rigorous treatment can be found in Harvey (1963).

Consider first the anode which has a set of vanes projecting radially inward forming slots between them which are approximately $\frac{1}{4}\lambda_g$ deep and therefore resonant at the operating microwave frequency. The slots are mutually coupled via the fringing field at their open ends and the whole structure forms a resonant circuit. In operation of the so-called π-mode the electric charges on adjacent vane tips are of opposite polarity (i.e., there is 180° phase shift in field between adjacent cavities). Starting at time zero assume that vane No. 1 has positive charge and vane No. 2 negative charge. On the next half-cycle vane No. 2 will be positively charged, and the next vane, No. 3, will be negatively charged, and so on. The positive charge can therefore be considered as rotating clockwise around the anode. Identical reasoning gives a counter-clockwise rotation of charge which is also valid and leads to a direct analogy with a single-phase induction motor with a pair of contra-rotating magnetic fields. It is evident, therefore, that in operation the electric charge on the vane tips of the magnetron anode can be resolved into a pair of contra-rotating sets of charges or field.

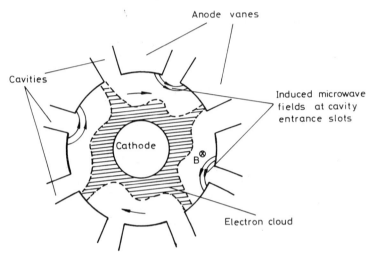

Fig. 9.5   *Interaction of electron cloud with induced cavity fields*
⊕ Magnetic field
B perpendicular to the plane of paper

Second, consider the motion of the electrons emitted from the cathode. The anode is supplied with a high voltage positive d.c. potential relative to the cathode and under its action electrons are attracted from the cathode radially towards the anode. A magnetic field is also provided with field lines parallel to the axis of the anode, and perpendicular to the electron path which introduces a force causing the electrons to travel in a quasi-circular path around the cathode. Clearly the actual path and velocity of the electrons is determined by the combined d.c. electric field strength and magnetic field strength in the space between the anode and cathode, and the rotational velocity they acquire around the cathode is the important feature. By choice of applied d.c. and magnetic field the cloud of electrons, as shown in Fig. 9.5, rotates synchronously with the induced microwave field in the anode structure: by increasing the d.c. voltage they rotate faster but are decelerated by the microwave field transferring some of the kinetic energy of their motion to the microwave field from the d.c. energy applied between anode and cathode since this is the source of motion for the electrons. Since all the cavities in the anode circuit are tightly coupled together, power can be extracted via a loop through only one cavity, as shown in Fig. 9.4. For medium to large power applications a coaxial to waveguide transition is usually employed.

The important characteristics of the operation of a magnetron evident from this descriptive action are:

(*a*) The power output rises with applied voltage since more electrons are attracted to the anode increasing the anode current.

(*b*) The power output rises with falling magnetic field. This superficially unexpected effect arises because at higher magnetic field the electrons are more constrained to paths looping close to the cathode, fewer reaching the anode. Anode current can readily be stopped altogether by raising the magnetic field strength further. Reducing the magnetic field strength allows the electron path to extend further and further from the cathode, and when the outermost electrons reach the anode, anode current starts to flow. In high power magnetrons the primary control of output power is often achieved by regulating the magnetic field rather than the applied voltage, as discussed later in this chapter.

(*c*) Heat is dissipated at the anode since the electrons reach it with a finite velocity and their associated kinetic energy is converted to heat. This heat represents the principal efficiency loss in the magnetron. The anode must be cooled by air or liquid to limit its temperature.

(*d*) To emit electrons the cathode is heated by an auxiliary electrical supply and it is usually necessary to allow time for the cathode to reach its operating temperature before applying the anode–cathode EHT. When the magnetron is operating some of the electrons fail to reach the anode and return to the cathode, an effect known as 'back-bombardment'. Their kinetic energy is dissipated as heat at the cathode, and to avoid excessive temperature rise it is necessary to reduce the power from the auxiliary cathode supply by a roughly equal amount. In some magnetrons the cathode heater current is reduced linearly with rising anode current, and others in one, two or three steps; whilst in some low power magnetrons the auxiliary supply is switched off altogether on application of EHT voltage.

9.2.3 Magnetron operating characteristics

A peculiar feature of magnetrons is the shape of their anode voltage–current characteristic and a typical one is shown in Fig. 9.6, and known as a performance chart. On raising the anode voltage from zero at fixed magnetic field very little current flows until a specific voltage is reached (called the π-mode voltage) when the outermost electrons reach the anode. Thereafter the anode current rises very rapidly reaching its maximum rated value with a further voltage increase of only 3–8%. It is therefore essential that the power supply be stabilised closely to prevent unacceptable changes in power output from the magnetron due to fluctuations in mains voltage.

Notice that a change in magnetic field produces an identical V–I characteristic, except at a different π-mode voltage, increasing magnetic field giving a proportional increase in π-mode voltage. This feature makes it possible to control a magnetron by adjustment of the applied magnetic field, a technique much used for high power magnetrons with electromagnets because only a low power adjustable supply is required to energise the electromagnet. Low power magnetrons usually have permanent magnets and are controlled solely by adjustment of anode voltage.

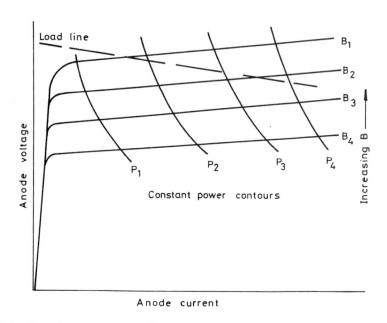

Fig. 9.6   *Typical performance chart for a magnetron*

In common with all oscillators the impedance of the load connected to the output affects the performance of a magnetron in both generated power and output frequency, and also to the d.c. anode voltage—current characteristic is affected. Although relatively small, these effects can be important in operation. Considering the load impedance at its point of connection to the magnetron, the reactive component causes a small change in output frequency because it represents an additional inductance or capacitance to the anode resonant structure, whilst the resistive component affects the power output. These characteristics are displayed on a Rieke diagram, shown in Fig. 9.7, in which contours of frequency and power output are plotted on an impedance (or admittance) circle diagram, a point on which represents the impedance of a load. Variations in frequency and power due to load variations are frequently called frequency pulling and power pulling respectively. Usually within the permitted load impedance range of the magnetron, frequency pulling does not exceed ± 0·2% of nominal frequency, and power pulling ± 15% of nominal output power, the nominal figures referring to matched load operation. In practice the load impedance of microwave applicators can vary over a wide range in operation and to minimise performance variations a ferrite circulator is frequently placed between magnetron and load to divert reflected power away from the magnetron. This arrangement also protects the generator from severely mismatched work loads, and by providing the magnetron with a good impedance match at all times results in a general improvement in its life.

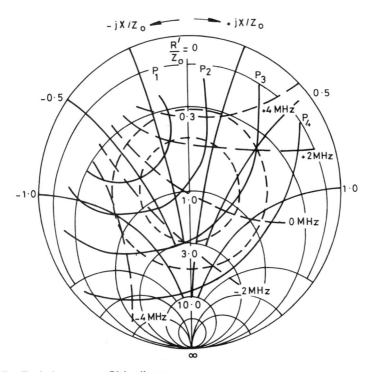

Fig. 9.7   *Typical magnetron Rieke diagram*

9.2.4 Magnetron moding

The most important consideration for the efficient operation of a magnetron, is the way in which the induced fields are set up between the anode and the cathode space. The fact that a magnetron consists of a number of cavities tightly coupled together gives rise to a number of possible field distributions, called modes, some of which may be close to each other in frequency. The mode described in the preceding sections is called the π-mode because it gives a phase difference of $180°$ between adjacent cavities. However, other phase differences are possible, the critical factor being that the electrical path length around the anode must be $2n\pi$ radians. The most efficient of all these modes is the π-mode and to ensure operation in this mode it is common practice to connect alternate anode segments by straps thus suppressing the other modes. Unwanted modes resonate at incorrect frequencies at which the magnetron efficiency is low with consequent excessive internal heating and therefore can result in damage. Moding is caused mainly by insufficient emission of electrons from the cathode and also by impedance mismatching in the output circuit. Moding can be observed by low power output and incorrect operating frequency. Modern design practise of magnetrons and their associated circuits has virtually eliminated moding as a malfunction factor.

9.2.5 Magnetron power supplies

Several methods exist for providing a high voltage (EHT) power supply of stable power for a magnetron. In effect the requirement is to provide substantially constant anode current independently of EHT voltage fluctuations or microwave workload impedance variations. For magnetrons with fixed magnetic field (i.e. with permanent magnet), the power unit must have adjustable EHT controlled to give constant anode current. Some of the methods of EHT control and their merits and disadvantages are as follows.

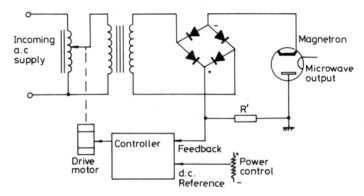

Fig. 9.8   *Schematic circuit diagram of basic magnetron power regulator using a motor driven variable ratio transformer with anode current feedback*

9.2.5.1 Variable ratio transformer: This is an obvious method in which a motor-driven voltage regulator is controlled by sensitive relays or electronic control circuits measuring the anode current to give servo-control against a preset reference, as shown diagrammatically in Fig. 9.8. Its technical disadvantage is slowness in response, but mechanical wear is also a feature. It is also an expensive control system and is infrequently used.

9.2.5.2 Thyristor control: In this system phase-controlled thyristors are used to control the a.c. supply to the EHT transformer in a servo system, as shown in Fig. 9.9. The required anode current is set by a d.c. reference which is compared with a d.c. voltage developed across R' proportional to anode current. The difference voltage is amplified and fed to a trigger circuit to control the firing angle of the thyristors.

This system is capable of good long-term stability and has a relatively fast response time limited by the firing of the thyristors to a few half-cycles of the supply frequency. However, the circuit can cause high peak instantaneous anode current which may exceed that allowed by the magnetron specification and also cause high peak instantaneous power (several times the mean power) resulting in possible microwave arcing. To minimise these effects it is usual for the EHT transformer to have high leakage reactance (30–40%).

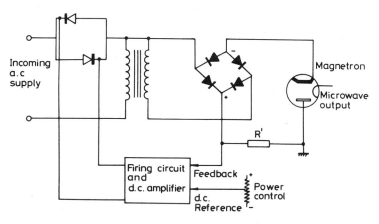

Fig. 9.9 *Schematic circuit diagram of basic magnetron power regulator using thyristor control of applied EHT with anode current feedback*

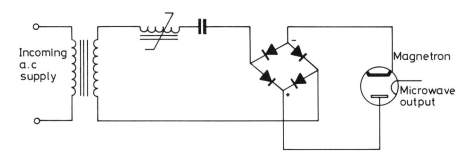

Fig. 9.10 *Schematic circuit diagram of resonant stabilised magnetron power supply using a saturable reactor*

9.2.5.3 Saturable reactor control: A much-used method of controlling low power magnetrons is to provide, as in Fig. 9.10, a non-linear inductor in series with the EHT transformer which is series-resonated with a capacitor at a frequency about 15% higher than the incoming supply frequency. At resonance the effective impedance is at its minimum value. The inductance value falls with rising voltage (resulting from an increase in anode current) so that the resonant frequency rises thereby increasing the impedance of the circuit at power frequency, and so limiting the increase in current. Excellent anode current stability can be achieved with this arrangement, a ± 10% change input voltage giving less than ± 1% change in anode current. Usually the leakage reactance of the EHT transformer forms the reactor and the transformer is arranged to saturate. Most domestic and small industrial ovens use this type of power supply on account of its low cost, simplicity and reliability since it has no 'active' components. Its disadvantages are that it is basically a fixed power circuit adjustment only being possible by switching transformer taps and that the output is sensitive to supply frequency variation.

9.2.5.4 Series resistor control: A simple inexpensive method of providing rough stabilisation to a load of low dynamic impedance is to place a resistor in series with it of such a value as to drop about half the supply voltage. Stabilised in this way a magnetron will give about 2% power change for 1% mains supply voltage change which is often adequate. The disadvantage is a severe reduction in overall efficiency. However, it is frequently necessary to provide conventional heating to a microwave oven and the power dissipated in this resistor can be used to heat air for blowing into the oven. Using a $1\frac{1}{2}$ kW magnetron of 60% efficiency, some 3 kW of hot air is available by this system making it very suitable for use with small ovens.

9.2.5.5 Variable magnetic field control: For high power generators considerations of efficiency, current waveform distortion, and power factor makes all the previous EHT stabilisation methods unattractive, particularly as a three-phase supply will be required. Control by magnetic field has already been discussed in principle. Magnetic field is a sensitive parameter in control of anode current, a 1% change of electromagnet current giving an anode current change of about 10%. Figures 9.11 and 9.12 show two basic schemes, the first being a series field arrangement, the second direct control of magnet current from a servo-loop in which anode current is regulated against a preset reference.

Series field operation is in principle simple and attractive but the detailed execution of a practical system results in a cost similar to the second system which is more generally used. The series-field system has two disadvantages. First, an auxiliary d.c. supply for the magnet is essential to provide a magnetic field at the instant of switch-on. Second, very high voltages can be induced across the winding of the electromagnet under fault conditions, requiring surge suppression and care in circuit design to protect semiconductors.

The separately excited field system suffers neither of these disadvantages. Typically a 30 kW magnetron can be controlled in power from zero to full power by an electromagnet power supply of only 300 watts. A control-loop gain of 30 dB is readily achieved giving an output power stability of about 3% for a supply voltage change of 1·7%.

Usually the closed loop control system regulates the magnetron anode current which is processed to provide a voltage which is compared against a set reference voltage, and following conventional systems the difference between these two is then amplified and controls the magnet current of the magnetron. Anode current is then regulated to a constant level proportional to the reference which is set to a

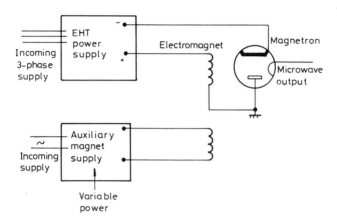

Fig. 9.11 *Anode current control of magnetron by series excitation of electromagnet*

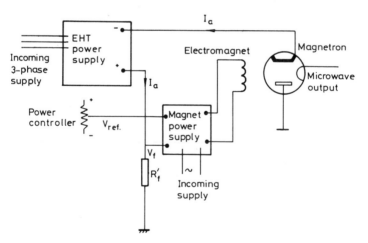

Fig. 9.12 *Closed-loop anode current control of magnetron by separate excitation of electro-magnet*

V_f = feedback voltage = $I_a R_f'$

V_{ref} = Reference voltage (manual set or set by a process parameter or computer)

level corresponding to the desired generator output. This simple method of power control assumes that the power output (and power dissipated in the workload) is in constant proportion to anode current, which is only approximately true but is adequate in the majority of applications. Departures from this assumption arise from variation in magnetron efficiency with load impedance (as indicated by the Rieke diagram) and reduction of power dissipated in the workload due to reflected power. Where more precise control of dissipated power is required it is possible to control power by measuring the forward and reflected power from the workload, process and difference as dissipated power and compare this parameter against the control reference instead of anode current. Because the accurate measurement of forward and reflected power requires some complex instrumentation this refinement is rarely adopted.

9.2.6 Magnetron protection systems
In relation to its physical size, the magnetron handles a large amount of power, and although having a high efficiency has a high power dissipation density. Air or liquid cooling is invariably required, high levels of reflected power must be avoided, and automatic high-speed trip systems must be provided to protect the magnetron from damage when adverse operating conditions arise from subsystem failure.

Typically a microwave generator has the following interlocks to protect the magnetron.

9.2.6.1 Water cooling: Flow switch of fail-safe pattern in water circuit. The circuit is often a closed circuit with pump and header tank to enable the quality of the water to be controlled, in which case a float switch for header tank level and thermostat alarm is also fitted.

9.2.6.2 Air cooling: Air flow pressure switch operating from pressure developed by cooling fan.

9.2.6.3 Anode temperature sensor: A thermostat switch in intimate thermal contact with the anode, which opens its contacts when excess temperature exists. Both low temperature fusible alloy and bimetal strip types are common. The anode temperature interlock has a slow response and must be considered as a second line of defence against the primary interlock failing; in a few cases of severe and sudden internal temperature rise the magnetron could be damaged before the anode temperature trip operates.

9.2.6.4 Anode over-current trip: This system uses a current-sensitive relay to trip the magnetron EHT if the anode current exceeds a preset level. Excess anode current usually arises from:

(a) an internal arc in the magnetron due to imperfect vacuum;
(b) badly matched radio frequency load;
(c) mains voltage surge;
(d) EHT short-circuit.

9.2.6.5 Reverse power trip: This consists of a microwave system to detect the presence of reverse power which usually operates either to limit the power output so that the reverse power does not exceed a preset specification limit, or to switch off the magnetron electronically (e.g., by control of EHT or magnetic regulator thyristors) and then to switch on again after a set delay. In the latter system the magnetron operates in a pulsing mode until the load impedance becomes normal, and the system has the advantage that the mean internal dissipation inside the magnetron during the fault is very low.

9.2.7 Circulator protection

The systems described in the preceding sections are aimed at offering some protection to the high power source from adverse power due to large mismatches of a terminating load. Complete protection of the power source from reflected power is desirable particularly in high power applications where large powers are fed to single mode resonant cavity applicators described in Chapter 7. This can be achieved by connecting a microwave ferrite device, called the junction circulator, between the power source and the resonant applicator, as shown in Fig. 9.1 (Helszajn, 1975). Although there are other types of microwave ferrite devices that could offer similar protection from reflected power, such as, for example, a Faraday rotation device, for reasons of cost and design the junction circulator is the most suitable protection device in high power microwave heating applications.

The junction circulator is a three-port waveguide device with all the three ports lying in a plane and a ferrite insert located at or around the centre of the junction. A magnetic bias field is applied to the ferrite through a permanent magnet, the direction of this bias field being perpendicular to the plane of the junction. Referring to Fig. 9.1 forward power applied to port 1 proceeds with negligible loss to port 2 while reflected power incident upon port 2 will now be deflected to port 3. This circulator action can be optimised in the design stage by careful choice of a number of parameters, such as the form and size of ferrite, its composition, the bias magnetic field, the dimensions of the centre of the junction, etc. Protection from large reflected powers is then accomplished by connecting a matched broadband water-load to port 3 and the complete assembly is then referred to as an isolator or iso-circulator. The water flowing through the load in port 3 is also passed around the centre junction to cool the ferrite element. The isolator could be connected inside the microwave power generator housing, as shown in Fig. 9.1.

Such junction devices have been built at 2450 MHz capable of handling 6 kW into a full short-circuit of any phase connected to port 2. Isolation is better than 27 dB over any 50 MHz band with a typical insertion loss of about 0·1 dB (Hudson, 1975). Similar devices have been built at 896 MHz and 915 MHz in WG4 capable of handling up to 30 kW c.w. power with once again a 0·1 dB insertion loss (Fumiaki Okada *et al.*, 1975). Although such ferrite devices are essential for protecting microwave sources feeding single mode resonant applicators, it is recommended that they are incorporated in high power applications to offer continuous protection to the source at all times and thereby to prolong the life of the magnetron or klystron tube.

9.3 High frequency breakdown phenomena

9.3.1 Introduction

It is well known that radio frequency and microwave electric fields of sufficient strength can break down a gas to its molecular constituents and produce a conducting medium which assumes various names, such as arc, plasma, corona, etc., depending upon many factors. In what follows we shall primarily use the generic term 'discharge' to describe some of these processes unless a specific type is under discussion. Fundamental analysis of these mechanisms at high frequencies is beyond the scope of this book and can be found in numerous authoritative treatises such as Francis (1960), Brown (1966), McDonald (1966) and von Engel (1965). Here we shall confine ourselves to some aspects which are relevant to microwave heating applications.

Broadly speaking there are two categories of high frequency (i.e., radio-frequency and microwave) instigated discharges which are pertinent to industrial heating applications, one beneficial and the other disadvantageous. In the beneficial category the plasma forms the reaction zone in which wanted chemical reactions take place in order to produce specific products. Here the processes may include the production of gaseous atoms, the synthesis of unsaturated hydrocarbons and fluorocarbons for acetylene production, converting waste chlorine to chlorine gas, coal pyrolysis, polymerisation of monomers, plasma–surface reactions and many other specific organic and inorganic synthesis reactions (Hamblyn and Reuben, 1975).

In the disadvantageous category, we include the effect of unwanted breakdown in the applicator zone. In processing under vacuum (see Section 9.4) where the air pressure in the reaction vessel is reduced below atmospheric value, the high frequency breakdown electric field is considerably lowered. As the air pressure increases towards its atmospheric value the likelihood of gas breakdown is reduced. However, given conditions favourable to breakdown such as moderately high electric fields of the order of $100\,kV/m$ ($1\,kV/cm$) in the presence of sharp metallic objects, corona or arc type discharges may also be obtained at atmospheric pressure even though the electric fields seem well below that required for breakdown between parallel plate electrodes of short separations (10^{-2} m), i.e., $3000\,kV/m$ ($30\,kV/cm$) (Craggs, 1978). Such effects are clearly disadvantageous because of possible damage to the product and applicator and because it represents loss of power.

9.3.2 Type of discharges

High frequency discharges at pressures in the range $0\cdot1\,Pa$ (about 10^{-2} torr) right up to atmospheric pressure can be most commonly obtained in a glass vessel by applying an oscillatory field between parallel plane electrodes placed inside or outside the vessel, as shown in Fig. 9.13(a) and (b) respectively (Metaxas, 1968), or by induction from a coil carrying high frequency current wrapped round the vessel, as shown in Fig. 9.13(c) (Francis, 1960). In the latter the discharge develops in two stages, initially a faint longitudinal discharge appears when the current is small and when the current is increased above a certain limit a transition to a bright ring discharge occurs showing the atomic lines in the case of hydrogen (Chandrakar and von Engel, 1965). At atmospheric pressures this induction discharge is also referred to as a plasma torch and has found many applications in growing single crystals and in spheroidising powders (Beguin *et al.*, 1967) (Hamblyn, 1977).

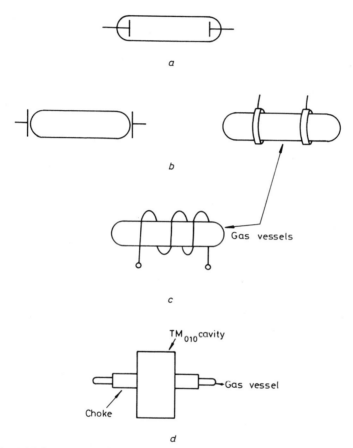

Fig. 9.13   *High frequency discharge vessels*
a Internal electrodes (radio frequency)
b External electrodes (radio frequency)
c Induction (radio frequency)
d Resonant cavity (microwave)

In the microwave regime resonant cavities, as shown in Fig. 9.13(*d*), or short wave couplers are used to feed the energy into the discharge vessel (Fehsenfeld *et al.*, 1965). Electrodeless discharges have the unique advantage of being independent of electrode effects and for this reason should be more suitable for industrial processing.

9.3.3 Rates of production and loss of charges

The mechanism for breakdown in a high frequency discharge depends upon many factors, such as the pressure range, the type of discharge, etc., and it is not really possible to produce a generic theory which effectively covers all the different types. The important common factor is the build-up of the electron concentration within the reaction vessel due to collision processes. This build-up can be studied by considering the continuity equation for electrons written in its most general form:

$$\frac{\partial n_e}{\partial t} = \sum_i P_i - \sum_j L_j + Q_i \qquad (9.1)$$

where $\sum_i P_i$ and $\sum_i L_j$ represent the summation of all processes of electron collision with atoms or molecules leading to the production and loss of charge per second respectively, Q_i is the rate of production of electrons from some external agency, i.e., cosmic rays, and n_e is the electron concentration. Electrons gain energy from the electric field by colliding elastically with gas molecules, since these collisions are responsible for changing their ordered oscillatory motion to a random motion. However, the energy gain due to the change of phase is initially more than the loss of energy due to collisions. Thus, energy is gained on the average after each collision until its mean energy is such as to make the gain from the electric field equal to the loss due to collisions.

Usually the main form of charge production processes is assumed to be by electron collision with ground state molecules, although it has also been demonstrated that in some gases electron collisions with excited or metastable atoms contribute significantly to the total rate of ionisation (Vidaud de Plaud, 1967; Metaxas, 1968). However, different loss mechanisms for charges could prevail in the gas phase such as diffusion to the walls, volume or wall recombination, attachment, etc., each of which may dominate at different pressure regimes. Each of these mechanisms, when substituted for L_j in eqn. (9.1), will produce a continuity equation giving a solution for n_e which is relevant to the particular type of process.

9.3.4 Diffusion controlled discharge

Let us examine in some detail a particular pressure regime, that between 13—4000 Pa (or about 0·1—30 torr) which has in the past been extensively used in plasma chemistry to instigate various plasma reactions. In ordinary vessels, where the mean free path and amplitude of oscillation of the electrons are much smaller than the dimensions of the vessel and where the collision frequency is larger than the applied frequency, the main loss mechanism is by diffusion of charges to the walls, recombination being usually small except when the concentration of charges is very large. Under these conditions the continuity equation for the electrons becomes

$$\frac{\partial n_e}{\partial t} = z_i n_e + D_e \nabla^2 n_e + Q_i \qquad (9.2)$$

where z_i is the rate of production per electron and D_e is the free electron diffusion coefficient. The approach to breakdown is assumed very slow so that before breakdown occurs $\partial n_e/\partial t = 0$ and the continuity equation (9.2) yields

$$z_i n_e + D_e \nabla^2 n_e + Q_i = 0 \tag{9.3}$$

The solution of this expression in one dimension, measuring x from the mid-plane between two parallel plate electrodes of separation d and assuming the electron concentration, n_e, to be zero at $\pm d/2$, is (Metaxas, 1968)

$$n_e(x) = Q_i/z_i \frac{\cos(z_i/D_e)^{1/2} \cdot x - \cos(z_i/D_e)^{1/2} \cdot d/2}{\cos(z_i/D_e)^{1/2} \cdot d/2} \tag{9.4}$$

Breakdown is normally defined by the condition that the electron concentration, n_e, rises to a very high value which occurs when

$$(z_b/D_e)^{1/2} \cdot d/2 = \pi/2 \tag{9.5}$$

or

$$z_b = D_e/(d/\pi)^2 = D_e/\Lambda^2 \tag{9.6}$$

where b refers to breakdown and Λ is known as the characteristic diffusion length. For a cylinder of radius R and height h the characteristic diffusion length is given by

$$\frac{1}{\Lambda^2} = \left(\frac{\pi}{h}\right)^2 + \left(\frac{2\cdot405}{R}\right)^2 \tag{9.7}$$

Since D_e and Λ^2 are easily determined, the D_e/Λ^2 is in fact a measure of the rate of ionisation per electron at the breakdown field, z_b. Once the discharge is struck, the electric field required to maintain it, E_m, is now very much lower than the breakdown field, E_b, and the motion of the charged particles (both electrons and ions) in the main body of the plasma, i.e., the positive column, is now controlled by ambipolar diffusion. This mechanism has the effect of considerably slowing down the electrons and somewhat increasing the velocities of the ions. Figure 9.14 shows a typical set of data for E_b and E_m in nitrogen at 16 MHz and $\Lambda = 8\cdot1$ mm. The electric fields are expressed in peak values.

9.3.5 Breakdown mechanism for the diffusion controlled discharge
Breakdown in the high frequency diffusion controlled discharge has been generally defined by the criterion that the electron density tends to infinity, which leads to the breakdown relation given by eqn. (9.6). It is a gross simplification of the true nature of the breakdown mechanism in the high frequency case to accept this statement as readily as in the case of the normal (Townsend) runaway breakdown mechanism occurring in ordinary d.c. discharges without pointing out the essential differences. This is because in the high frequency case there is no feedback to

the multiplication condition as with the positive ions in the d.c. case. The latter impinge upon the cathode releasing secondary electrons which ionise the gas molecules in the same way as primary electrons and the ions they produce return to the cathode and release more electrons.

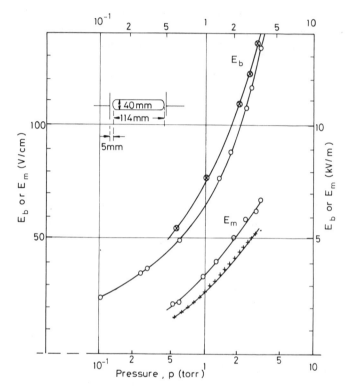

Fig. 9.14   *Theoretical and experimental responses for breakdown and maintenance (peak values) in nitrogen at 16 MHz and $\Lambda = 8 \cdot 1$ mm*
(Metaxas, 1968)
(1 torr $= 133 \cdot 3$ Pa)
ooo experiment
⊕⊕⊕ theory using data of Deas and Emeleus (1949)
xxx Schottky theory

A possible mechanism for the high frequency breakdown may be as follows. Assuming that the rate of change of n_e before breakdown is slow, that is, $\partial n_e / \partial t = 0$, the continuity equation (9.3) yields a trigonometric solution for n_e which is satisfied for fields below breakdown, e.g., $E < E_b$, as shown in Fig. 9.15(a). Ionisation in the gas equals the rate of charge loss by diffusion. Excitation is also occurring but the number of excited particles which emit visible light as they decay is very small indeed since n_e is very small and hence no glow is observed. Q_i determines the magnitude of n_e under these circumstances but as seen from eqn. (9.6) it does not affect E_b. The electron concentration just before breakdown n_{e0} is given by eqn. (9.4) which after expanding the cosine term in the denominator and rearranging gives with $z = z_b$

$$1 - Q_i(\text{cosine terms})/n_{e0}z_b = z_b d^2/D_e 8 \tag{9.8}$$

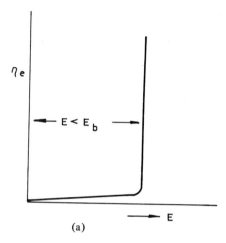

(a)

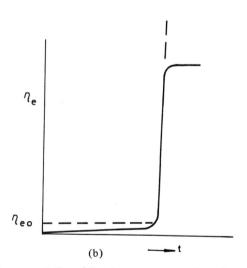

(b)

Fig. 9.15 *Schematic representation of the electron concentration prior to and at breakdown*

For N_2 at about 133 Pa (i.e., 1 torr), $z_b \approx 10^6$ s^{-1}. From the above expression it can be seen that the ratio $z_b d^2/D_e 8$ is always equal to unity at $E = E_b$ unless

$$10 Q_i(\text{cosine terms}) \geqslant n_{e0}z_b \tag{9.9}$$

or

$$Q_i \geqslant 10^5 n_{e0}/(\text{cosine terms}) \tag{9.10}$$

Since Q_i is normally very small, eqn. (9.10) shows that when breakdown field is reached the critical condition for breakdown is always satisfied irrespective of both the initial electron density and the magnitude of Q_i.

When the electric field just exceeds the breakdown value the rate of production per electron z_b exceeds the loss by diffusion, causing the electron concentration to increase rapidly. The diffusion loss is sufficiently slow that it does not influence the growth of n_e. Thus during the multiplication the electron concentration is governed by the continuity equation

$$\frac{\partial n_e}{\partial t} = z_b n_e \tag{9.11}$$

giving

$$n_e = n_{e0} e^{z_b t} \tag{9.12}$$

where n_{e0} is the electron density just before the growth process. As soon as the electric field just exceeds the breakdown value rapid growth develops which causes n_e to rise exponentially to a high value. To give an example, with $z_b \approx 10^6 \text{ s}^{-1}$, $n_{e0} \approx 10^8/\text{m}^3$ and assuming breakdown occurs at $n_e \approx 10^{13}/\text{m}^3$, eqn. (9.12) yields a time for the growth process of 11×10^{-6} seconds. In practice the electron concentration is limited by the maximum current that the external circuit can supply. After breakdown the excitation is greatly enhanced due to the high value of n_e which results in a visible glow. Figure 9.15(b) gives the likely variation of n_e with time. The solid curve represents the variation of the electron concentration in a practical circuit with time.

To summarise, eqn. (9.4) describes the distribution of electrons for $E = E_b$, n_e increases rapidly by multiplication and eqn. (9.12) controls its growth, diffusion loss now being unimportant. After breakdown the motion of the charges is controlled by ambipolar diffusion.

At pressure above about 1.3 kPa (i.e., a few tens of torr) the dominant loss mechanisms are by volume recombination and attachment which are much larger than diffusion losses. On the other hand at pressures below 1 Pa (or below about 10^{-2} torr) in ordinary vessels the mean free path is much larger than the vessel dimensions and the dominant loss mechanism is recombination at the walls, volume diffusion, recombination and attachment losses being now of secondary importance (Francis, 1960). At sufficiently low pressures (i.e., below 0.1 Pa or about 10^{-3} torr) the main form of charge production is by secondary wall emission (Gill and von Engel, 1948).

9.3.6 Effects of high frequency discharges in processing plants
Having described some of the processes which take place in the gas volume leading to breakdown and subsequent maintenance of the fully developed discharge and

which are relevant to high frequency plasma industrial applicators used in chemical processing, let us now briefly turn our attention to the deleterious effects of discharges instigated by high frequency electric fields developed in industrial applicators of moderate to high powers. Such unwanted discharges have been observed in three distinct types of application the origin of the problem being highlighted in Fig. 9.16, which shows the required breakdown electric field in air as a function of the pressure for two frequencies. The form of this curve is typical of that with most other gases in that the breakdown field reduces with reducing pressure until a minimum is reached beyond which the electric field rises once again with reducing pressure. At the minimum breakdown field the electron collision frequency is approximately equal to the applied frequency of the source.

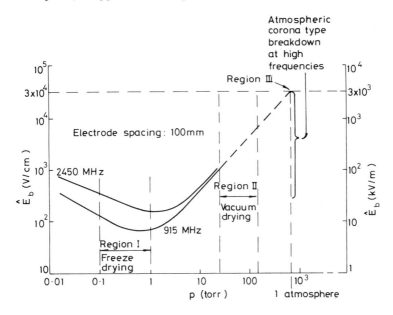

Fig. 9.16   *Typical breakdown electric field responses in air (peak values) as a function of the pressure for two frequencies*
(1 torr = 133·31 Pa)
(Adapted from Parker, 1968, by kind permission of Academic Press, Inc.)

The other parameters which are normally used to completely specify such breakdown characteristics are the type of gas, the vessel dimensions (yielding Λ) and the frequency of operation. For a given gas, characteristic length Λ and pressure the required breakdown field reduces as the operating frequency is reduced. Thus it appears, judging by the data shown in Figs. 9.14 and 9.16, that frequencies near the industrial allocated bands at 13·56 MHz and 27·12 MHz may be more prone to breakdown than either the 900 MHz or 2450 MHz, the latter being the least likely to instigate discharges, all other considerations being equal. It has also been observed that, for a given gas and characteristic length Λ, the responses at different frequencies tend to converge towards the same set of values at pressure near atmospheric. It should also be recalled from eqn. (4.11) that the power dissipation is proportional to the frequency and square of the electric field strength which favours the higher frequencies for vacuum processing.

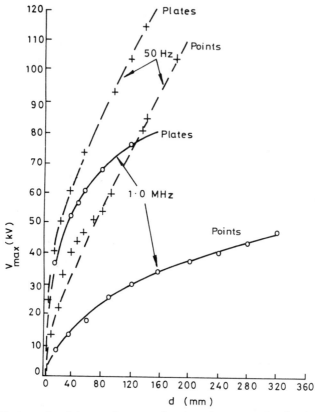

Fig. 9.17   *Maximum breakdown voltage as a function of gap separation for air at atmospheric pressure and for various electrode configurations*
(After Misere, 1932, Copyright by Springer-Verlag.)

The three pressure regions where the effects of gas breakdown must be avoided by careful design of the processing equipment are depicted in Fig. 9.16, region I, at around 13–133 Pa (0·1 to 1 torr) in microwave freeze drying application, region II at around 1·33–26·66 kPa (10–200 torr) in microwave vacuum drying application and region III at atmospheric pressures in applications involving high intensity electric fields in the vicinity of sharp metallic objects. Gas breakdown effects in regions I and II will be discussed in Section 9.4 under vacuum processing.

As far as the effects in region III are concerned the data of Fig. 9.16 show that, as the pressure increases beyond the minimum of the response, the required high frequency electric field to initiate breakdown increases and tends towards 3000 kV/m which is the value required for d.c. atmospheric pressure breakdown in air between parallel plate electrodes of small gaps ($\leqslant 10^{-2}$ m). However, Misere (1932) has shown that the combination of larger gaps, pointed electrodes and elevated operating frequencies is such as to reduce considerably the required breakdown field from the 3000 kV/m value at atmospheric pressure. Figure 9.17 shows that at 1 MHz with pointed electrodes and a gap of 0·1 m the required electric field for gas breakdown is only about 280 kV/m. Similarly in point to plane discharge gaps at atmospheric pressure the onset potential for high frequency coronas, V_b, relative to the onset potential for streamers, V_{str}, reduces with increasing frequency, as shown in Fig. 9.18 (Prokofiev *et al.*, 1959; Aints *et al.*,

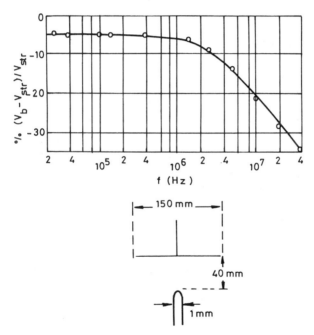

Fig. 9.18 *Percentage reduction of the high frequency corona voltage V_b, relative to the onset potential of streamers, V_{str}, as a function of the frequency in atmospheric air*
a % reduction of the voltage
b Dimensions of electrodes and discharge gap
(Data up to 1·4 MHz after Korge and Kudu, 1973 — Data above 1·4 MHz after Prokofiev *et al.*, 1959. By kind permission of Prof. Kudu, Tartu State University.)

1971; Korge and Kudu, 1973). At 27 MHz the onset potential for high frequency corona reduces to 5·6 kV relative to a streamer potential of 8 kV, which gives an onset electric field strength of 140 kV/m (1·4 kV/cm). This value is certainly within the range of electric field strengths established in radio frequency applicators.

Therefore, the possibility of electrical breakdown, in the air which may include dust or solvent vapour, surrounding a radio frequency applicator as transient discharges, coronas or sustained arcs needs to be carefully examined and taken into account in the initial applicator design. Arc detectors are used to switch off the source of power in the event of arcing. However, these detectors are primarily used in protecting the source rather than the processed material. There is a need, there-fore, to study factors such as carbon deposits on conveyor belts, inclusion of foreign metallic objects with processed material as in some textiles, electrode insulating material, etc., which influence breakdown of the surrounding air and to optimise the applicator design so as to minimise the likelihood of such discharges (Tomlinson and Barber, 1981). No data have been published for onset potentials for point coronas in the region of the industrially allocated microwave frequencies 400–3000 MHz.

9.4 Vacuum processing

9.4.1 Introduction
Many industrial processes are performed at reduced pressure to take advantage of the reduction in boiling temperature of solvents that occurs with falling pressure, or to exploit the difference in diffusion rates of volatiles which also increases. In all vacuum processes requiring heat, transfer of heat to the workload becomes difficult because convection, which usually is the predominant heat transfer process at atmosphere pressure, is ineffective at low pressure. Most conventional vacuum ovens therefore rely on conduction heat transfer from hot plattens, and in a few specialist applications by the injection of low pressure steam which provides heat transfer by condensation on the workload. Conduction heat transfer is slow and consequently difficult to control and requires a large surface area making for bulky plant. In most cases steam condensation is unacceptable.

Vacuum heating is well known in the metallurgical industries for the processing of high purity alloys where the same conventional heat transfer problems exist but which can be overcome by low frequency induction heating because of the relatively high electrical conductivity of the workload. It is apparent that micro-waves can provide a substantial increase in the rate of heating in a vacuum environ-ment compared with other methods but may be limited not only by the usual restraints of mass transfer but by the onset of electric field breakdown of the low pressure gas.

9.4.2 Pressure and boiling temperature of solvents
Fundamental to the operation of a vacuum drying system is the operating pressure to restrain the temperature of the workload below a prescribed limit. Often the temperature limit is a function of dwell time, and the limits which apply in a conventionally-heated vacuum system where the dwell time is long may be unnecessarily low for a microwave installation with more rapid process time. In this respect each application must be individually evalutated for optimum

conditions for the process duration involved. A further point also to be considered is that the boiling point of solvents is increased with rising concentration of dissolved solids (e.g., a sugar–water solution) which may require the process to operate at a lower pressure than indicated by the boiling point-pressure characteristic of the pure solvent. Figure 9.19 shows the boiling temperature of water vs. pressure in the region 1–133 kPa (or about 0·01 to 1000 torr). Note that the curve continues below 0°C (freezing point changes with vacuum pressure by a negligible amount) which indicates the region of sublimation of ice to steam (water vapour). Freeze drying operates in the 13–133 Pa (or about 0·1 to 1 torr), shown as region I in Fig. 9.16.

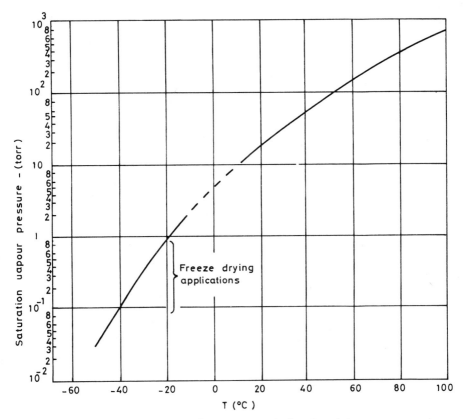

Fig. 9.19 *Saturation vapour pressure of ice and water as a function of the temperature* (1 torr = 133·31 Pa)

9.4.3 Microwave freeze drying

Microwave heating under vacuum was first explored in the specific application of freeze-drying of food, which is a sublimation process where the moisture is removed from the frozen material by direct conversion into vapour without having to go through the liquid phase (Copson, 1975; Parker, 1968). Conventional freeze drying is a slow process taking many hours to complete and the latter part of the freeze–dry cycle can be considerably speeded up by using a heat source such as microwaves which accelerates the energy absorption through the dried layers by effectively acting as a volumetric heat source to the receding ice layer (Gould and Kenyon, 1971).

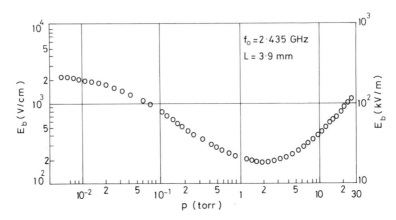

Fig. 9.20   *Continuous wave (c.w.) breakdown electric field (r.m.s.) as a function of the pressure of water vapour*
(1 torr = 133·31 Pa)
(After Tetenbaum and Weiss, 1978.)

The major problem has been microwave breakdown of the air and/or water vapour residing in the atmosphere surrounding the frozen material. This is because the optimum pressure for the sublimation process is between 13–133 Pa (or about 0·1 and 1 torr), a pressure range which, as the data of Figs. 9.20 and 9.21 clearly show, lies close to the minimum breakdown electric fields in dry air or water vapour. Operating at 2450 MHz, and at a pressure no lower than is necessary for sublimation, will reduce the likelihood of discharge formation. In addition to gas breakdown, the maximum power density that can be used is also limited by material melting only if the microwave energy is dissipated at a faster rate than can be conducted away by mass transfer leading to an increase of the pressure at the ice interface above the triple point. The formation of pockets of water causes preferential absorption of the microwave energy and may lead to thermal runaway, as discussed in Section 3.5.2. These constraints must, therefore, be taken into consideration when determining process throughputs that can be achieved in a particular installation. Whether the limiting factor is breakdown of the surrounding air or breakdown of the water vapour is difficult to predict and depends on the distribution of the electric field in the evacuated chamber.

Although intensive research was undertaken on freeze drying in which the fundamental data were established, at the time (1960–65) the technology was insufficiently developed to provide economically satisfactory installations. Since then, processing with microwaves in vacuum but at higher pressures, as described below, has been successfully applied in several food industry processes for the drying of heat-labile products (Meisel, 1979; Magnetronics Ltd. 1976, 1980, 1981) and low temperature thawing of meat and fish (Meredith, 1980).

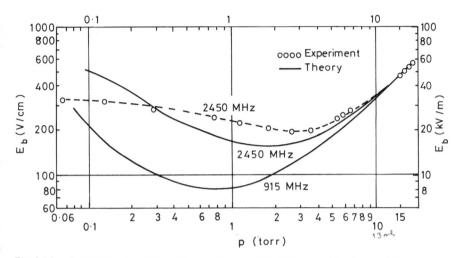

Fig. 9.21 *Experimental and theoretical breakdown data (E in r.m.s.) in air, L = 6·7 mm* (1 torr = 133·31 Pa) (After Gould and Kenyon, 1971, Copyright *The Journal of Microwave Power*, 1917.)

9.4.4 Microwave vacuum drying

Having a higher operating pressure for vacuum drying within the range 1·33–26·66 kPa (or about 10–200 torr) brings us to a more successful application of vacuum processing with microwaves, for drying heat sensitive materials such as pharmaceutical products and other chemicals. The drying must be carried out at temperatures below 100°C to avoid damaging the material and this requires processing under partial vacuum in order to reduce the boiling temperature. The data in region II of Fig. 9.16 shows that in the pressure range of 1·33–26·66 kPa (or about 20 to 200 torr) the electric field required for breakdown in air has now considerably increased and is many times higher than that required at the minimum of the breakdown response. The critical electric field in the processing applicator sets a limit to the dissipation power density for a particular set of conditions of pressure, temperature and dielectric properties of the dielectric material. The electric field breakdown appears as a localised glow discharge and its onset occurs at high-field points resulting from standing waves and field concentration due to the configuration of applicator walls and workload. It is important to avoid glow discharges because they waste power, cause local overheating and are liable to transfer to the microwave window through which the energy is introduced into the oven, with a risk of damage. For these reasons, coupled with the statistical nature of breakdown electric field strength and the process system tolerances, it is necessary to operate with a factor of safety of at least times two for the working peak instantaneous electric field.

In common with glow discharges at d.c. and low frequency a.c. the breakdown electric field strength is much higher than that required to maintain the discharge, so that a power reduction usually in excess of 10 dB is needed to extinguish a discharge once formed.

9.4.5 Applicator design

It is clear that microwave processing under vacuum carries the additional constraint of liability to electric field breakdown, which must be given careful consideration at the conceptual stage of applicator design. In industrial plant where continuous uninterrupted operation may be required for periods of weeks, the effect of build-up of deposits of workload from spillage and dust is important, particularly in the region of the microwave feed windows.

Calculation of performance characteristics commences with estimation of the maximum permissible temperature of the workload and hence the working vacuum pressure. The maximum safe working electric field stress can then be determined and, using eqn. (4.11), the maximum power dissipation density within the workload can be calculated. In performing this calculation due allowance must be made for the waveform of the generated microwave power which is likely to have a ripple content yielding a peak instantaneous power significantly exceeding the mean value.

Knowing the required mass flow of the process the total power can be calculated from eqn. (4.37), and hence the minimum volume of material which is required within the applicator. Thus the minimum size of the heating applicator is determined for the particular application and chosen frequency; it may well indicate that the most economic frequency should be 2450 MHz rather than 915 MHz. However, the *E*-field limitation may not be the limiting factor, mass transfer of steam at low pressure can cause foaming of a wet product which is usually unwanted, but in certain specific cases the control of the density of foam can be achieved by control of power density and pressure.

From microwave consideration there is no restriction in the type of applicator which may be used under vacuum within the limitations of electric field breakdown of the gas at the operating pressure. However, the applicator body usually forms the vacuum envelope, resulting in a structure to be designed and manufactured to pressure vessel codes of practice (e.g., British Standard Specification No. 5500). Large scale applicators have been built at 896 MHz (915 MHz in the U.S.A.) based on axial waveguide and serpentine configurations where the requisite stiffness has been achieved by a supporting structure of rigid beams. At 2450 MHz multi-mode ovens have been used successfully.

9.4.6 Pressure windows

In industrial applications the power is such that the energy must be introduced to the applicator via a microwave transparent window. Only in low power systems is it possible to feed the power directly from the output window of the magnetron generator because of the high field intensity surrounding it. Moreover, most magnetrons require some air cooling of the output window by natural convection if not forced draught.

The essential features of a microwave vacuum window are:

(*a*) At the peak instantaneous power to be transmitted the electric field stress at the window surface (including fringing field and standing-wave effects) must be less than the critical breakdown electric field by a margin of safety appropriate to the application. In most cases this means that it should not exceed half the critical value but special circumstances may demand a smaller factor. This factor determines the area of the window.

(*b*) The loss factor of the material of the window must be low so that overheating does not occur. Apart from wasting power, the hot surface causes a reduction in the critical electric field for breakdown.

(*c*) The window must be of a material which is resistant to high temperature and does not suffer surface tracking. If a glow discharge does occur in an oven it is liable eventually to stabilise across the face of the window causing excessive heating.

(*d*) The window material must have good mechanical properties:
   (i) completely gas-tight;
  (ii) high tensile strength to resist fracture;
 (iii) smooth surface for easy cleaning;
  (iv) resistant to fatigue cracking to withstand pressure cycling;
   (v) good thermal conductivity;
  (vi) mechanically stable;
 (vii) chemically inert.

(*e*) The window must have a good broadband VSWR (typically 1·25 over 2% of the band).

The above requirements exclude plastics (PTFE, polypropylene and ABS) for obvious reasons, and the materials most frequently used are alumina and quartz, both of which have very low tan δ_{eff} and good mechanical properties.

9.5 Hybrid systems

9.5.1 Introduction

Most industrial heating processes have several distinct phases, and whilst high frequency heating is generally able to perform most of them, the optimum system is usually one in which microwave energy is used solely for the exploitation of its unique characteristics because it is invariably more expensive in capital cost than conventional heating plant. It is becoming, therefore, more and more apparent that many successful high frequency applications require a degree of hybridisation with conventional drying or heating systems, as shown in Fig. 9.22. Industrial processes which comply to that shown in Fig. 9.22(*a*) include large scale drying where the bulk of drying is carried out using hot air followed by end drying or levelling by high frequencies. Similarly, the hybrid system shown in Fig. 9.22(*b*) is a typical example of microwave preheating of rubber prior to vulcanisation with hot air. It is impossible to give a general code of procedure for optimisation and each process must be considered individually, but the principles involved are illustrated by the following examples.

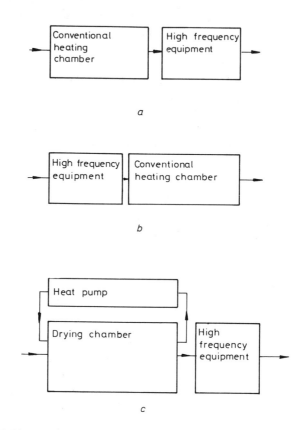

Fig. 9.22 *Hybrid systems*
  a Final drying or levelling with high frequency energy
  b Preheating with high frequency energy
  c Combination drying with a heat pump and high frequency

9.5.2 Processing of moulded rubber parts

This is an example of a combined system in which microwave energy is used as a pre-heater. In this process a block of unvulcanised rubber is inserted in a press-mould and is heated to about 150°C with the mould closed. Depending greatly on the size and thickness of the rubber, thermal equilibrium will occur after a period of several minutes to a few hours. If, instead of using blocks at ambient temperature, they are heated to, say 85°C, the press time can be shortened by some 30% to 50%. Microwave plant can provide this pre-heating, and the economic viability is immediately apparent in the increase in production throughput in relation to the capital costs of the microwave equipment and the press.

9.5.3 Continuous vulcanisation of extrusions
This is a technique where rubber profiles are extruded through a die and have then to be vulcanised by raising the temperature to about 200°C as quickly as possible, and maintaining this temperature for a time long enough to permit the chemical reaction of vulcanisation or polymerisation to take place.

Because of its volumetric heating properties, microwave energy is able to give the required temperature rise through the entire cross-section of the profile in a short time compared with the vulcanisation time. It is necessary then to provide a conventionally-heated dwell tunnel with the sole purpose of preventing the rubber from cooling before the reaction has taken place. No heat is to be transferred to the rubber by the dwell tunnel, and its efficiency is controlled entirely by heat losses from its external surface which can be made small. Because of the high temperatures required in the rubber, serious surface heat loss can occur unless precautions are taken, which consist of heating the environment applicator to a temperature comparable to that of the rubber.

Ideally in this system all the heat energy in the product is provided by microwave energy; however, in practice, the wide range of speeds, types of rubber and cross-sectional area results in some cases where only part of the energy is of microwave origin; the balance being conventional. Evidence for this is of the mass flow and temperature rise of product requiring an energy input in excess of that actually applied by microwaves.

9.5.4 Drying processes

9.5.4.1 End drying or levelling with high frequency energy: Because of its high latent heat of evaporation, water requires a relatively large amount of energy to effect drying, and therefore high frequency energy should only be used where conventional drying methods cease to be efficient.

In a very wet material as much water as possible should be removed by mechanical methods such as suction, squeezing or centrifuging. This should then be followed by conventional hot air drying ovens in which the latent heat for evaporation is provided by the heated air blowing across the surface of the workload. In most cases the efficiency of this drying technique falls away as the drying proceeds, largely because the rate of surface evaporation exceeds the rate of migration of water to the surface (see also Section 4.7). A condition then arises in which the surface is substantially dry, with a rising moisture content towards the centre of the workload. Often the thermal conductivity of the dry matter is low so the rate of heat transfer to the wet inner regions is reduced. Drying then becomes very slow resulting in an extended length of drying oven with associated costs of floor space and housing and energy wastage from heat lost from the external surfaces of the drying oven.

It is near this point where drying rate begins to fall that electromagnetic energy can be applied efficiently. The thermal barriers of the dried surface are transparent to the radio frequency or microwave energy so that extremely rapid evaporation of the residual water is possible. In practice the rate of evaporation is, in the limit, determined by the diffusivity of the dried workload which in turn determines the internal steam-pressure developed. Such pressure must obviously be maintained well below a point at which damage might occur.

Substances dried industrially vary so widely in their physical and electrical properties that it is not possible to provide general data on the optimisation of such a hybrid system, which has to be designed to suit the parameters of each specific application. However, the necessary information can readily be obtained from dielectric property measurements (see Sections 3.4.2 and 3.8), from tests in the laboratory and on existing drying ovens, to establish in the first instance where the falling rate point occurs in the dryer, and what then is the moisture content profile of the workload. Scaled laboratory trials can then be performed to determine the limits of drying which could be applied by radio frequency or microwave energy.

Hot air is usually supplied to a high frequency dryer to transfer the evaporated moisture away from the dryer to the exhaust or for feeding back into the preceding conventional dryer. The rate of air flow and its temperature are crucial for optimum performance of the dryer.

9.5.4.2 Combination drying using a heat pump and high frequency dryer: In the case of the hybrid drying system discussed above, and typically shown in Fig. 9.22(a), the conventional drying chamber which precedes the high frequency equipment can be made more efficient by recycling some of the exhaust vapours and using the sensible heat to heat up the incoming air. Such is the heat and mass balance that most of the latent heat of the evaporated moisture is lost. However, by recycling the exhaust vapours through a heat pump, as shown in Fig. 9.22(c), a large part of this latent heat can be transferred to the incoming air with substantial gain in overall efficiency of the preceding drying chamber (Hodgett, 1976; Baxter *et al.*, 1977; Lawton, 1978).

In essence the heat pump is based on the principle that a gas cools when it is expanded and heats up when it is compressed. There are basically two types of heat pumps used for drying very wet workloads essentially characterised by the operating temperature and the nature of the working fluid; that of the close cycle type and that of the open or steam recompression cycle, as shown in Figs. 9.23(a) and (b) respectively.

9.5.4.2.1 Closed cycle heat pump: In Fig. 9.23(a) the drying air is separate from the working fluid of the heat pump and heat exchange between the two media occurs only in the evaporator and the condenser. The heat pump fluid

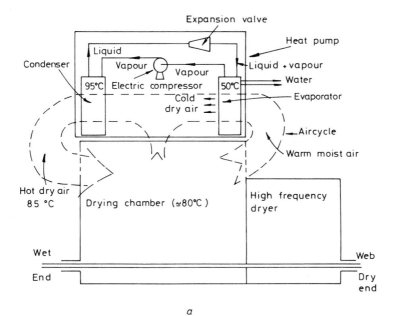

Fig. 9.23 *Hybrid drying systems using a heat pump and a high frequency dryer*
  a Close cycle heat pump
  b Steam recompression heat pump

enters the evaporator and is boiled off at low temperature and at a pressure of about 350 kPa (50 lb in^{-2}). The vapour is then passed through a compressor which increases its temperature and pressure to say about 1400 kPa (200 lb in^{-2}). The resulting high temperature vapour condenses as it passes through the condenser liberating heat at about 95°C which is used to heat the cold dry air. The high pressure condensate is cooled by evaporation as it goes through the expansion valve. The mixture of cool low pressure liquid and vapour passes once more through the evaporator to complete its cycle. As far as the air cycle is concerned warm moist air extracted from the drying chamber is cooled well below its dew point by passing it over the evaporator, water condenses and heat (primarily latent) is passed over to the heat pump working fluid. The cooled dried air is circulated over the condenser with make-up air from the drying chamber while the water is drained. Finally hot dry air from the condenser is admitted back into the drying chamber to complete the cycle.

The performance of such a heat pump dryer is best characterised by the amount of water extracted per unit of input energy and values of 3 kg/kWh are quite attainable at present, where the kWh refers to the mechanical energy supplied to the compressor. The best steam heated conventional dryers extract about 0·55 kg/kWh – where kWh now refers to the primary fuel consumption. Assuming a primary fuel to main electricity efficiency of about 30%, a heat pump dryer from mains electricity would therefore reduce the primary fuel requirement by about 40% (Lawton, 1978).

9.5.4.2.2 Steam recompression heat pump: The limitation of the close cycle heat pump is that the operating temperatures reached in the drying chamber are relatively low for some drying operations in the manufacturing industry, for instance, paper or some textiles where temperatures of the order of 140 or 150°C are required. An alternative type of heat pump will have to be used, shown in Fig. 9.23(*b*), where the evaporated steam itself is the heat pump fluid. The chamber is filled with superheated steam slightly above atmospheric pressure. Drying occurs in such a chamber by heat transfer from the steam raising the temperature of the product above the boiling point of water thereby creating a pressure gradient directed away from the surface. The water evaporated from the product leaves as high temperature liquid and contains the bulk of the mechanical energy supplied to the compressor. This energy having been converted into high grade heat can be used further in allied processes such as dyeing. Once again the extraction rates with this heat pumping arrangement are of the order of 4–4·5 kg/kWh.

The hybrid systems described above represent the ultimate in optimising the efficiency of a particular drying line where conventional, heat pumping and electro-magnetic techniques can be used wisely to produce a very efficient drying system of the future. The capital costs of such a system will be high and will only be at present economically justified in a few special cases. Heat pumps have been used on their own without the need to include a radio frequency or microwave dryer at the end of the line.

9.5.5 Other hybrid systems

So far we have discussed hybrid systems comprising conventional energy, a heat pump and high frequencies. Other combinations are possible which include direct electrical $I^2 R'$ heating of the product, infra-red or ultra-violet techniques. The first of these techniques is suitable for electrically conductive materials while the other two are essentially surface treatment techniques. Infra-red energy is, for example, currently used in the production of foam-backed tufted carpets to obtain a skin of the uncured Styrene Butadiene Rubber (SBR) foam prior to its vulcanisation by hot air or as future systems might employ a combination of hot air and microwave energy (Metaxas, 1981). Both infra-red and ultra-violet techniques suffer due to the lack of penetration into the processed material but can be used in combination with high frequency energy for thicker materials if both a surface treatment and volume heating are required.

9.6 Automatic control of the process

Having an extremely rapid heating capability microwave energy is particularly suitable for control by feedback techniques where a process is operated at the optimum condition by maintaining a particular parameter (e.g., temperature or moisture content) at a pre-determined level. Departures from the desired level can be data processed and used to control the microwave power to readjust to the preset point, either by a closed-loop feedback circuit, or by an open-loop correction based on the known sensitivity of the system. Combined control is also possible in which changes in parameters such as mass flow, or input temperature can be measured to make open-loop corrections of microwave power, with a closed-loop control used to stabilise the final output parameter.

The design of closed-loop control systems is well-established. However, there are a number of aspects of such a system incorporating a microwave generator which need consideration:

(a) Magnetron generators exhibit a frequency change when their power is changed. If the applicator has a high Q-factor, its impedance match may change to give a secondary but not insignificant change to the power dissipated in the workload.

(b) The principle time delay in the closed-loop system is in the transport delay between the product leaving the heating chamber and arriving at the sensing head. This is a pure time delay, not an exponential factor, and may cause severe overswing of the control parameters.

(c) Other exponential time delays arise from heat flow to the surface of the workload, and the fixed delay time of the workload in the heating zone.

(d) The response time of the generator is generally limited by the firing of control thyristors working at power-line frequency. However in ultra-fast control systems series-regulator triodes have been used to control magnetron power via the applied EHT, and in this case response of the order 10^{-6} seconds is possible.

9.7 Automatic tuning and matching of resonant applicators

9.7.1 Automatic tuning

Many microwave heating systems require the use of a resonant cavity to provide a sufficient E-field strength to obtain the required power dissipation density in the workload. Often the Q-factor is such that the bandwidth of the system response is comparable with or even less than the natural frequency excursions arising from frequency drift with temperatures, load variations, frequency change with power of the oscillator, mains voltage variations, etc. It is therefore necessary to provide means for tuning the cavity to hold the frequency difference, δf, between the generator frequency and cavity resonant frequency at a small fraction of the bandwidth, typically $\delta f < 0.1 f_0 / Q_0$. Moreover, during start-up the uncertainty of frequencies is such that the initial frequency separation may be outside the 'capture' range of an automatic tuning system. Operation with a high Q system therefore requires a control system with two principle roles, one to 'search' the frequency spectrum for coincidence between cavity and generator frequencies following initial switch on, the second to close a servo-loop to maintain frequency tracking.

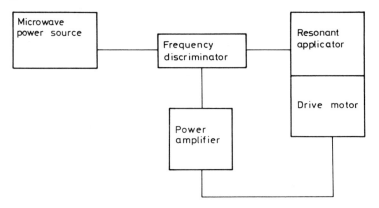

Fig. 9.24   *Schematic diagram of automatic tuning system for resonant applicators*

A block diagram of the principle elements of a frequency tracking system is shown in Fig. 9.24. The resonant cavity has a tuning plunger driven by a servo-motor which is controlled from a power servo-amplifier which has its input connected to a microwave frequency discriminator. This discriminator is a circuit assembly which gives a voltage output proportional to frequency error between cavity and generator, with a polarity reversal when the frequency error reverses, i.e., the cavity frequency changes from being less than to being greater than the generator frequency. The frequency discriminator may take a variety of forms, one of which is described by Wyslouzil and VanKoughnett (1973), but all are based on the measurement of the input impedance of the cavity which is a known function of frequency.

9.7.2 Automatic matching

In addition to frequency control of the cavity heater it is important to maintain a good impedance match at resonance for maximum efficiency, achieved by adjusting the effective aperture of the coupling hole (see Chapter 7). This requires a variable susceptance which may consist of a sliding inductive or capacitive aperture plate, a capacitive tuning screw, or a T-stub with adjustable short-circuit piston. Choice of system requires care because heavy surface currents are likely in the heater giving rise to sparking and overheating of the component parts. In many cases where load variation is not great a manual preset adjustment is adequate but a servo-control system is possible, deriving the necessary control signal from a second output from the frequency discriminator.

Wyslouzil and VanKoughnett (1973) have described an automatic triple stub matching system of high power handling capacity, capable of transforming a load VSWR of 20 or less down to less than 1·5 and possessing a fast response time (less than 2 seconds) for automatic operation. Its operation as a matching device is based on the principle of being able to match an arbitrary load by introducing series reactance at three fixed points, say $\frac{7}{8}\lambda_g$ apart. Basically the motion of the tuning stubs is controlled by the signal levels from a pair of probes within a frequency discriminator device. This matching device can be optimised by feeding the discriminator outputs into logarithmic amplifiers thus achieving good performance which is independent of the absolute signal levels from the discriminator device.

9.8 References

AINTS, M., VEIMER, V., and KUDU, K., 'Onset potentials of h.f. point discharge phenomena at lower pressures', *X Int. Conf. of Ionisation Phenomena in Gases*, Proceedings, Oxford, p. 144 (1971).

BAXTER, W. R. S. B., HODGETT, D. L., KOLBUSZ, P., and LAWTON, J., 'Drying by heat pumps', *Conf. on Electricity for Mat. Proc. and Conservation*, IEE Conf. Pub. No 149 (1977).

BEGUIN, C. P., EZELL, J. B., SALVEMINI, A., THOMPSON, J. C., VICKROY, D. G., and MARGRAVE, J. L., 'Chemical synthesis in radio frequency plasma torches', in *The Application of Plasmas to Chemical Processing* (Edited by R. Baddour and R. S. Timmins), pp. 35–53. MIT Press (1967).

BROWN, SANBORN C., *Introduction to Electrical Discharges in Gases*. John Wiley & Sons, New York (1966).

CHANDRAKAR, K., and VON ENGEL, A., *Proc. Soc.* 284, 442 (1965).

COPSON, D. A., *Microwave Heating*, 2nd Edition. AVI Publishing Co. In. (1975).

CRAGGS, J. D., *High Frequency Breakdown of Gases in Electrical Breakdown of Gases* (Edited by J. D. Craggs and J. M. Meek). John Wiley & Sons, New York (1978).

DEAS, H. D., and EMELEUS, K. G., *Phil. Mag.* 40, 460 (1949).

FEHSENFELD, F. C., EVENSON, K. M., and BROIDA, H. P., 'Microwave discharge cavities operating at 2450 MHz', *Rev. Sci. Instrum.* 36 (3), 294 (1965).

FRANCIS, G., *Ionisation Phenomena in Gases*. Butterworths Scientific Publications (1960).

FUMIAKI OKADA, KOICHI OHWI, and MAKOTO MORI, 'The development of a high power microwave circulator for use in breaking of contrete and rock', *J. Microwave Power* 10 (2), 171 (1975).

GILL, E. W. B., and VON ENGEL, A., *Proc. Roy. Soc.* A192, 446 (1948).

GOULD, J. W., and KENYON, E. M., 'Gas discharge and electric field strength in microwave freeze drying', *J. Microwave Power* 6(2), 151 (1971).

HAMBLYN, S. M. L., 'Plasma technology and its application to extractive metallurgy', *Mineral Sci. Enging.* 9, no. 3, July, pp. 151–176, 1977.

HAMBLYN, S. M. L., and REUBEN, B. G., 'Uses of RF plasma in chemical synthesis', *Advances in inorganic and Radio chemistry*, 17 (Edited H. J. Emeleus) New York, Academic Press, pp. 90–115, 1975.

HARVEY, A. F., *Microwave Engineering.* Academic Press, New York (1963).

HELSZAJN, J., *Non-reciprocal Microwave Junctions and Circulators.* Wiley Interscience Publications, New York (1975).

HODGETT, D. L., 'Efficient drying using heat pumps', *Chem. Engnr.* (July/August), 510–512 (1976).

HUDSON, A. S., 'Ferrite devices for magnetron protection in microwave power systems', *J. Microwave Power* 10(3), 257 (1975).

LAWTON, J., 'Drying: the role of heat pumps and electromagnetic fields', *Physics in Technology* 9, 214 (1978).

KORGE, H., and KUDU, K., 'The onset potential of h.f. corona in the frequency range from 25 kHz to 1·4 MHz in the point to plane discharge gap', *XI Int. Conf. of Phenomena in Ionised Gases*, Proceedings, Prague, p. 200 (1973).

MALONEY, E. D., and FAILLON, G., 'A high power klystron for industrial processing using microwaves', *J. Microwave Power* 9(3), 231 (1974).

MCDONALD, A. D., *Microwave Breakdown in Gases.* John Wiley & Sons, New York (1966).

MEISEL, N., 'Microwave energy applications', *Newsletter XII*, No. 6, pp. 3–6 (1979).

MEREDITH, R. J., Magnatronics Ltd., Unit A, St Mary's Mills, Evelyn Drive, Leicester, England (1980).

METAXAS, A. C., 'Reignition in electrodeless discharges', PhD Thesis, Imperial College, University of London (1968).

METAXAS, A. C., 'The future of electrical techniques in the production of printed tufted carpet', *J. Microwave Power* 16(1), 43 (1981).

MISERE, F., 'Luftdurchschlag bie Niederfrequenz und Hochfrequenz an verschiedenen electroden', *Arch. Electrotech.* 26, 123 (1932).

OKRESS, E. C., *Microwave Power Engineering*, pp. 19–135. Academic Press, New York (1968).

PARKER, W. N., *Freeze Drying in Microwave Power Engineering* (Edited by E. C. Okress), pp. 38–64. Academic Press, New York (1968).

POHL, W. J., *Power Klystrons and Related Devices in Microwave Power Engineering* (Edited by E. C. Okress), Vol. 1, pp. 84–104. Academic Press, New York (1968).

PROKOFIEV, A., KABARDIN, O., and KUDU, K., 'The onset of the h.f. point-discharges in the atmospheric air', *Proc. Acad. Sci., U.S.S.R.*, TXXIII, No. 8, Physics Section, 1959, pp. 1004–1006 (1959) (In Russian).

TETENBAUM, S. J., and WEISS, J. A., 'Microwave breakdown in water vapour', *Microwave Power Symposium*, Ottawa, pp. 56–58 (1978).

TOMLINSON, P. and BARBER, H., 'Discharges in radio frequency drying equipment', Electricity Council Research Centre/M1405, January (1981).

VIDAUD DE PLAUD, P. H., D.Phil Thesis, University of Oxford (1967).

VON ENGEL, A., *Ionised Gases* Second Edition. Clarendon Press, Oxford (1965).

WYSLOUZIL, W., and VANKOUGHNETT, A. L., 'Automated matching of resonant microwave heating systems', *J. Microwave Power* 8(1), 89 (1973).

Chapter 10

Hazards, leakage and safety

10.1 Introduction

Due to the invisible, silent nature of high frequency energy, there is understandable concern from the general public about safety standards, and because of the high technology of the subject there is a great deal of conflict of opinion of results of experiments, and therefore of proposed local standards of safety. Not the least problem is the apprehension caused by the media handling the subject in a fickle way without in-depth knowledge of the facts and tending to focus on instances out of context, thereby creating alarmist episodes.

In assessing from first principles the possible hazards of radio frequency and microwave energy there are two broad types of risk that might be incurred. The first is thermal, the simple heating of the human body as a result of exposure. The second group is the non-thermal risk which comprises generally effects on the nervous system, and genetic changes; these may be further sub-divided into possible temporary or permanent effects.

Considerable research has been done in many countries with the objective of establishing a code of practice and eventually legislation on permissible levels of exposure to radio frequency and microwave energy. Reaching agreement on a world standard has been made extremely difficult because of a few reports, mostly in the early 1970s from the Soviet Union, on effects at levels of exposure on order of magnitude less than those deduced from extensive experience and experiments generally in the West. Attempts to reproduce the effects reported appear invariably to have failed, and evidence is accumulating, backed by experience that the codes of practice adopted in the West are satisfactory. What is desirable is to set a low level of permissible exposure which affords a safety-factor to a known hazard level and yet is not such as to exclude, unnecessarily, a significant number of installations.

An excellent review of the health aspects of radio frequency and microwave energy exposure has been published by the Environmental Health Directorate of Canada (Anon., 1978).

10.2 Exposure standards

Human tissue is highly receptive to microwave radiation and must be protected against excess exposure to it, establishing an exposure standard for personnel working within a microwave heating environment (Stuchly, 1978). For many years an upper exposure standard of $10\,mW/cm^2$ ($100\,W/m^2$) was used (Anon., 1971).* The basis of this power density is that it is the same order of magnitude as the heat flux from the human body in a normal sedentary state so that the body will be able to dissipate the heat easily. It is also roughly one third of the flux density which can just be detected as a sensation of warmth. Let us represent an average human body as a cylinder of $1\cdot7\,m$ in height and $1\,m$ in average circumference as shown in Fig. 10.1, weighing $80\,kg$ and made of water-like substance, i.e., with specific heat of about $4180\,J/kg°C$. This gives a total body surface area of $1\cdot7\,m^2$. Medically the body can withstand average temperature changes of a few $°C/hour$, giving a power density of about $11\,mW/cm^2$ ($110\,W/m^2$) assuming a change of $2°C/hour$. Thus, according to this rough calculation the body can withstand this dosage for long periods without any adverse effects.

It is important to appreciate that the body cannot be described as a homogeneous dielectric. It is, in fact, highly heterogeneous with some specific parts, such as the eyes and testicles being more sensitive to microwave radiation since the heat generated cannot be rapidly taken away and dissipated in other parts of the body because of blood-flow restriction. However, the body's thermoregulatory system could easily stabilise against such a slow rate of rise of temperature, and if the power density were increased until the temperature regulation failed it would be at a level many times $11\,mW/cm^2$ ($110\,W/m^2$). A further feature is that microwave energy will generally be arising from a single source so only one half (i.e., one side) of the body would be exposed to the full intensity and moreover the field intensity is most unlikely, in an industrial environment, to be uniform over an area as great as that of the human body. Thus with a maximum level of $10\,mW/cm^2$ ($100\,W/m^2$) the average is likely to be considerably less. All these points lead to the conclusion that from thermal considerations a limit of $10\,mW/cm^2$ ($100\,W/m^2$) for continuous exposure is safe and is supported by the absence over some fifteen years of injury or reports of discomfort from operators or personnel working with industrial microwave equipment designed and meeting these standards; in the U.K. this represents some 10^7 operator man-hours without incident.

In addition to the above, there have been some publications on the damaging effects of prolonged exposure to low level radiation ($< 1\,mW/cm^2$) which, unfortunately, is difficult to confirm without a systematic research programme over a long period of time (Fröhlich, 1980). Many authorities and committees have been formed worldwide to establish a universally acceptable limit. Despite

---

* The non-SI unit of radiation flux, mW/cm^2, is quoted first since it is very well established while the SI unit is that shown within the brackets.

these efforts, there remains much controversy, depicted in Fig. 10.2, which shows the state of art regarding exposure standards in some countries over short and prolonged periods of time (Dodge and Glaser, 1977). What is clear though, is

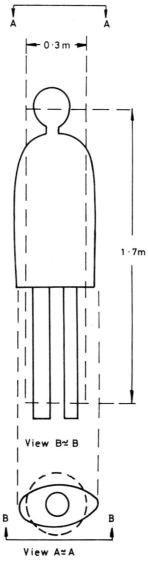

Fig. 10.1 *Human model for power density calculation*

that there is a positive trend towards more strict limits and as a result the upper limits are being progressively reduced. However, as many r.f. and microwave equipment are part of on-line systems with input and output ports to take the product in and out and as the recommendations become stricter there is additional pressure to design ports exhibiting adequate attenuation and reduce the leakage below the new limits. In turn this means that attenuation ducts become longer and the equipment more expensive. Even at moderate power levels of a few tens of kW leakage levels below $1 \, mW/cm^2$ $(10 \, W/m^2)$ will be quite difficult to attain unless sophisticated elaborate choking systems are employed.

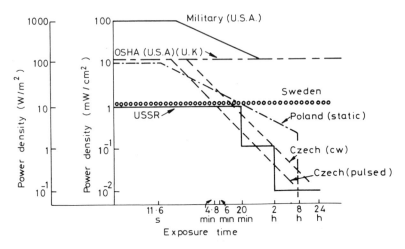

Fig. 10.2 *Microwave personnel exposure standards*
(Simplified from Dodge and Glaser, *The Journal of Microwave Power*, 1977.)

A review of selected biological effects and dosimetry data useful for developing high frequency safety standards for human exposure concludes that above $1 \, GHz$ there can be no conceivable damaging thermal stress at the presently accepted guideline in the Western world, of $10 \, mW/cm^2$ $(100 \, W/m^2)$. At lower frequencies particularly near the radio frequency regime the review stresses that unacceptable temperatures in localised tissue is possible during exposure for several hours at the accepted level. On present data the lowering of the permissible level for continuous whole body exposure to perhaps $1 \, mW/cm^2$ $(10 \, W/m^2)$ may seem justified at these lower frequencies (Tell and Harlaen, 1979).

10.3 Emission standards

In some countries manufacturers of microwave equipment such as power supplies, ovens, etc., are compelled by law to keep radiation leakage below a given level, which establishes an emission standard. The U.S. regulations state that the emission from a domestic oven must not exceed $1 \, mW/cm^2$ $(10 \, W/m^2)$ before its acquisition by the purchaser and $5 \, mW/cm^2$ $(50 \, W/m^2)$ thereafter. The tests are carried out by a standard $0.275 \, dm^3$ $(275 \, cm^2)$ water load placed in the oven.

10.4 Communication and industrial frequency bands

Throughout this book we have been concentrating on two frequency bands, i.e., 896/915 and 2450 MHz and reference was made to 27·12 MHz in the radio frequency regime. In fact there are many more bands in the range 6·78 MHz to 245 GHz, shown in Table 10.1, which have been allocated for Industrial, Scientific and Medical (ISM) use. The frequency bands centred at 13·56 MHz, 27·12 MHz, 896 MHz (in U.K.) 915 MHz (in U.S.A.) and 2450 MHz are the principle ones, for which industrial equipment can be readily purchased. These frequency band allocations were primarily made by selecting unused frequencies within the very congested telecommunication range.

Table 10.1 *ISM allocations in revised international frequency allocation table agreed at the World Administrative Radio Conference (WARC) in 1979*

Bandwith authorised	Centre frequency	Nature of use (see text)
6765–6795 kHz	6 780 kHz	Subject to agreement
13 553–13 567 kHz	13 560 kHz	Unconditional
26 957–27 283 kHz	27 120 kHz	Unconditional
40·66–40·70 MHz	40·68 MHz	Unconditional
433·05–434·79 MHz	433·92 MHz	Unconditional in FRG, Austria, Liechtenstein, Portugal, Switzerland, Yugoslavia
433·05–434·79 MHz	433·92 MHz	Subject to agreement except for countries shown above
902–928 MHz	915 MHz	Unconditional (Region 2 only)
2400–2500 MHz	2 450 MHz	Unconditional
5725–5875 MHz	5 800 MHz	Unconditional
24–24·5 GHz	24·25 GHz	Unconditional
61–61·5 GHz	61·25 GHz	Subject to agreement
122–123 GHz	122·5 GHz	Subject to agreement
244–246 GHz	245 GHz	Subject to agreement

It will be noticed that alongside these ISM allocated frequency bands the terms 'unconditional' or 'subject to agreement' have been included. The unconditional use simply means that there is no interference based radiation limit within these frequency bands, the only restriction being that of personnel safety discussed in the preceding Section. Radio communication services operating within these bands must accept harmful interference, if any, which may be caused by these applications. However, care must be taken with spurious signals and harmonics of the fundamental frequency since, although their levels are usually small, they may fall within a telecommunication band, generally having far stricter levels, quoted in μV/m at a certain distance from the radiating source.

Table 10.2   *Frequency allocations for industrial, scientific and medical purposes (existing allocations, 1980)*

Frequency MHz	Frequency tolerance (+/−)	Area permitted
0·07	10 kHz	USSR
13·56	0·05%	World-wide
27·12	0·6%	World-wide
40·68	0·05%	World-wide
42, 49, 56, 61, 66	0·2%	Great Britain
84, 168	0·005%	Great Britain
433·92	0·2%	Austria, Netherlands, Portugal, W Germany, Yugoslavia, Switzerland
896	10 MHz	Great Britian
915	13 MHz	North and South America
2 375	50 MHz	Albania, Bulgaria, Hungary, Roumania, Czechoslovakia, USSR
2 450	50 MHz	World-wide except where 2375 MHz is used
3 390	0·6%	Netherlands
5 800	75 MHz	World-wide
6 780	0·6%	Netherlands
24 150	125 MHz	World-wide
40 680		Great Britain

The term 'use subject to agreement' infers that use of an ISM equipment within a particular frequency band be subject to special authorisation by the administration concerned in agreement with other administrations whose radio communication services might be affected. In applying this provision, administrations should have due regard to the latest relevant CCIP (Comité Consultative Internationale de Radio Communication) recommendations. The existing allocations (1980) for ISM applications are shown in Table 10.2.

10.5 Leakage from industrial microwave equipment

Having briefly introduced the standards for the use of industrial microwave equipment we now turn our attention to possible sources of leakage of microwave radiation and offer practical solutions which go a long way towards satisfying the recommendations stipulated in the preceeding Sections. Health hazard apart, there is an incentive to reduce leakage by designing systems which waste very little power and therefore increase the overall utilisation efficiency. Broadly speaking industrial equipment can be divided into two categories: batch and on-line systems.

10.5.1 Batch systems

The only source of leakage from a batch system, usually comprising an oven-type applicator and its magnetron source attached close to it, are the door seals. Quarter-wavelength choke systems have been used (Ohkawa *et al.*, 1978; Püschner, 1966) which reduce the leakage to well below the acceptable limits. These chokes were considered in some detail in Section 6.8. The safety regulations require interlocks on the oven doors which switch off the power source if the oven door is not fully closed and clamped or locked. It is mandatory that such interlocks fail to safety.

10.5.2 Continuous flow systems

Continuous flow high-frequency systems are more prone to leakage than batch systems. This is because an on-line system must incorporate in its design an entry and an exit port through which the processed material enters and leaves the applicator. These ports must be adequately choked to absorb residual energy radiated out from within the applicator. Generally these ports are as small as the dimensions of the material to be processed allow. Processing textile webs at 2450 MHz may require ports of a metre wide and up to 5 mm in height. Applicators operating at the 900 MHz band have greater processing slot heights, up to 100 mm with modest choking systems.

Leakage of energy from open apertures can be controlled by the following techniques singly or in combination thus forming a structure generally referred to as a choking system.

(*a*) Restriction of the aperture size to dimensions less than those of a cutoff waveguide at the operating frequency.

(*b*) Restricting the aperture to a rectangle with one dimension significantly less than a half-wavelength so as to control the leakage to one plane of polarization only.

(*c*) For wide, low height apertures, designing the oven so that the internal wall currents are polarized so as to have zero component normal to the aperture slot, as in a serpentine applicator.

(*d*) Fitting a reactive choking system of quarter-wave slots or structures to present an open circuit to energy propagating in the aperture passage.

(*e*) Providing a resistive absorbing system to attenuate the leakage energy and to maintain residual leakage within prescribed limits.

10.5.2.1 Cutoff waveguide chokes: The cutoff waveguide choke is the simplest and most effective means of limiting leakage but the restricted size makes it suitable only for small cross-section workloads (e.g., extrusions) or, if it is in the form of an array or grating of tubes, suitable only for the processing of particulate materials. The principle of operation of such a choke is based on the attenuation of micro-wave energy propagating in a cutoff waveguide if its dimensions are less than some critical figure which is a function of the free-space wavelength of the energy.

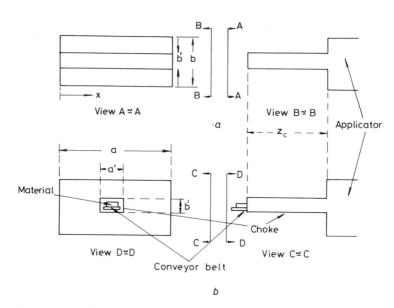

Fig. 10.3   *Principle of cutoff waveguide choke*
a Waveguide $(a \times b')$ at the feed end not acting as an effective choke
b Waveguide $(a' \times b')$ at the feed end acting as a choke

Consider the feed end of an axial travelling wave rectangular TE_{10} applicator, as shown in Fig. 10.3(*a*). A rectangular waveguide of dimensions x and b' is used to feed the conveyor belt and material into the applicator. Let us examine the properties of this feed waveguide for a fixed b' and variable x. When $x = a$, the attenuation per metre due to wall losses for the TE_{10} mode in the axial applicator $(x = a, y = b)$ and feed waveguide $(x = a, y = b')$ are shown qualitatively in Fig. 10.4 by solid curves exhibiting the same cutoff frequency, f_c. With such dimensions of the feed waveguide all that is achieved is to attenuate the dominant waveguide mode propagating through the feed waveguide. However, as x is made much smaller, then a condition is reached, say at $x = a'$ in Fig. 10.3(*b*), where the cutoff frequency in the feed waveguide, f_{cc}, has now shifted well beyond the operating frequency in the axial applicator f_a, and therefore propagation of the dominant mode through the feed waveguide ceases. Under cutoff conditions the characteristic wave impedance of the feed waveguide will be pure reactance for the dominant mode and no energy will propagate along it. This is a reactive attenuation in that the energy is not dissipated in the walls but reflected back into the applicator. The attenuation per metre due to wall losses for the TE_{10} mode at frequencies $f > f_{cc}$ through the feed waveguide is shown dotted in Fig. 10.4.

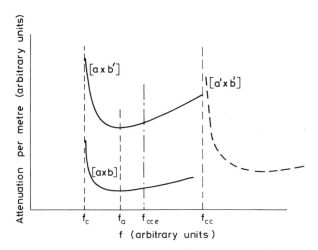

Fig. 10.4   *Attenuation per metre due to wall losses as a function of the frequency for the dominant TE_{10} mode in rectangular waveguides*

  f_a   operating frequency in TE_{10} applicator
  f_c   cutoff frequency for TE_{10} mode in applicator
  f_{cc}   cutoff frequency for TE_{10} mode in choke
  f_{cce}   cutoff frequency for TE_{10} mode in a choke partially filled with processed material

Care must be taken when designing a cutoff waveguide to allow for the dielectric constant of the processed material. For a completely filled feed waveguide the cutoff frequency now becomes $f_{cce} = f_{cc}/\sqrt{\epsilon'}$, so for a practical partially filled choke, the shift of the cutoff frequency to the right will be smaller but nevertheless it must be ensured that $f_{cce} > f_a$, otherwise the effectiveness of the choke will be significantly reduced.

The propagation constant γ, of the wave passing through the feed waveguide $(a' \times b')$ under cutoff condition, is real for $f < f_{cce}$ and the attenuation of the power follows $P_0 e^{-2\alpha z_c}$, where z_c is the length of the smaller waveguide or choke and α is its attenuation per unit length which for an empty choke, and assuming negligible wall losses, attains the form (Johnk, 1975)

$$\alpha = \omega(\epsilon_0 \mu_0)^{1/2} \left[\left(\frac{f_{cc}}{f} \right)^2 - 1 \right]^{1/2} \text{Np/m} \qquad (10.1)$$

Substitution of $c = 1/\sqrt{\epsilon_0 \mu_0} = \lambda_0' f$ in eqn. (10.1) yields, after rearrangement,

$$\alpha = 2\pi \left(\frac{1}{\lambda_c^2} - \frac{1}{\lambda_0'^2} \right)^{1/2} \text{Np/m} \qquad (10.2)$$

Laslo and Newman (1980) used this cutoff principle to design a choke for tobacco grains constructed by an array of small rectangular tubes operating beyond cutoff at 2450 MHz and 915 MHz. A typical choke assembly for 2450 MHz is shown in Fig. 10.5 and achieves theoretical power reductions of about 10^6. The length of the choke was made about 0·25 m to compensate for the loading effects of the dielectric. At 915 MHz the dimensions of the cutoff tubes were 83 × 56 mm achieving equally high power reductions.

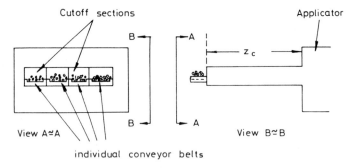

Fig. 10.5   *Choke assembly constructed fo an array of small cutoff sections*
(After Laslo and Newman, 1980, Copyright *The Journal of Microwave Power*, 1980.)

Circular tubes can also be used as chokes. For example, for a 0·15 m diameter tube acting as a choke at 915 MHz ($\lambda_0' = 0·328$ m) and with a cutoff wavelength for the dominant TE_{11} mode of $3·41 R$, where R is the tubes' radius, eqn. (10.2) yields an attenuation of 133 dB/m. In practice a cutoff wavelength not more than $0·7\lambda_0'$ is used because the attenuation falls rapidly as $\lambda_c \rightarrow \lambda_0'$ and the sensitivity to manufacturing tolerances and the loading effect of the work load becomes excessive. It should also be noted that the above cutoff characteristics apply to a tube in the regions well removed from its ends, and less attenuation is achieved in the end regions. Other modes will have smaller values for their cutoff wavelengths and therefore would exhibit higher attenuation and can be safely ignored.

10.5.2.2 Quarter-wavelength stubs: One of the most common reactive choking techniques is to use a series of short-circuited stubs along the broader sides of the waveguides at the input and output ports of the continuous processing applicator, as shown schematically in Fig. 10.6, each stub adjusted in length to be $\frac{1}{4}\lambda_g$ of the particular mode it is intending to suppress. The input impedance of the stub of length d is

$$Z = jZ_{0s} \tan (2\pi d/\lambda_g) \qquad (10.3)$$

where Z_{0s} is the characteristic impedance of the stub. For $d = \frac{1}{4}\lambda_g$ this impedance becomes infinite and therefore the stub presents an open circuit to the particular mode. Such stubs are effective in blocking the TE_{mn} modes in the rectangular

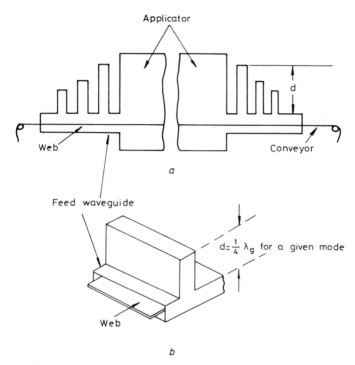

Fig. 10.6   *Quarter-wavelength stub choke for input and output ports*
    a Complete applicator
    b Details of a single stub

feed waveguide but not the TM_{mn} modes. The attenuation of such filters can be about 20 dB. However, they exhibit a narrow bandwidth (about 1 MHz) which may require tuning screws for operation at the centre frequency of the microwave source. Alternative schemes based on this filter principle have been described by Püschner (1966) where, for example, mode transducers have been used between the applicator aperture and a short-circuited stub for mode transformation with subsequent blockage by a single filter. The main disadvantages of such filters are the narrow bandwidth, the critical dimensions of the stub assembly and its dependance on web width.

10.5.2.3 Corrugated chokes: Some of the shortcomings of quarter-wave stub chokes can be overcome by a corrugated choke which is another reactive type filter (VanKoughnett and Dunn, 1973), shown schematically in Fig. 10.7(a). Essentially it consists of a series of equal length stubs which periodically load the output waveguide and exhibit pass and stop bands characteristic of similar periodic structures (Harvey, 1963). The frequency characteristics of such structures depend critically on the dimensions of the structure illustrated in Fig. 10.7(b)

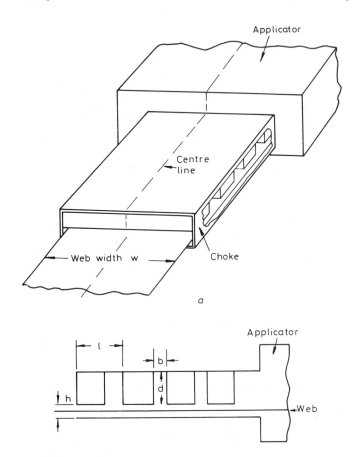

Fig. 10.7   *Principle of a corrugated choke*
(After VanKoughnett and Dunn, 1973, Copyright *The Journal of Microwave Power*, 1973.)

such as the length and width of the stubs, d and $(l-b)$ respectively, the distance between adjacent stubs, b, and height, h. The attenuation constant α in Np/m can be derived from the dispersion characteristic and takes the form (Harvey, 1963)

$$\cosh(\alpha l') = \cos(2\pi l'/\lambda_g) - \frac{b}{2h}[\tan(2\pi d'/\lambda_g)][\sin(2\pi l'/\lambda_g)]$$

$$(10.3)$$

where the primes refer to equivalent electrical lengths. The insertion loss for n stubs is $8 \cdot 686 \, \alpha l' n$ dB. The design of the choke assembly follows from eqn. (10.3) by carefully choosing the various constants such as, for example, $h < 0 \cdot 2 \lambda_g$, $l' = \lambda_g / 4$, $d' = \lambda_g / 4$ and $b = l' - h$.

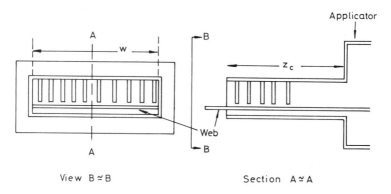

Fig. 10.8   *Doubly corrugated choke constructed with screws in a waveguide*
(After VanKoughnett and Dunn, 1973, Copyright *The Journal of Microwave Power*, 1973.)

The main drawback of this type of corrugated choke is that its design can be optimised for only one particular mode in the waveguide. To attenuate other propagating modes as well, this design is extended to doubly corrugated chokes which is essentially similar to that shown in Fig. 10.7 but with slots cut in the direction of material travel forming individual blocks connected to the broad side of the output waveguide. A variation of this, formed by cylindrical screws, is shown in Fig. 10.8. The insertion characteristics of such a structure comprising ten rows of $6 \cdot 35 \times 29 \cdot 2$ mm ($\frac{1}{4} \times 1 \cdot 15''$) screws spaced on $1 \cdot 73$ mm ($0 \cdot 68''$) centres across the output waveguide is shown in Fig. 10.9 for the TE_{10} and TE_{20} modes. The insertion loss reduces somewhat with the passage of material on the web through the corrugated choke. To achieve operating gaps of more than $0 \cdot 2 \lambda_g$, in order to process thicker materials, two-sided corrugated chokes have been designed, as shown in Fig. 10.10, with screw assemblies at the top and bottom of the broader waveguide faces. In this case special attention should be given to suppressing the E fields parallel to the web plane by rods extending the width, w, of the output waveguide.

10.5.2.4 Hybrid chokes: The chokes referred to in the sections above are of the reactive type which ensures a high degree of mismatching of the outgoing radiative energy causing it to be reflected back towards the applicator where part of it is reabsorbed by the processed material while the remainder is dissipated in the walls of the waveguide and applicator. However, any radiation not suppressed by the reactive choke can be further attenuated by materials which readily absorb

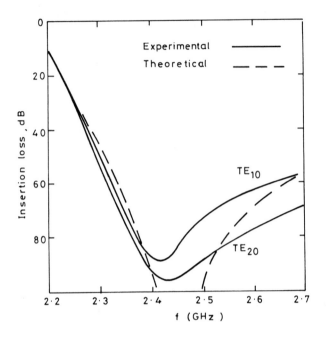

Fig. 10.9 *Insertion loss of doubly corrugated choke constructed with screws in a wave-guide*
(After VanKoughnett and Dunn, 1973, Copyright *The Journal of Microwave Power*, 1973.)

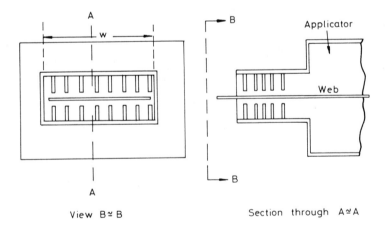

Fig. 10.10 *Two-sided doubly-corrugated choke constructed with screws*
(After VanKoughnett and Dunn, 1973. Copyright *The Journal of Microwave Power*, 1973.)

microwaves, incorporated at the end of the reactive choking structure, thus forming a hybrid choke comprising resistive and reactive elements. A variety of hybrid choke schemes is shown in Fig. 10.11.

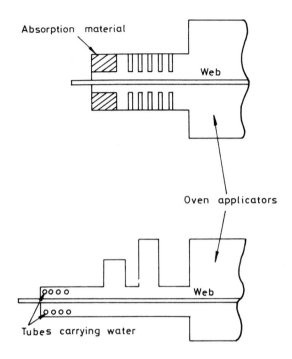

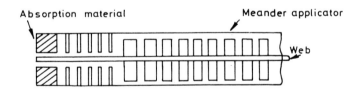

Fig. 10.11 *Typical hybrid choke systems made up of both resistive and reactive components*

Examples of materials for resistive chokes include carbon, eccosorb (Emerson and Cumming) and water. The latter can be very effective in acting as a radiation shield, as has been demonstrated in a large volume plasma reactor (Bosissio *et al.*, 1972). Essentially, the plasma reactor was powered from a slow wave structure and to protect personnel against stray radiation a double walled rectangular box made of plexiglass and carrying water was very successfully used. In fact cooling water is always available in a high frequency installation so part of it can be easily branched out, assuming adequate flow exists, and be used as the absorbing element.

Although such combination choking systems are more or less always necessary for on-line systems, they may not offer sufficient protection. This was the case in a development of a twin-cavity carpet drying applicator (Holme and Metaxas, 1979) which, of necessity, had a complete gap separating the upper and lower parts of the cavities. Additional protection was obtained by incorporating the applicator in a Faraday cage complete with interlocked doors and other safety measures, as shown in Fig. 10.12. The Faraday cage can be made simply of metal sheet or of a double metallic shield separated by a gap, possibly filled with a microwave absorbing material. The internal walls of the Faraday cage could be lined with a microwave absorbing material to reduce stray radiation reaching the entry and exit ports due to multiple wall reflections.

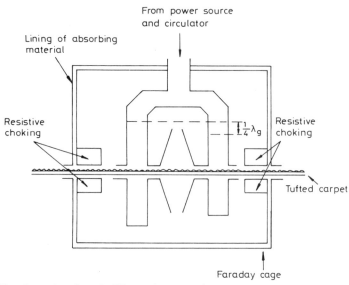

Fig. 10.12 *Screening of a twin TE_{10} cavity carpet dryer*

10.5.2.5 Choking of large apertures: Frequently a large aperture is required for a continuous-flow process in which the cross-section of workload has dimensions in both directions which are large compared with a half-wavelength. In this case leakage can occur in any polarisation and with a large number of modes. Such an aperture is impossible to choke by pure reactive means because with the surface impedance of the walls of the aperture approaching infinite value, solutions exist to Maxwell's equations allowing propagation to occur. These solutions to Maxwell's equations satisfy the boundary condition of $H_t = 0$ and can be visualised as a set of sinusoidal space harmonics (i.e., normal waveguide modes) in which the vector sum of the H_t components is zero at the boundary. As the energy propagates along the choke duct, changes in relative amplitude occur to maintain the same boundary condition in spite of phase changes due to the differing cutoff wavelengths of individual modes. At the other extreme the surface impedance can be zero (i.e., a metal wall) in which case the aperture behaves as a normal waveguide. Periodic loading is another alternative (i.e., alternating zero and high impedance sections), designed to attenuate specific modes, but energy interchange between modes tends to nullify this effect.

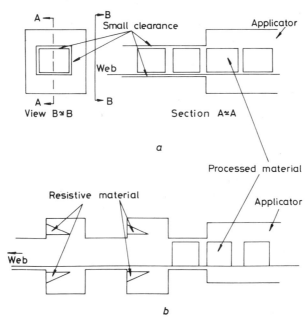

Fig. 10.13   *Choking of large apertures*

Resistance choking is the only effective method of limiting leakage from a large aperture and takes one of two forms, or a combination of them. In the first shown in Fig. 10.13(*a*) the workload of high tan δ_{eff} material passes along a closed waveguide duct with minimum clearance so that its absorption provides a high value of attenuation which is readily calculable to all modes. Depending on the tan δ_{eff} of the material such a choke may typically be fully effective over a length of about three times its largest cross-section dimension. However, it has the serious shortcoming that it is totally ineffective in the absence of a workload, and it must be supported by other means to guard against this potentially hazardous condition, which usually includes a sensor to switch off the microwave power. Such sensors may be microwave leakage detectors, but the overall system must essentially be fail-safe. Probably the safest scheme is a microwave detector system incorporating fuses through which a small pilot current flows. The fuses form part of a microwave circuit detecting leakage and 'blow' if it exceeds a safe level, interrupting the pilot current and switching off the microwave power. This system is the final safety stop and at least one other system is also used for normal operation to avoid unnecessary fuse-blowing.

In the second system, illustrated in Fig. 10.13(*b*), one or more large chambers are fitted in the choke duct containing absorbant material or a water load system. Leakage from the duct diffuses outwards into the chamber reducing the power density at the entrance to the next section of duct. By this means some 10 dB attenuation can be achieved per section, which is augmented by the attenuation due to the tan δ_{eff} of the workload. Again this system is usually inadequate to protect against total absence of workload and an additional fail-safe facility must be included. It also has the disadvantage of wasting power in the absorber, though in the case of highly absorbing workloads such as meat blocks, this power loss is within acceptable limits.

10.6 Safety precautions

Whatever the industrial application, all possible precautions must be taken to ensure the lowest possible leakage from a high power source and its associated applicator structure. Following a comprehensive survey carried out by the Bureau of Radiological Health in the U.S.A. of industrial microwave systems it is recommended that the following safety measures are observed in order to protect personnel using such equipment:

(*a*)   Adequate reactive and absorptive loads at the entry and exit ports to reduce stray radiation to well below the maximum allowable limit.

(*b*)   Additional screening should be used in special circumstances such as a Faraday cage, singly layered or doubly layered, sandwiching an absorptive material.

(*c*)   Well designed cleaning doors.

(*d*)   Interlock and safety equipment incorporating door and power unit interlocks as well as water flow monitors, arc detectors and quick access for fire-extinguishing equipment in case of processing flammable materials. Regular checking of these interlock and safety features must be carried out.

(*e*)   Labels placed at most hazardous locations, such as entry or exit ports, warning personnel from putting foreign objects or hands, etc., in entry ports while the system is operating.

(*f*)   Warning lights must be placed around the zone of the microwave equipment to avoid personnel other than the operators approaching the hazardous areas.

The survey showed that the most common causes of excessive radiation leakage were defective cleaning doors, due to warping, burning, loose hinges and inadequate choking at the conveyor input or exit ports.

10.7 References

ANON, Medical Research Council, Press Notice MRC/70/1314 updating of GPO safety precautions relating to Intense Radio Frequency Radiation issued by HMSO No. 48/82, 1960 (1971).

ANON, Health aspects of radio frequency and microwave exposure, Parts I and II, Reports 77-EHD-13 and 78-EHD-22, Information Directorate, Dept of National Health and Welfare, Brook Claxton Building, Ottawa, KIA OK9 (1978).

BOSISSIO, R. G., WEISSFLOCH, C. F., and WERTHEIMER, M. R., 'The large volume microwave plasma generator. A new tool for research and industrial processing', *J. Microwave Power* 7(4), 325 (1972).

DODGE, C. H., and GLAZER, Z. R., 'Trends in non-ionising electromagnetic radiation bio-effects research and related occupational health aspects', *J. Microwave Power* 12(4), 319 (1977).

EMERSON and CUMMING, Europe NV, DEVEL, Belgium

FRÖHLICH, H., 'The biological effects of microwave and related questions', *Advances in Electronics and Electron Physics* 53, 85 (1980).

HARVEY, A. F., *Microwave Engineering*, p. 439. Academic Press, New York (1963).

HOLME, I., and METAXAS, A. C., 'Microwave drying of tufted carpets. III. Field trials', *J. Microwave Power* 14(4), 367 (1979).

JOHNK, T. A., *Engineering Electromagnetic Fields and Waves*. John Wiley & Sons Ltd., New York (1975).

LASLO, T. S., and NEWMAN, T. A., 'The application of the cutoff principle to the end seal of microwave conveyor systems', *J. Microwave Power* 15(3), 173 (1980).

OHKAWA, S., WATANABE, M., and KANETO, K., 'High performance door seal for microwave oven', *Proc. IMPI Conference*, Ottawa, pp. 2–4 (1978).

PÜSCHNER, H., *Heating with Microwaves*. Philips Technical Library (1966).

STUCHLY, M. A., 'Microwave bio-effects and radiation safety', *Trans. IMPI* 8 (1978).

TELL, R. A., and HARLAEN, F., 'A review of selected biological effects and dosimetry data useful for development of radio frequency safety standards for human exposure', *J. Microwave Power* 14(4), 405 (1979).

VANKOUGHNETT, A. L., and DUNN, J. G., 'Doubly corrugated chokes for microwave heating systems', *J. Microwave Power* 8(1), 101 (1973).

Chapter 11

Industrial applications

11.1 Introduction

Microwave energy has been used in industrial processing for many years. Its adoption against competition from more conventional heating methods has been its appeal to special advantages, such as faster throughputs, space and energy savings and quality improvement. In the early stages of the evolution of microwave heating these advantages were often difficult to justify against the relative cheapness of fossil fuel heat. This, together with the natural reticence of many industrialists to change existing but often inefficient and obsolete conventional systems for microwave systems, has resulted in the well-documented slow growth of the microwave heating industry. The 1960s are primarily characterised by the obstinate opportunism of many newly formed manufacturers of microwave equipment to capitalise during an era of economic expansion but who unfortunately did not possess the technical expertise and after-sales service necessary to ensure customer satisfaction.

After the failures of many of these early endeavours, these trends were slowly reversed during the seventies following a determined effort to form design teams having the necessary wide range of professional engineering capability and to provide an appropriate after-sales service. The disappearance from the scene of many of the initial firms, consolidated the remainder in the field. It is, however, difficult to predict the market trends for microwave industrial heating in the post-recession period of the early eighties; suffice to say that public perception of the impending shortage of fossil fuel is growing sharper all the time. Also it is argued that in order to survive in the eighties firms have to invest now in new plant and machinery. What the microwave manufacturing equipment sector and those who support it must ensure is that it is ready to meet the opportunities that will arise in the eighties following the realisation that conventional energy is getting scarcer. This requires that there are proven and successful microwave applications such as exist in the rubber and food sectors of the manufacturing industry.

Taking a global view of the future of microwave industrial heating is more difficult because the various countries have different forms of energy available and have different central government energy policies. Certain countries do not possess much indigenous fossil fuels and are committed to electricity from hydro-electric or nuclear sources and consequently manufacturing firms tend to select electrical methods, which gives wider opportunities for microwave techniques. However, in countries rich in fossil fuel, the penetration of high frequency techniques will be slower.

This chapter will briefly summarise the past, highlighting mistakes and changing attitudes, will give an up-to-date account of industrial applications world-wide and will examine the economic and other factors in the installation of equipment.

11.2 The last three decades

What has become apparent during the past three decades is the overriding need for close cooperation between the user and the manufacturer of microwave equipment. Often the manufacturer of microwave equipment had little knowledge of the industrial process being broached, while on the other hand, the user did not understand the basic principles of microwave heating and apparatus. It is only the convergence of these two interests that can result in successful applications.

A pre-requisite for this is clearly that each must be master of his discipline. The past decade or so has brought in more reliable microwave equipment. However, equally important is the appreciation by the user of the energy costs of his existing process. This is because when cheap fossil fuels were readily available energy costs played a relatively unimportant role in investment decisions. Although the energy crisis has alerted manufacturers to a greater awareness of energy costs it is often difficult to separate the running costs of individual processes from the collective total, which is a necessary precursor to making an investment decision.

For the equipment manufacturer expertise must range far beyond microwave technology to embrace thermodynamics of heat flow and mass transfer and extensive capability in mechanical and electrical power and control engineering; moreover a considerable depth of understanding is required of the process such as has been acquired, for example, in the rubber industry, where rubber compound ingredients have a prime effect on performance.

Although the lack of user—manufacturer cooperation has in the past hampered many potential applications it is by no means the only reason for the slow adoption of microwave processes by industry (Freedman, 1972, 1973). Amongst other influential factors, Assinder (1974) lists some typical reasons that led to the failure of early installation:

(*a*) Technical problems with equipment generate disillusionment.
(*b*) The problem changes. The original processing problem no longer exists due to a variety of reasons ranging from changes in process technology to management policies, or market demands.
(*c*) Improvements in conventional machinery: not monitored carefully.

(*d*) No thorough economic analysis was undertaken, so economically unviable microwave solutions were pursued.

(*e*) Conservatism within the processing industry, reluctance to change.

According to a critical review of the ingredients for successful microwave applications Bedrosian (1973) advises the equipment manufacturer to question deeply the merits of using microwaves for a proposed system before experimental work begins. One must consider the application from all possible angles, such as the technology of the product, the thermodynamics of the system as well as the views of the plant engineer and company accountant.

Of prime importance is the precise definition of the processing problem which should preferably originate from the user. Following this definition the next question to be asked is whether microwave energy offers cost effective advantages which cannot readily be achieved by improved conventional equipment. A thorough analysis of the costs must be made at an early stage of the investigation and should be compared with the costs of the most up to date conventional equipment. Finally, a detailed plant specification and process analysis must be prepared by the machinery designer and agreed with the user: this often reveals a serious lack of data of the existing process which may cause the plant specification to be written in terms of heating power dissipation rather than of throughput mass flow capacity.

Some of the well established advantages of microwave (or radio frequency) energy over conventional techniques are well documented (Perkin, 1979):

(*a*) More efficient in drying within the falling rate period — this may reduce running costs.

(*b*) Drying times may be reduced — this allows increased throughputs.

(*c*) The system is more compact than conventional systems.

(*d*) Energy is transferred in a clean manner.

(*e*) Selective energy absorption by lossy constituents — this may result in moisture levelling of web materials.

(*f*) Puffing of a material such as tobacco, doughnuts and food products by internal pressure.

(*g*) Heat transfer is independent of air stream and mass transfer is increasingly independent of air stream as solids temperature increases — high velocity air flows are not required which may result in fan power energy savings.

(*h*) Energy dissipated rapidly throughout the volume of the material — deeper penetration.

(*i*) Avoids overdrying.

(*j*) Substitution of an expensive raw material with a cheaper one — as in bread baking, thus reducing raw materials costs.

(*k*) Relatively low maintenance costs.

To these must be added the important advantages of vacuum and pressure processing.

11.3 The present position

Metaxas (1981*b*) has carried out a literature survey of all work known to be going on in various Government laboratories, universities and private or nationalised industries. The work has been broadly divided into main industrial processing areas. A further subdivision, as shown in Table. 11.1, is made in terms of full scale industrial installations (I), prototype industrial installations (P) and laboratory work (L). In such a voluminous list there is bound to be a considerable overlap between the three subdivisions with applications progressing from the laboratory model straight into a full production model without going through the prototype stage. Alternatively there are applications at a stage between laboratory and scaled up prototype stages particularly in processes where large powers would be required, such as 100 kW or more. Upon close examination of these processes it becomes apparent that the ultimate power levels required will not make a good criterion for categorising the applications under I, P or L since a particular industrial heating application may only require 15 kW of microwave power whereas a laboratory foam-backed carpet dryer could quite easily need 25 kW.

From the extensive range of potential applications studied there has emerged a core of processes which do possess a sound technical and economic base. Some of these will be mentioned briefly.

11.3.1 Full scale industrial microwave heating processes
A clear example of the uniqueness of microwave energy is demonstrated in the baking industry where microwaves are used for proving doughnuts raised with yeast and, more interestingly, where a combination of microwave energy and deep frying are used to process chemically leavened doughnuts with unique quality results — unequalled, it is claimed, by conventional techniques. One installation led to increased profits due to a 15% reduction in the ingredients, increased the production capacity due to shorter frying time, improved eating quality and shelf life and resulted in a 20% increase in actual sales. In doughnut proving, microwave energy reduced the proving time and transformed a cumbersome batch process into a continuous one (Schiffman, 1976).

Conventional tempering of frozen products is a long and arduous process taking from several hours to many days to complete depending upon the size, type, etc., of the package (Swift and Tuomy, 1978). Microwave units have been successfully developed to increase the temperature of the product to $-4°C$ or $-2°C$ for ease of further processing. Invariably, it is imperative that the temperature of the product remains below the freezing point in order to avoid thermal runaway due to the enhanced loss factor of water droplets. The operating costs of a typical 25 kW tempering unit in the U.S.A. processing 0·28 kg/s (1000 kg/h) of frozen meat amounts to about 2 p/kg assuming a 10% rate of depreciation and electricity unit costs of 0·6 p/kWh. The microwave equipment can be justified in terms of higher throughputs, savings in floor space and handling, improved quality and sanitation and particularly greater floor flexibility. Similar commercial applications

Table 11.1 *Industrial microwave applications.* * I = Full production equipment.
P = Pilot scale. L = Laboratory models, Patents (Pt), etc.

Process	Material/Process	Scale
Baking	Doughnut proofing and frying	I
	Bread rolls, meringues, potatoes, cakes	(P)L
Blanching	Vegetables	P
Chemical reactions	Neutralisation of toxic substances, plasma heating	P
	Ferromagnetisation of FeS_2 (pyrite), contamination in natural kaolin, devulcanisation of scrap rubber, recovery of water from urine, cracking of waste plastics, electroplasmolysis of plant tissue, processing of starches, pyrolysis of wastes, coal desulfurisation, devulcanisation of sulphur vulcanised elastometers, detoxification of hazardous wastes, dehydrochlorination of PVC and synthesis of ion exchange resin from residue, carbon fibre production from waste resins, disintegration of toxic gases, oxidation of sulphur to sulphuric acid, regeneration of spent activated carbon	L
Cooking	Chicken, bacon, meat (sausages), herring, potato crisps, fillets, meat pie, rice	I
	Potato puree, gravy, grain	P
Curing/Hardening	Polyvinylchloride belts, urathene foam, polyester, epoxy resin, foundry core	I
	PVC coated gloves	P
	Polystyrene foam, plaster of Paris, asphalt road surface, cement, cellylosic membranes, bricks, precision moulds and cores, plastics, wire coatings, paints and laquers, solid propellants (non-metal), fibreglass, epoxy laminates, wood gluing with urea based adhesive, core binders and coatings, alkaline phenolic resins, cellular plastic laminates (phenolic, polyurathene), hardening of cement mortar, concrete sodium silicate bonded cores, post-curing of formaldehyde-phenol polymer, varnish on transformer armatures	
Dye fixation	Yarns (wool, acrylic, cotton, nylon, polyester), loose stock, fabrics, cotton polyester fibres	L(P)

Table 11.1   *Continued.*

Process	Material/Process	Scale
Drying	Paper, film, noodles, pasta, onion, printing ink, paint laquer, casting moulds, plastic coatings, lumber adhesive, pharmaceutical (fluidised), pastes and washes (foundry), photographic silver halide coating, profiling of biscuits, leather, vitamins, nuts, slip cast crucibles, textiles	I
Atmospheric, fluidised, or spray drying	'Ceramic products', latex sponge, polyamide pellets (F/B)[1], granules, flour (F/B), solvents, carpets	P
	Prints, manure (Pt), ceramics, carpets, tea, egg yolk, foam products, sand moulds/core washers, asbestos containing materials, cotton seed, fibrous sheet (Pt), plastic pellets, thermoplastics, acrylamide polymer, polystyrene flakes (F/B), polygels, pharmaceutical powders, cotton fabric, wheat gluten and agents, pine wood disks, fertilizer powders, clay products, moulds, and cores made of plaster-water blends, yarns, coal, mandarin orange juice (S/D)[2], molecular sieves, fibre bundles, plaster, alumina slurry, refractory bricks, coke, tobacco, sewage, gypsum and plaster, moulds, beechwood, silica fillers for isocyanate sealants, detergent compositions, fireproofed cotton textiles, coconut flesh (Pt), MnO_2 for battery cells	L
Vacuum drying	Beverage, fruit juices, harvested crops	I
	Labile products, soya beans, corn, rice, peanuts	P
	Heat sensitive products, powders, foodstuffs	L
Freeze drying	Sliced beef – silica gel	P
	Apple sauce, spun yarn under vacuum or at atmospheric pressure (Pt), carrots	L
Heating	Polyurathene foam, tobacco, high nitrile reforms, cigarettes, blood, press-setting in pencils – plasmas	I
	Tar, nylon monofilaments, cellulose fibres (for paper making), egg yolk paste	P
	Plastic waste for recycling, plastic sheet sealing, cellusosic tow for cigarette making (Pt), rope,	L

Table 11.1   *Continued.*

Process	Material/Process	Scale
Heating cont.	polymer film between wood laminates, wool, adhesive for plywood jointing, crosby durum wheat, foamable thermo-plastic resin beads, coal, foaming of polyamide resins, eggs (boiling of), hazardous waste (RCB's, kepone, CCl_4, phenyl-mercuric acetates), anti-corrosive coatings, epoxy powder coatings, fabric/polyurathene foam/thermoplastic laminates for embossing, heat sealing of laminated plastic tubes, foaming of silicon rubber	L
Melting	Lipid, oil fats, refractory dielectrics (oxides), dewaxing in investment casting,	I
	Polyurathene foams, nylon flakes, sludge (sewage/radioactive waste), bomb explosives, ice cream and chocolates, metal plating sludge, combustion ash	L(P)
Pasteurisation	Bread, peeled potatoes	I
	Ham — precooked food, crabfish, beverage, milk	L
Rendering	Residues from meat industry	P(I)
Roasting	Laver (seaweed)	I
	Beans, nuts	P
	Cacao bean, beef	L
Sterilisation (mould inhibition)	Japanese straw mats, bread, pizza, peeled potatoes, yogurt	I
Purification Disinfectation/ Sanitation/ Pasteurisation	Pre-cooked foods, bread, tobacco shreds, raw sewage, medicine ampules, cosmetics poultry droppings, clothes, nitrile-styrene thermoplastic bottle mouldings, domestic water supply, vinyl chloride polymers, cosmetic colour additives, heat resisting spores and micro-organisms, pea flour and protein concentrate, grain, soil (partial sterilisation/deactivation of fungi), gel worming bulbs, coconut flesh (Pt)	L
Tempering–Thawing	Meat and various materials (rubber, fish, fruits, butter, etc.)	I
	Triglycerides (fats)	P(I)
	Tissue cultured cells	L

Table 11.1   *Continued.*

Process	Material/Process	Scale
Vulcanisation (curing)	Natural or synthetic rubber products (extruded, blanks, tyres, bales, injection moulded), elastomers and foams	I
Specials	Coagulation of fish protein, puffing or rice cakes	I
	Opening oysters, road maintenance, carbon production, vitrification of high-level radio-active materials	P
	Moulding of reinforced plastics, polystyrene pattern removel, converting grass to hay, machining of ceramics, concrete/rock crushing, producing laminated paper (Pt), annealing of plastic bottles (Pt), making elastomer sheet material, soil stabilisation, germination of seeds, production of oil from cashew nut, reclamation of scrap rubber, softening/stiffening material via carbon black additives (Pt), migration of flame retardant formulation to surface in cotton packages, improvement of ore concentrates (minerals, pyrites and sphalerites), reconstituting chicken waste, embossing of polyurathene foam containing laminate sheets, polymerisation of fibreglass laminates, selective destruction of wild oats when mixed with cereal seeds, decomposition of laboratory animals carcasses, recovery of petroleum from petroleum-impregnated media	L

* *Sources*: (Metaxas, A. C., 1981*b*) (Some data on foodstuffs taken from Bengtsson and Ohlsson, 1974, Copyright, 1974, The Institute of Electrical and Electronics Engineers, Inc.)
[1] Fluid bed;
[2] Spray dried.

have been reported for other foodstuffs such as butter and fruits and for tempering frozen rubber bales. For example a 90 kW, 915 MHz microwave unit thaws 0·3 m³ blocks of butter, passing on a conveyor belt through a horn applicator, from − 20°C to about + 2°C. Butter contains relatively small amounts of water, about 15% dry basis, and is therefore less prone to thermal runaway compared with other foodstuffs which contain much larger amounts of water. A more recent application involves the development of a combination system for tempering and thawing meat and fish products (Schiffman, 1976).

A proven application is a combination of microwaves and hot air to pre-cook bacon (Smith, 1979). The microwaves are used to supply the sensible heat to the bacon, to render the fat and to coagulate the proteins while the hot air removes the moisture from the surface. A 50 kW unit is capable of processing 0·07 kg/s (250 kg/h) of bacon to a 40% yield. Present developments aim at producing bacon products with very little nitrate content.

Upgrading the quality of otherwise unsaleable meat scrap has been achieved by microwave energy prior to dicing and canning. Specifically, microwave energy is used to coagulate the meat in order to obtain a coherent product (Swift and Tuomy, 1978). Similarly, meat is cooked with microwaves in the preparation of meat pies from small extruded samples. A combined system of short frying and in depth cooking by microwaves has been introduced for the preparation of meat patties prior to freezing. This replaces a much longer frying technique. The new process is cheaper because of higher yield and less fat usage (Bird, 1979). Otherwise, microwave heating applications involve processing of chunks of meat, sardines and skinless frankfurters (Ohlsson, 1977).

A combined system of microwaves and steam has improved the quality and yield of cooked chicken parts (Schiffman, 1976). The problem here has been to identify a real commercial advantage by using microwave energy. The idea initially was to reduce cooking losses by using microwaves thereby increasing the final weight of a given chicken. However, the final product is sold on a piece basis so that increasing the final weight of a given number of pieces by the use of microwaves does not confer a real advantage. Processing of small birds might have been the answer but unfortunately the unit costs of purchasing smaller birds increases. Despite these difficulties there are still a number of units operating in the U.S.A. However their growth has been relatively slow.

Continuing with heating processes in the food industry there are some additional applications of microwave/conventional systems which have reached full industrial exploitation but it is too early to judge their real commercial performance. Microwave bread pasteurisation is one such process which enhances the destruction of micro-organisms and heat labile spores which infect bread during the cooling, slicing and packaging operations. Two 40 kW and one at 80 kW units have been installed in Sweden. Similarly, successful installations in Europe have been reported for the preservation of peeled potatoes, pizzas and yogurt (Olhsson, 1980).

Finally, further successful microwave applications from Japan have been reported, such as heating nuts to liquefy the inherent fats and aid slicing, the opening of oysters, the roasting of laver, coagulation of fish protein and puffing of rice cakes.

Outside the food industry, a positive impact has been made in the rubber industry where vulcanisation with microwave energy has now become more widely used on account of its convenience and high efficiency yielding worthwhile savings in energy and cost (Meredith, 1976). Due to the volumetric heating of microwave energy the rate of rise of temperature is no longer dependent on the heat flow through the surface of the rubber as in conventional vulcanisation and is very much faster. Depending upon the thickness of the product, microwave energy can heat rubber up to a hundred times faster than with conventional heating methods and reduces the process time significantly with consequent improvement in productivity. In continuous vulcanisation of extruded rubber this improved productivity is realised in greatly increased line speed for a given size of plant with less consumption of energy. Moreover the extrudate is processed clean and free of contaminants arising in liquid curing media which are difficult to remove completely. In addition extrudates of large cross-sectional area and complex shape can be processed continuously where previously batch curing was the only practical method. There are also many batch microwave curing installations offering significant improvements in quality, efficiency and costs over conventional methods (Shute, 1971).

Amongst special applications, outside the food and rubber industries, there are a few worth noting which enjoy partial commercial success in that a few units are still operating in industry. The first process involves microwave dewaxing of the ceramic shells used in investment casting. Conventional methods of dewaxing, such as in an open furnace, results in high breakage rate and the wax is lost, whereas in a microwave process the shells remain completely intact and the wax may be reclaimed. The second process originates from Japan for killing insects and nits in straw mats using microwave energy to rapidly raise their temperature to 60°C or more.

11.3.2 Full scale industrial microwave drying processes

Up to 1981 the most successful industrial applications have been in the rubber and food industries as evidenced by the continuing installation of further plant. It is interesting to note that the particular applications within these industries generally fall into two main categories; first, the addition of sensible heat by rapid volumetric heating and, second, the final drying of a product from a modest moisture content below which conventional dryers are slow and inefficient. This pattern of usage underlines the generality that microwave energy should be confined to applications where its unique properties can be exploited to full advantage, and that a combined system of conventional and microwave heating often represents the optimum installation in terms of capital cost, operating cost and energy usage (Metaxas *et al.*, 1979). Each industrial plant must be considered in complete detail in establishing its economic case, taking particular care not to overlook obvious factors such as that the cost of conventional heating plant includes contributions not only from the heating chamber but also possibly from a boiler house and extensive steam mains, or a fuel oil store. Also likely future trends must be considered, such as the relative costs of electricity to other energy sources.

Table 11.1 lists a number of drying applications reported in the literature. A notable continuing success in the U.S.A. is pasta drying where the unique features are sanitation, product control, time and space savings. Conventional drying of pasta is an exceedingly slow process taking 10–20 hours at 40°C in long hot air recirculating ovens. A typical combined system of microwaves and hot air in the

U.S.A. uses 60 kW of microwave power and processes 0·42 kg/s (1500 kg/h) of pasta at a cost reduction of operation, electricity and maintenance of up to 30%. The combined system reduces the overall drying time to about 15 minutes, involves shorter dryers and presents a product bacteriologically more acceptable in that better control over microbial contamination and insect infestation is achieved (Forwalter, 1978).

Similarly, final drying of quick-cooking noodles with microwave/hot air has been reported from Japan. The system employs 5 × 25 kW, 915 MHz sources handling 0·17 kg/s (600 kg/h) of noodles where the moisture content is reduced from 30% to 13%. The applicator is of a meander type with hot air maintained at about 80°C.

A recent microwave/hot air installation for the final drying of onions has resulted in substantial benefits in terms of moisture levelling of the output product, a 30% reduction of the energy costs in the final drying and a reduction of the bacteria count by 90%. The system represents the ideal usage of an expensive form of energy in that it is only used to dry the onions from 10% to 5% moisture and a conventional hot air dryer reduces the moisture level from 80% to 10% (Smith, 1979).

The drying of laver, a seaweed product with a fine roast flavour, is another successful application emanating from Japan. The product is dried as paper thin 0·3 m square shapes, passing through the dryer on low loss conveyors, in typical 12 kW units handling about six sheets per second and evaporating the last 5% of moisture (Forwalter, 1978).

The drying processes cited above are clear examples of the potential of microwave energy in dehydration processes so long as the benefits are clear and the economics viable (Smith, 1979). Particular care is needed where the product to be dried is seasonal (onions) and a large capital equipment might remain idle for a large proportion of the year.

One of the recent technological advances in the penetration of microwave energy into the industrial scene is that of vacuum drying. Some materials are temperature sensitive and cannot be dried at 100°C. To reduce the boiling point the pressure is lowered to between (2·7–27) kPa (about 20–200 torr) and microwaves applied producing a unique drying system which offers great advantages in processing intrinsically expensive temperature sensitive materials such as pharmaceutical powders, and foods such as beverages, fruit juices or harvested crops, etc. A real problem with vacuum drying is that of corona due to the reduction of the breakdown electric field from its high value at atmospheric pressure to values between 90–800 kV/m at pressures in the range (2·7–27) kPa (or about 20–200 torr) respectively. For foodstuffs and pharmaceutical powders adequate powder dissipation can be achieved with 20 kV/m which allows a safety factor of four times at the lowest pressure (see also Section 9.3.6).

A 48 kW, 2450 MHz microwave vacuum dryer is reported to dry orange juice concentrate at a rate of 0·014 kg/s (49 kg/h) in a cycle which takes 40 minutes while the cost/kg is lower than the equivalent freeze or spray drying techniques.

The combined action of vacuum and a microwave field dries and foams the original concentrate down to a few per cent moisture content, the product is then shaped and brushed from a belt into a hopper and through an air lock into a collecting vessel (Meisel, 1979).

Similarly a 60 kW vacuum dryer followed by a rapid cooling zone has just been commissioned by a British firm to dry a beverage powder. The wet slurry feeds into the dryer and is transported on four parallel belts running down the length of four WG4 waveguides operating at 896 MHz (Meredith, 1980; Lefort, 1981). Another 60 kW vacuum dryer with a multiple feed multimode oven is operating in the U.K. at about 5·3 kPa (40 torr), finish drying a heat sensitive food product.

A number of relatively small microwave drying equipment have also been installed in industry over the years processing food, various textiles, foundry cores and moulds, etc., with power handling capability between 2–10 kW (Shute, 1980). Very little publicity has been given to these applications and it is therefore exceedingly difficult to follow up their progress.

Drying by high frequencies has passed through an evolutionary period which has demonstrated that optimisation of these systems necessitates a degree of hybridisation with other techniques. A clear example of this is in drying applications where economic criteria point towards combined systems in which the electromagnetic energy is used to evaporate residual water near the dry condition whilst the bulk of dehydration is carried out by conventional, or as a recent advance has shown, by heat pumping techniques (Witt, 1980; Lawton, 1978) (see Section 9.5). Also, in an effort to reduce costs, there are moves towards using high frequency heating techniques for supplying only sensible heat, i.e., inducing diffusion of the moisture to the surface so that cheaper conventional techniques may be employed to extract and carry the moisture thereby avoiding the use of microwave energy in supplying the latent heat of evaporation, which, particularly for water, represents a high proportion of the total energy required (Lefeuvre *et al.*, 1978). In most of the successful applications cited above there has been a mixing of techniques in order to achieve the best technical and economic compromise.

11.3.3 Pilot systems

This category cites applications where a considerable amount of work has been carried out and is either being at present introduced to industry or is awaiting the right economic climate. Some well-publicised examples will be briefly described.

A recent development has been the microwave rendering of meat, bone and fat residues to produce tallow and animal feed. Preliminary tests in the U.S.A. with prototype microwave rendering units point to better grade tallow which is used in the production of margarine and quality soaps. As soon as the grease or tallow starts to melt and is liquified, it runs away from the microwave applicator in direct contrast with conventional pressure cookers where the tallow or grease keeps on being processed until it is taken out of the equipment, a process which downgrades the ensuing tallow. Microwave rendering is a very economical way of using electromagnetic energy where practically no evaporation of the initial moisture occurs, the microwave energy being used solely to separate the fat from the bone. A 30 kW unit has been developed capable of processing 0·4 kg/s (1500 kg/h) of fat (Bird, 1979).

Another microwave vacuum development uses a vertical feed applicator for drying grain. The working pressure is about 5 kPa (about 38 torr) which reduces the boiling point of water to about 40°C, thus minimising material damage which has been the major problem of conventional drying machinery processing temperature sensitive materials. The unit is powered by two 6 kW magnetrons operating at a frequency of 2450 MHz. The quality of the final product may now become a major selling point by U.S.A. exporters who compete world-wide. An interesting feature of this device is the extraction of vaporised moistures which in the first instance condense on the outer walls of the drying column, the walls being kept cool by external water spray. The condensate is drawn up from the bottom of the unit as clear water. A scaled up unit up to 100 kW, operating with a column diameter of about 3 m and 20 m tall is under development. The products to be dried include soya beans, rice, wheat, corn, rye, cotton seed, pecans and peanuts (Forwalter, 1978).

A major experimental investigation on microwave final drying of tufted carpets has been undertaken in England (Holme and Metaxes, 1979). The objective of the work has been to design a special microwave applicator to overcome the arcing problem which may occur with radio frequency due to corona discharges on the stenter pins which transport the carpets through the various stages of manufacture. A pilot dryer has been designed comprising two TE_{10n} resonant cavities which are displaced by $\frac{1}{4}\lambda_g$ in order to produce the necessary field displacement in the two cavities to ensure uniform drying. The unit utilises a 25 kW source operating at 896 MHz and is capable of up to 20% evaporation at about 0·028 kg/s (100 kg/h) throughput (see Section 8.2).

A potential application which has received considerable attention in the last decade, with active support from the Flour Milling and Bread Baking Association in England, has been in microwave bread baking (Chamberlain, 1975; Driscoll, 1976). The whole concept was based on the economic advantage that British bakers would be able to use up to 100% of the cheaper British or European soft wheat in a microwave/hot air baking process instead of the more expensive mixture of 50% British wheat and 50% harder Canadian wheat. The savings in raw material costs would offset the capital expenditure on the microwave equipment. The high α-amylase content of the soft wheat breaks down the starch during baking and tends to 'gum up' slicer blades and yields a loaf which becomes stale fairly quickly. Also the low protein content of soft wheat makes a dough which produces less gas and is more permeable during the early stages of baking resulting in a low quality loaf. The tests carried out at FMBRA have shown that baking in an oven which includes microwaves and hot air, the dough temperature rises rapidly through the critical temperature region of 55–90°C, thus allowing less time for the enzymes to break the starches down. In this way the initial composition of the dough mixture may be made of practically 100% softer European wheat. The major technological barrier still focusses on identifying a suitable dielectric pan for use in the microwave/hot air process.

Another long standing potential application for microwave energy has been in vegetable blanching. The objective here is to inactivate the enzymes which would otherwise reduce the quality during storage (Hill, 1979). The advantage of microwave energy is rapid temperature rise compared with conventional techniques which also suffer from the major disadvantage of loss in weight caused by leaching out of the solids and nutrients during prolonged exposure to water or steam. However the main problem in the past has been to justify the high capital cost of the microwave equipment for such a seasonal and low cost product. The best compromise appears to be a combined system wherein the vegetables are heated to about 60°C conventionally and microwave energy is used to heat rapidly to 90°C in a steam environment (steam in preference to hot water in order to avoid leaching out solids). A 20 kW microwave vegetable blancher with steam injection has recently been installed in a British factory to assess its performance and it is capable of processing brussel sprouts at up to 0·084 kg/s (300 kg/h), with the steam supplying 20% of the heat required. The combined system produces a more uniformly blanched product with reduction in the overall processing times compared with the conventional technique.

Freeze drying has been extensively studied in the past twenty years (Copson, 1975). Essentially freeze drying is a sublimation process where the ice is vaporised from the frozen product without passing through the liquid stage (see also Section 9.4.3). The distinct advantage of freeze drying is that the product after processing retains its natural quality without the need for refrigeration. A number of pilot systems were designed around 1965 but it seems that the major problems still ahead are three-fold. First, owing to the much reduced pressures required for the sublimation process to proceed, i.e., pressures about a few tens of Pa or less (0·1 torr), the probability of arcing is enhanced due to much lower threshold electric fields required. Second, due to non-uniformity of the microwave electric fields set up within the drying chamber, hot spots could melt parts of the product which would then readily absorb a substantial proportion of the available power and damage the product. Third, the economics are as yet not favourable against conventional systems.

The vitrification of high level radioactive waste by microwave power is the subject of continuous research and development. The principle entails the evaporation and calcination of solutions containing uranyl salts and radioactive waste into which glass forming additives are present. After evaporation of the liquids the mixture continues to absorb microwave energy to the point where it fuses and vitrifies. This has been demonstrated in the laboratory and prototype equipment is being developed for eventual transfer into the nuclear industry (Gayler and Hardwick, 1979).

Experiments are in progress in the U.K. and abroad which examine the possibility of using radio frequency and microwave energy for heating purposes in thermo-nuclear plasma. Three main areas of interest exist at present (Mourrier, 1979):

(*a*) Ion cyclotron resonance heating where the frequency of the incident radio frequency energy is matched to ion resonance in a magnetic field. Present power levels are typically about 1 MW peak pulse at between 5 to 50 MHz.

(*b*) Lower hybrid resonance heating where the incident energy is matched to plasma resonance and typically lies between 500 to 2000 MHz. Powers involved are of the order of 1 MW peak pulse.

(*c*) Electron cyclotron resonance heating where the incident microwave energy is coupled to the electron resonance in a magnetic field. Present power levels are about $\frac{1}{4}$ MW at frequencies between 20 to 100 GHz.

11.3.4 Laboratory models

Referring back to Table 11.1 this is perhaps the most voluminous of all categories with a multitude of reported applications. Most will perhaps never pass beyond this preliminary stage ending up as interesting publications in scientific journals. Of the remainder so far very little has been heard. A brief insight into activity in this sector will be given for some of the better publicised topics of research.

A possible application for microwaves is in the field of sterilisation of foodstuffs, which are conventionally preserved by subjecting the containers to heat treatment by thermal conduction in pressure retorts. From the knowledge of the thermal death properties of botulinum, the spore forming organism usually considered in process time calculations and the heating characteristics of the food being studied, it is possible to develop a process which will ensure the sterility of the contents of the container. For botulinum the statutary thermal death time at 121°C is 2·78 minutes.

If microwaves were used to rapidly raise the temperature to 121°C this temperature being maintained for about 3 minutes followed by rapid cooling, there might result a significant improvement in the quality of food. A resonant cavity operating in the 900 MHz band has been used to rapidly raise the temperature to 121°C, in an experiment to study High Temperature Short Time (HTST) treatment of meat particulates in gravy (Metaxes, 1976). The temperature variation within the heating zone was found to be about 7% of the final temperature which would probably be too large for effective treatment. It is hoped that variations could be largely reduced during the subsequent holding time to complete the HTST process. Ohlsson (1977) has reported satisfactory results for microwave sterilisation of solid food samples, if these were immersed in liquid during irradiation to equalise the temperature variations. The treated foodstuffs are of high quality, but the remaining problem is to find a good packaging material for long storage.

Similarly, the application of microwave techniques to the fixation of dyes has been the subject of a number of investigations (Metaxas *et al.*, 1978). This entails the rapid preheating of textiles, in the form of yarns, loose stock, tow, sliver, etc., in batch or continuous processes, using high frequency energy to near fixation temperatures, which depends on the type of fibre. The fixation process is then completed by conventional steaming although there has been the suggestion, largely unsubstantiated, that under certain circumstances the microwave or radio frequency energy acts more than merely to rapidly raise the temperature. The possibility that field effects might enhance diffusion of the dye into the fibres is now being studied, although it is readily appreciated that very large electromagnetic fields may be necessary for this effect to take place. The overriding advantage of microwave assisted dye fixation would be an improvement in quality, e.g., in the elimination of 'frosting', and the improvement of dye penetration. As yet this process, promising as it may appear, has not been widely exploited in industry, apart from a small scale experimental installation at radio frequency (27·12 MHz) for dying loose wool stock where it is claimed that internal pressure effects due to the generation of steam trapped in a glass envelope containing the loose stock, plays a major role in the mechanism of dye fixation, by raising the boiling point of the water present. Many small scale microwave oven type field trials have been carried out in the past ten years but no commercial installation is reported.

Selective inactivation of fungi has been recently reported (Diprose *et al.*, 1979) during soil sterilisation by microwave techniques. Such treatment allows minimal inactivation of the more useful bacteria population in the soil. Small scale laboratory tests have so far been carried out and a 1 kW continuous unit is being developed for soil treatment at a rate of about 0·01 kg/s (40 kg/h) which will be suitable for greenhouses.

The treatment of agricultural products with microwave or radio frequency techniques has been studied for many years, particularly in the U.S.A., by the Department of Agriculture (Nelson, 1976). The areas of research have been, for example, control of insects in stored products, improving seed germination, induction germination of impermeable seed, the effects on dormancy and longevity of seeds, etc. The prime reason for the reluctance of the seedsman to accept the new techniques is that of the cost compared to much cheaper treatments by chemical fumigants. Moreover it appears that no premium can be charged for seeds with reduced content of hard seed, since although the hard seeds take much longer, they do eventually germinate. Competition from cheaper infra-red techniques together with a reluctance to change to the more sophisticated radio frequency and microwave techniques seems to have sealed the fate of these methods.

There is still a considerable effort devoted to multifarious drying applications which are tabulated as L in the atmospheric, fluidised or spray drying section. The justification for research in this is hard to find in the light of cheaper and more viable electrical techniques. However, such efforts have to be viewed in the context of individual company needs, government energy policies, availability of fuel, etc.

Allied to the microwave vulcanisation of rubber there are many other potential applications in curing or hardening of various chemicals such as glues, resins, cement mortar, fibreglass epoxy laminates, sodium silicate cores, polystyrene foam, cellulosic membranes, etc. (Metaxas, 1981*a*). There have been many reports of success in the laboratory with occasional references to small industrial units.

A sector which has largely remained at the laboratory scale is that of using microwave energy to induce chemical reactions and processing in microwave plasma. This topic was extensively studied particularly at MIT in the U.S.A. during the nineteen sixties (Baddour and Dundas, 1968). The processes examined ranged from the production of gaseous atoms of hydrogen, oxygen, chlorine, etc., to synthesis of unsaturated hydrocarbons and fluorocarbons for acetylene production. Microwave radiation was used to produce carbon black and hydrogen from methane or to produce 1-butane and propylene from reactions in cyclohexane. Other examples include converting waste chlorine into chlorine gas, oxidising silicon, and the production of sulphuric acid through microwave oxidation of sulphur dioxide. In certain cases microwave radiation can be used as a substitute for expensive catalysts in kinetically controlled gas phase reactions such as, for example, vanadium pentoxide in the oxidation of sulphur dioxide to manufacture sulphuric acid. In such a case the capital cost of microwave equipment can be partly offset by the omission of expensive catalysts and their maintenance.

Pyrolysis of coal in a microwave discharge has been extensively studied with H_2, CO_2, CO and acetylene being the main products (Bodily *et al.*, 1973). Pyrolysis of propane in an argon-H plasma jet at atmospheric pressure gave C_2H_2, C_2H_4, CH_4, C, H and small amounts of C_3H_6 (Sato Kazuo *et al.*, 1973). The reaction of coal with nitrogen in a microwave discharge results mainly in the production of HCN and C_2H_2. First the active nitrogen interacts with the coal molecules to cause rapid volatilisation of gaseous products and then the active nitrogen slowly gasifies the residual char (Fu *et al.*, 1972). Other reactions studied include the synthesis of hydrazine from a microwave discharge in ammonia (Weisbeck, 1971) and the synthesis of biphenyl from benzene similarly in a low pressure microwave discharge (Barker, 1971).

Microwave and radio frequency polymerisation of monomers has also been studied in the past. In such studies monomer gases or volatile liquids react in a microwave discharge to produce specific polymer films on solid substrates (Thompson and Mayhan, 1972). Comparison between the two techniques on the polymerisation of benzene has been recently published where the findings indicated that the polymer structure and plasma composition were significantly affected by plasma energy levels and discharge frequency (Duval *et al.*, 1975). Similar comparative studies have been previously reported on the radio frequency polymerisation of vinyl chloride. In this study the optimum polymer deposition rate, determined by adjustment of the monomer flow and discharge power, was independent of the geometry and dimensions of the system (Brown and Copsey, 1972).

Akin to such experiments is the study of atom—solid reactions, where in a particular case it was observed that water contact angles of polymer films decreased significantly after exposure to excited atomic species from microwave discharges in NO, SO_2, N_2 and H_2O (Books and Wightman, 1975). A bibliography of earlier work with microwave plasma can be found in Johnston (1970*a*, *b*).

A severe technological barrier in trying to translate these laboratory tests to useful industrial processes is that of limitations in throughput, although working at radio frequencies larger processing volumes can be accommodated. However, Bosisio *et al.* (1972) have developed a new large volume microwave plasma generator (LMP) system where the energy is coupled to the reactor via a slow wave structure resulting in a plasma processing volume well in excess of 10^{-3} m^3 (1000 cm^3). They have presented experimental data for the dissociation reaction of diatomic gas such as oxygen and nitrogen. The atoms or free radicals produced in the LMP can be used in subsequent chemical reactions or in pumping lasers. Other possible applications include specific organic synthesis reactions or plasma surface reactions in textile chemistry. There appears to be no fundamental limitation to scaling up the LMP to industrial process proportions. However, as yet there seems to be no industrial microwave plasma reactor in operation.

11.4 Economics

11.4.1 Introduction
The processing of rubber products, either for thawing or vulcanisation, the tempering of meat, the drying of pasta, onions or food particulates, the cooking of chicken and bacon, and doughnut proving, are all established microwave techniques in their own right (Metaxas, 1981*a*). They each offer some unique advantage over the equivalent conventional technique and for this have found their rightful place within the various industrial environments. High frequency equipment, however, is capital intensive and the economics of a particular application must be thoroughly examined before any equipment is installed in industry. Furthermore, taking the efficiency of converting prime fuel to electricity to be about 30% and a mains electricity to heat into the product efficiency of 65% gives an overall fuel efficiency of about 20%. However, many conventional heating drying processing plants exhibit efficiencies considerably less than 20%.

As was described in Section 11.2 high frequency industrial processing offers some unique advantages over conventional methods. Heat can be rapidly generated within the volume of the material creating faster heating rates or moisture movement and optimising the heating or drying process particularly in the falling rate period. A conventional drying process can indeed be very inefficient at the last stages, as Fig. 11.1 shows qualitatively, where it takes more than half the length of the dryer to reduce the moisture from M_2 to M_3 which is a comparatively small amount compared with the initial drop from M_1 to M_2 achieved in the first half of the dryer length. It is at such a final stage of drying that high frequency techniques can be fully justified economically when they are a substitute for conventional processes which might be running at an overall efficiency of a few per cent.

Before introducing a microwave system as part of forming the complete process, it is necessary to carry out a thorough economic analysis (Ishii Kozyu, 1974;

Jolly, 1972, 1976). Below we shall examine briefly the economics of two ways normally considered in the first instance, that of energy savings or increased throughput.

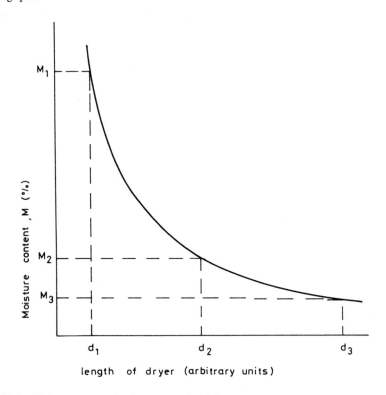

Fig. 11.1   *Moisture evaporation in a conventional dryer*

11.4.2 Energy savings

The justification for purchasing a microwave processing unit to achieve energy savings is based on the fact that many conventional heating processes are so inefficient that part or all of the equipment can be replaced by a more efficient microwave unit whose capital costs are repaid after a number of years through savings in energy usage. Figure 11.2(b) shows a typical case in a continuous drying line, where the high frequency equipment dries the last 10% moisture. If the running costs per hour of the microwave dryer and the original conventional dryer are g_{mw} and g_{con} respecitvely, we can equate the total savings in y_n years to the capital costs incurred. This gives

$$(g_{con} - g_{mw})6000y_n = g_1 \cdot P_{sc} - C \tag{11.1}$$

where g_1 is the capital cost of the microwave dryer in £/kW (source), P_{sc} is the source kW, C is any recoverable capital due to resale of the conventional equipment and the number of operational hours in a year is assumed to be about 6000.

The theoretical power required, P_{th}, to dry the last 10% moisture can be calculated, from

$$P_{th} = \frac{M_{ad}}{t}\left[c_p \Delta T + \frac{M}{M_{ad}}(\Delta T + L_h)\right] \text{W} \qquad (11.2)$$

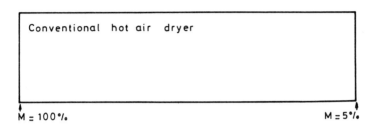

a

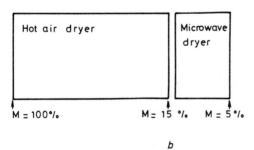

b

Fig. 11.2 *Drying the last stage with microwaves*
a Conventional plant
b Combination plant

where M_{ad}/t is the mass throughput in kg/s and M/M_{ad} is the fractional moisture content on dry basis or fractional regain (0·1 in this case). The specific heat c_p and the latent heat L_h are expressed in J/kg°C and J/kg respectively. The power of the microwave unit, P_{sc}, and the mains electricity drawn power, P_m, can then be found from the expressions

$$P_{sc} = P_{th}/n_u \qquad (11.3)$$

and

$$P_m = P_{th}/n_u n_g = P_{th}/n_0 \qquad (11.4)$$

where n_u and n_g are the microwave utilisation and mains electricity to microwave generation efficiencies. Therefore, the running costs of the microwave dryer, g_{mw}, are given by $g_0 P_m$, where g_0 is the unit of electricity in p/kWh. Assuming that the relevant figures for the application in question are: $g_{con} = £20/hour, g_0 = 4$ p/kWh, $g_1 = £1500/kW$, $P_{th} = 100$ kW, $C = £40\,000$ and $n_u = n_g = 0.7$, then $P_{sc} = 143$ kW and through eqn. (11.4), $g_{mw} = £8.17/hour$. Substitution in eqn. (11.1) yields a repayment period of about 2·4 years. The reduction in energy costs in this case amounts to nearly £71 000 per annum. Tube replacement and maintenance costs are included in the figure for g_0.

11.4.3 Increased throughput

If the prime requirement of a particular process is to increase its throughput, this could be achieved by installing a high frequency unit at the end of an already existing line and running the line at faster speeds (see Section 9.5). Referring to Fig. 11.3, the conventional hot air dryer, assuming it is running to capacity already, cannot cope with the increased throughput and instead of drying to 2% it now dries only down to 8%. The high frequency unit would now need to evaporate the remaining 6%. The profits due to the extra, say 20%, increase in throughput should now pay for the capital costs, the running costs and any interest charges on the borrowed capital. First we would have to estimate the yearly profits.

Say the throughput of a drying line, which at present processes a material at 5 m/min, would at an increased speed of 6 m/min, be about 3000 kg/h. This would give a total annual production of 18 million kg assuming 6000 hours operation. At, say, 2 kg/m² base weight, this represents 9 million m² of material. Making a tentative assumption of a net profit of 20 p/m², the extra revenue due to the 20% increase in production, i.e., an extra 1·8 million m², is £360 000/year.

The microwave dryer would require 150 kW, giving a source of 300 kW which assumes an overall utilisation efficiency of only 50%. The electrical input power would now be 365 kW. At 4 p/kWh the yearly running cost would be £87 600. At £1 500/kW (source) the purchase of the microwave equipment would be £450 000. The yearly interest on the borrowed capital (say 15%) would amount to £67 500. Balancing the profits against the expenditure yields (Holme and Metaxas, 1979)

$$450\,000 + 67\,500 y_n + 87\,600 y_n = 360\,000 y_n \qquad (11.5)$$

where y_n is the number of years. Equation (11.5) yields a repayment period of just over two years. The unit costs, defined as yearly running costs/total production, would be about 0·5 p/kg.

11.4.4 Other factors

The simple calculations described above should only serve as guidelines towards the justification for purchase of a high frequency installation. If the running costs of the conventional equipment are lower than the running costs of the microwave equipment for carrying out the same part of the process there must be some unique features of the microwave heated product such as improved shelf life or flavour to

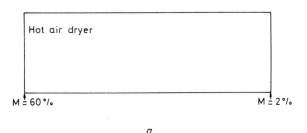

a

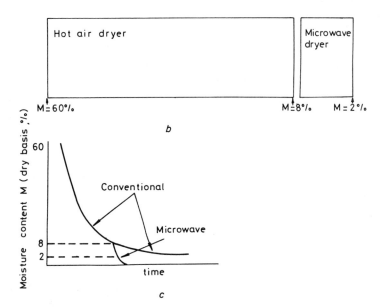

b

c

Fig. 11.3 *Increased throughput with microwave energy*
  a Conventional process
  b Combination process
  c Drying curve

outweigh the extra cost. For example microwave bread baking is said to be economically viable because a less expensive raw material can be used and the savings accrued justify the capital purchase of the microwave equipment (Chamberlain, 1975).

Another important consideration in making a decision is related to the available space that exists in a factory for the new equipment. Conventional equipment tends to be longer and heavier than microwave equipment and there are instances where part of the factory had to be extended or the existing foundations reinforced in order to accommodate the conventional equipment. If the already available space can accommodate the more compact and lighter microwave system this may represent a major financial consideration towards which of the two systems to purchase.

Often conventional heating plant has particularly low overall efficiency due to such factors as falling rate drying or the need to maintain temperature during breaks in production because of a long or difficult warm up procedure. Slow response to adjustment also results in inefficiency as does operation at throughput substantially different from the design rating. In such cases the quick response, volumetric heating characteristics and rapid warm up of microwave plant is likely to give substantial cost savings.

Equally, in the surface treatment of thin coatings or in drying thin web materials it is often prudent to consider the suitability of ultra-violet or infra-red techniques because of the lower capital costs involved compared to either radio frequency or microwave techniques. Having considered alternative modern conventional or other electrical techniques such as 50 Hz a.c. energy in the case of highly conductive coatings, infra-red, ultra-violet, etc., and had their unsuitability firmly established on technical grounds or possibly fuel unavailability, it is then prudent to consider high frequency techniques.

There is no set code of practice as to when to use high frequency energy. Following the general guidelines outlined so far each application must be considered for its own merits and what is more must be examined from within its own environment and company constraints, availability of fuel, outlook of management and workforce and other similar factors which characterise the essence and individuality of each company.

11.5 Future trends

With little doubt the strongest single factor controlling the future development of industrial microwave heating as a market is the state of the world economy, dominated by the cost of fuel in its various forms. In products where energy forms a major component of their manufacturing cost, energy-efficient processing is likely to become of even greater importance, bringing into prospect a steadily growing range of processes in which microwave energy can play an important role in energy saving.

At the same time there is likely to be some reductions in the purchase price of industrial microwave equipment arising first from a greater volume of manufacture to cover design and overhead costs, and second, from a likely trend in certain types of equipment towards the use of large numbers of low-cost mass-produced power generators of the order of 1 kW output instead of a few high power generators. Clearly this must be a slow development for the market will develop as prices fall, but until the market develops prices cannot fall, a classic stalemate position which will only be resolved over many years if the rate of evolution of the technology over the past two decades does not increase.

In the technology of microwave heating the trend will be towards equipment giving high power density so as to realise the ultimate advantage of microwave energy over lower frequency systems. Such systems, based on automatic tuning of resonant cavity applicator will result in extremely compact drying systems and heating devices able to give acceptable performance on materials hitherto considered unsuitable because of their low values of loss factor. Moreover, these devices are of great importance in the food industry in providing a process tool of intense power density capable of performing unique functions. In this respect the development of the microwave market is in part dependent on the perception and inventiveness of process engineers in having the vision to apply these techniques in the development of new products.

Clearly, also, there will be a growing integration of microwave with conventional heating methods. This places a burden on microwave equipment manufacturers in that the process machinery involves a greater quantity of non-microwave apparatus.

The full potential of micorwaves as a process tool in industry has yet to be realised. There is a general trend for more sophistication, automation and control. Industrial microwave systems will be operated from microprocessors and there will be an emphasis on the provision of transducers to measure the parameters to be controlled. Looking further into the future the combination of microwave sources and robotics offers unique opportunities which will be translated to practical applications for the benefit of industry only if enterprising and inventive scientists and engineers are prepared to take up the challenge.

11.6 References

ASSINDER, I., 'Microwaves for food processing – industrial applications of microwave energy' (Edited by R. Smith), *IMPI Transactions* 2, 92 (1974).

BADDOUR, R. F., and DUNDAS, P. H., *Microwave Power Engineering*, 2, Applications, Chapter 5.5.2 'Chemical processing in a microwave discharge' (Edited by E. Okress), pp. 329–346 (1968).

BARKER, R., 'Synthesis of the hydrazine from a microwave discharge in ammonia', *J. Microwave Power* 6(1), 37 (1971).

BEDROSIAN, K., 'The necessary ingredients for successful microwave applications in the food industry', *J. Microwave Power* 8(2), 173 (1973).

BENGTSSON, N. E., and OHLSSON, T., 'Microwave heating in the food industry', *Proc. IEEE* 62, No. 1, pp. 44–55, January (1974).

BIRD, L., 'Microwave rendering', *Proc. IMPI Conf.*, Monaco, pp. 63–65 (1979).

BODILY, D. M., CHE, S. C. L., and WISER, W. H., 'Microwave pyrolysis of coal and related hydrocarbons', *Am. Chem. Div. Fuel Chem.* 18(2), 221 (1973).

BOOKS, J. T., and WIGHTMAN, J. P., 'Reactions of polymer films with active species produced in a microwave discharge', *J. Microwave Power* 10(1), 71 (1975).

BOSISIO, R. G., WEISSFLOCH, C. F., and WERTHEIMER, M. R., 'The large volume microwave plasma generator (LMP): a new tool for research and industrial processing', *J Microwave Power* 7(4), 325 (1972).

BROWN, K. C., and COPSEY, M. J., 'Polymerisation in radio frequency glow discharges', *European Polymer J.* 8, 129 (1972).

CHAMBERLAIN, N., 'Microwave energy in baking of bread', *Food Trade Review*, pp. 8–12, September (1975).

COPSON, D. A., *Microwave Heating*, Second Edition. AVI Publishing Co. Inc. (1975).

DIPROSE, M. F., HACKMAN, R., and BENSON, F. A., 'The use of microwave radiation for soil partial stabilisation', *Colloquium of Techniques and Applications of Microwave Power*, IEE Digest No. 1979/65, pp. 3.1–3.3, 15 November (1979).

DRISCOLL, J. L., 'Suitable pan materials for baking bread in a microwave hot air oven', Electricity Council Research Centre, Capenhurst, Chester, U.K., /M898, February (1976).

DUVAL, M., and THEORET, A., 'Comparative study of microwave and radio frequency plasma polymerisation of benzenes', *J. Electrochem. Soc.* 122(4), 581 (1975).

FREEDMAN, G., 'The future of microwave power in industrial applications', *J. Microwave Power* 7(4), 353 (1972).

FREEDMAN, G., 'The future of microwave heating equipment in the food industry', *J. Microwave Power* 8(2), 161 (1973).

FORWALTER, J., *Food Processing*, November (1978).

FU, Y. C., BLAUSTEIN, B. D., and SHARKLEY, A. G. Jr., 'Reaction of coal with nitrogen in a microwave discharge', *Fuel* 51(4), 308 (1972).

GAYLER, R., and HARDWICK, W. H., 'The vitrification of high level radio-active waste using microwaves', *Colloquium on Techniques and Applications of Microwave Power*, IEE Digest No. 1979/65, p. 21, 15 November (1979).

HILL, M., 'Treatment of foodstuffs by microwave energy', *Colloquium on Techniques and Applications of Microwave Power*, IEE Digest No. 65 pp. 61–64, 15 November (1979).

HOLME, I., and METAXAS, A. C., 'Microwave drying of nylon tufted carpets. III. Field trials', *J. Microwave Power* 14(4), 367 (1979).

ISHII KOZYU, T., 'Theoretical basis for decision to microwave approach for industrial processing', *J. Microwave Power* 9(4), 354 (1974).

JOHNSTON, D. A., Bibliography. II. Microwave plasmas Part I, *J. Microwave Power* 5(1), 17 (1970*a*).

JOHNSTON, D. A., Bibliography. II. Microwave plasmas, Part II, *J. Microwave Power* 5(3), 192 (1970*b*).

JOLLY, J. A., 'Financial techniques for comparing the monetary gain of new manufacturing processes such as microwave heating', *J. Microwave Power* 7(1), 5 (1972).

JOLLY, J. A., 'Economics and energy utilisation aspects of the application of microwaves: a tutorial review', *J. Microwave Power* 11(3), 233 (1976).

LAWTON, J. L., 'Drying: the role of heat pumps and electromagnetic fields', *Physics in Technology* 9, 214 (1978).

LEFEUVRE, S., PARESI, A., MANGIN, B., and REZUAN, Y., 'Industrial material drying by microwave and hot air', *Proc. IMPI Conf.* Ottawa, Ontario, pp. 65–67 (1978).

LEFORT, J., 'Experience in the drying of foodstuffs using microwave energy', *Seminar in processing using r.f. and m.w. Techniques*, BNCE, 30 Millbank, London (1981).

MEISEL, N., Microwave-vacuum drying by the Gigavac process for continuous manufacture of instantly soluble fruit powders, 'Microwave energy applications', *Newsletter XII*, No. 6, pp. 3–6 (1979).

MEREDITH, R. J., 'Microwave energy for high speed efficient vulcanisation of extruded rubber', *J. Elastomers and Plastics* 8, 191, April (1976).

MEREDITH, R. J., Magnetronics Ltd., Unit A, St Mary's Mills, Evelyn Drive, Leicester (1980).

METAXAS, A. C., 'Design of equipment for microwave sterilisation of food particulates', Electricity Council Research Centre, Capenhurst, Chester, England, /M921, July (1976).

METAXAS, A. C., 'Radio frequency and microwave techniques in the manufacturing industry', *Speciality Chemicals* **1**, No. 1 pp. 27–30 February (1981*a*).

METAXAS, A. C., 'Industrial microwave heating – the past, present and future trends', Electricity Council Research Centre, Capenhurst Chester, England, Internal Note N/1437, October (1981*b*).

METAXAS, A. C., CATLOW, N., and EVANS, D. G., 'Microwave assisted dye fixation', *J. Microwave Power* **13**(4), 341 (1978).

METAXAS, A. C., MEREDITH, R., and HOLME, I., Boost your shift production with microwaves, *Carpet Review Weekly*, pp. 30, 41 July 19 (1979).

MOURRIER, G., 'Micorwave power for thermonuclear fusion on machines', *9th European Microwave Conference, Proceedings*, pp. 54–63 (1979).

NELSON, S. O., 'Use of microwave and low frequency r.f. energy for improving Alfalfa seed germination', *J. Microwave Power* **11**(3), 271 (1976).

OHLSSON, T., 'Possibilities and limits of microwave sterilisation', *Proc. Int. Symposium, Karlsruhe, Germany* (Edited by K. Paulus), pp. 105–114. S. Karger (1977).

OHLSSON, T., *Newswaves* **1**, No. 2, p. 6 (1980).

PERKIN, R. M., 'Prospects of drying with radio frequency and microwave electromagnetic fields', *J. Separation Process Technology* **1**(1), 14 (1979).

SATO KAZUO, NISHIMURA YUKIO, TAKESHITA KENJIRO, and SAKAI WATARU, 'Pyrolysis of propane in a microwave plasma jet', *Sekiyu Gakkai Shi* **16**(1) 30 (in Japanese) (1973).

SCHIFFMAN, R., 'An update of the applications of microwave power in the food industry', *J. Microwave Power* **11**(3), 221 (1976).

SHUTE, R. A., 'Industrial microwave systems for the rubber industry', *J. Microwave Power* **6**(3), 193 (1971).

SHUTE, R. A., Microwave Heating Ltd., Skimpot Industrial Estate, Luton, Beds. (1980).

SMITH, F. J., 'Microwave – hot air drying of pasta, onions and bacon', *Microwave Energy Applications, Newsletter XII*, No. 6, pp. 6–12 (1979).

SWIFT, J., and TUOMY, J. M., *Microwave Energy Applications, Newsletter XI*, No. 1, pp. 3–10 (1978).

THOMPSON, L. F., and MAYHAN, K. G., 'The plasma polymerisation of vinyl polymers', *J. Appl. Polym. Sci.* **16**, 2291 (1972).

WITT, A., *Proc. UIE Conf., Cannes* (1980).

WEISBECK, R., 'Synthesis of biphenyl from benzene in a microwave low pressure discharge', *J. Microwave Power* **6**(1), 31 (1971).

Definitions of moisture content

The proportion of water in a wet substance can be defined on either a wet or dry basis. In wet basis the moisture content, M', is defined as the weight of water W_w over the weight of water W_w plus the weight of dry matter W_d:

$$M' = \frac{W_w}{W_w + W_d} \times 100 \tag{I.1}$$

The dry basis definition (referred to in industry as percentage regain)* is given by

$$M = \frac{W_w}{W_d} \times 100 \tag{I.2}$$

The relationship between M' and M follows from eqn. (I.1) and (I.2):

$$M = \frac{M'}{1 - M'} \tag{I.3}$$

Another definition involves the percentage pick-up, M'', defined as the weight of water picked up by the material W_{wp} over the total weight at equilibrium. The relationship between M and M'' can be derived as follows. The equilibrium percentage regain M_0 is given by

$$M_0 = (W_{we}/W_d) \times 100\% \tag{I.4}$$

where the subscript e refers to equilibrium weights. The total weight at equilibrium W_{te} is given by

$$W_{te} = W_d + W_{we} \tag{I.5}$$

Substitution of W_{we} from eqn. (I.4) yields

$$W_{te} = W_d (1 + M_0/100) \tag{I.6}$$

---

* In some industries (textile and paper) the dry basis definition is also described as moisture regain percentage.

By definition the percentage pick-up, M'', is given by

$$M'' = (W_{wp}/W_{te}) \times 100\% \qquad (I.7)$$

and since pick-up water + equilibrium water = total water, we obtain

$$\frac{M''}{100} W_{te} + \frac{M_0}{100} W_d = \frac{W_d}{100} M \qquad (I.8)$$

Substitution of W_{te} from eqn. (I.6) yields

$$M = M_0 + M'' \left(1 + \frac{M_0}{100}\right) \qquad (I.9)$$

When the equilibrium moisture content M_0 is negligible the moisture contents M and M'' are equal. Finally another expression for the moisture content is that on a percentage solids basis, M''', defined as

$$M''' = \frac{W_d}{W_d + W_w} \times 100 \qquad (I.10)$$

or after substitution of eqn. (I.2) into eqn. (I.10) yields

$$M''' = \left(1 + \frac{M}{100}\right)^{-1} 100 \qquad (I.11)$$

As an example a material with a moisture content on dry basis equal to $M = 10\%$ and with negligible M_0, the amount of moisture can also be expressed as $M' = 9 \cdot 09\%$ and $M''' = 90 \cdot 9\%$ and $M = M''$. This book adheres mostly to the definition of dry basis.

Conversion of various units to SI units

Other Units	SI units

Distance

1 angström	$= 10^{-10}$ m
1 thou (0·001 inch)	$= 25·4 \, \mu$m
1 inch	$= 25·4$ mm
1 ft	$= 0·3048$ m
1 yd	$= 0·9144$ m
1 mile (U.S. statute)	$= 1·609$ km
(U.S. nautical)	$= 1·852$ km

Area

1 in²	$= 645·16$ mm²
1 ft²	$= 0·0929$ m²
1 yd²	$= 0·836$ m²
1 acre	$= 4046·86$ m²

Volume

1 U.K. gal $= 277·42$ in³	$= 4·546$ dm³ (1 litre)
1 U.S. gal $= 231$ in³	$= 3·785$ dm³ (1 litre)

Force and weight

1 dyne	$= 10^{-5}$ N
1 pdl	$= 0·138\,255$ N (1 N $= 1 \, \text{kg ms}^{-2}$)
1 lbf (i.e., the weight of lb mass)	$= 4·448$ N
1 lbm (Avoirdupois)	$= 0·4536$ kg
1 ton (long)	$= 1016·05$ kg
1 tonne	$= 1000$ kg

Density

1 lb/in³	$= 2·767\,99 \times 10^4$ kg/m³
1 lb/ft³	$= 16·0185$ kg/m³
1 lb/UK gal	$= 99·7763$ kg/m³

Pressure

1 dyne cm^{-2}	$= 0.1$ Pa
1 mbar	$= 100$ Pa $= 100$ N/m^2
1 torr $= 1$ mmHg	$= 133.31$ Pa
1 inch of H$_2$O $(39.2°$F$) = 2.49$ mbar	$= 249$ Pa
1 inch of Hg $(32°$F$) = 33.86$ mbar	$= 3386$ Pa
1 lbf/in^2 $= 68.95$ mbar	$= 6895$ Nm^{-2} $= 6895$ Pa $= 6.895$ kPa

Energy and heat

1 erg	$= 10^{-7}$ J
1 cal	$= 4.1868$ J
1 BTU $= 2.93 \times 10^{-4}$ kWh $= 252$ cal	$= 1.05506 \times 10^3$ J
1 hph $= 2544$ BTU	$= 2.684 \times 10^6$ J
1 kWh	$= 3.6 \times 10^6$ J
1 Therm $= 29.3$ kWh $= 10^5$ BTU	$= 1.05506 \times 10^8$ J

Specific heat

1 BTU/lb°F $= 1$ cal/g°C	$= 4.1867$ kJ/kg°C

Power

1 ftlbf/s	$= 1.356$ W
1 hp (U.K.)	$= 745.7$ W $(1$ W $= 1$ J/s $= 1$ Nm/s$)$

Some useful trigonometric functions and formulae

$$\sin x = \frac{1}{\operatorname{cosec} x}, \quad \cos x = \frac{1}{\sec x}, \quad \tan x = \frac{1}{\cot x}$$

$$\sin^2 x + \cos^2 x = 1$$

$$\sin 2x = 2 \sin x \cos x, \quad \cos 2x = \cos^2 x - \sin^2 x = 2 \cos^2 x - 1$$
$$= 1 - 2 \sin^2 x$$

$$\sin 3x = 3 \sin x - 4 \sin^3 x, \quad \cos 3x = 4 \cos^3 x - 3 \cos x$$

$$\tan 2x = \frac{2 \tan x}{1 - \tan^2 x}, \quad 1 + \tan^2 x = \sec^2 x, \quad 1 + \cot^2 x = \operatorname{cosec}^2 x$$

$$\sin x = \cos (\pi/2 - x) = \sin (\pi - x), \quad \cos x = \sin (\pi/2 - x) = -\cos (\pi \pm x)$$

$$\sin (\tfrac{1}{2}x) = \pm \sqrt{\frac{1 - \cos x}{2}}$$

$$\cos (\tfrac{1}{2}x) = \pm \sqrt{\frac{1 + \cos x}{2}}$$

$$\sin (x \pm y) = \sin x \cos y \pm \cos x \sin y$$

$$\cos (x \pm y) = \cos x \cos y \mp \sin x \sin y$$

$$\tan (x \pm y) = \frac{\tan x \pm \tan y}{1 \mp \tan x \tan y}$$

$$\sin x \pm \sin y = 2 \sin \tfrac{1}{2} (x \pm y). \cos \tfrac{1}{2} (x \mp y)$$

$$\cos x + \cos y = 2 \cos \tfrac{1}{2} (x + y). \cos \tfrac{1}{2} (x - y)$$

$$\cos x - \cos y = -2 \sin \tfrac{1}{2}(x+y) . \sin \tfrac{1}{2}(x-y)$$

$$\sin x = \frac{1}{2j}(e^{jx} - e^{-jx}), \quad \cos x = \frac{e^{jx} + e^{-jx}}{2}$$

circle (radius R),   perimeter $(2\pi R)$,   Area (πR^2)

sphere (radius R),   volume $(\tfrac{4}{3}\pi R^3)$,   Area $(4\pi R^2)$

cone volume $\tfrac{1}{3}$ (area base) × (height)

Some useful series and Bessel functions

Binomial

$$(x + y)^n = x^n + nx^{n-1}y + \frac{n(n-1)}{2!}x^{n-2}y^2$$

$$+ \frac{n(n-1)(n-2)}{3!}x^{n-3}y^3 + \ldots \qquad (y^2 < x^2)$$

$$(1 \pm x)^n = 1 \pm nx + \frac{n(n-1)x^2}{2!}$$

$$\pm \frac{n(n-1)(n-2)x^3}{3!} + \ldots \qquad (x^2 < 1)$$

Trigonometric

$$\sin x = x - \frac{x^3}{3!} + \frac{x^5}{5!} - \frac{x^7}{7!} + \ldots \qquad \text{(all real values of } x\text{)}$$

$$\cos x = 1 - \frac{x^2}{2!} + \frac{x^4}{4!} - \frac{x^6}{6!} + \ldots \qquad \text{(all real values of } x\text{)}$$

$$\tan x = x + \frac{x^3}{3} + \frac{2x^5}{15} + \frac{17x^7}{315} + \frac{62x^9}{2835} + \ldots \qquad (x^2 < \pi^2/4)$$

$$\tan^{-1} x = x - \frac{x^3}{3} + \frac{x^5}{5} - \frac{x^7}{7} + \ldots \qquad x^2 < 1$$

$$= \frac{1}{2}\pi - \frac{1}{x} + \frac{1}{3x^3} - \frac{1}{5x^5} + \ldots \qquad x^2 > 1$$

Exponential

$$e^x = 1 + x + \frac{x^2}{2!} + \frac{x^3}{3!} + \frac{x^4}{4!} + \ldots \qquad \text{(all real values of } x\text{)}$$

Power

$$a^x = 1 + x \log_e a + \frac{(x \log_e a)^2}{2!} + \frac{(x \log_e a)^3}{3!} + \ldots$$

Logarithmic

$$\log_e x = 2 \left[\frac{x-1}{x+1} + \frac{1}{3} \left(\frac{x-1}{x+1} \right)^3 + \frac{1}{5} \left(\frac{x-1}{x+1} \right)^5 + \ldots \right] \qquad (x > 0)$$

$$\log_e (1+x) = x - \frac{x^2}{2} + \frac{x^3}{3} - \frac{x^4}{4} + \ldots + \frac{(-1)^{n-1} x^n}{n} \qquad -1 < x < 1$$

Taylor's

$$f(x+h) = f(x) + hf'(x) + \frac{h^2}{2!} f''(x) + \frac{h^3}{3!} f'''(x) + \ldots$$

$$= f(h) + xf'(h) + \frac{x^2}{2!} f''(h) + \frac{x^3}{3!} f'''(h) + \ldots$$

Maclaurin's

$$f(x) = f(0) + xf'(0) + \frac{x^2}{2!} f''(0) + \frac{x^3}{3!} f'''(0) + \ldots$$

Bessel function

$$J_0(x) = 1 - 2\cdot249 \left(\frac{x}{3} \right)^2 + 1\cdot265 \left(\frac{x}{3} \right)^4 - 0\cdot316 \left(\frac{x}{3} \right)^6$$

$$+ 0\cdot044 \left(\frac{x}{3} \right)^8 - 0\cdot004 \left(\frac{x}{3} \right)^{10} + \ldots \qquad (-3 \leqslant x \leqslant 3)$$

$$Y_0(x) = \left(\frac{2}{\pi} \right) \log_e (\tfrac{1}{2} x) J_0(x) + 0\cdot367 + 0\cdot606 \left(\frac{x}{3} \right)^2$$

$$- 0\cdot743 \left(\frac{x}{3} \right)^4 + 0\cdot253 \left(\frac{x}{3} \right)^6 - 0\cdot042 \left(\frac{x}{3} \right)^8$$

$$+ 0\cdot004 \left(\frac{x}{3} \right)^{10} - \ldots \qquad (0 < x \leqslant 3)$$

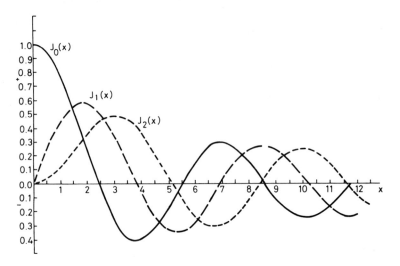

Fig. IV.1   *Bessel functions of the first kind and of order 0, 1 and 2*

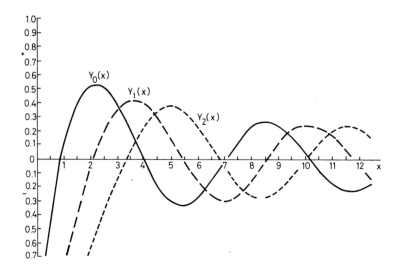

Fig. IV.2   *Bessel functions of the second kind and of order 0, 1 and 2*

$$J_{n+1}(x) = \frac{2n}{x} J_n(x) - J_{n-1}(x), \quad J_{-n}(x) = (-1)^n J_n(x)$$

$$Y_{n+1}(x) = \frac{2n}{x} Y_n(x) - Y_{n-1}(x), \quad Y_{-n}(x) = (-1)^n Y_n(x)$$

Figures IV.1 and IV.2 show some Bessel functions of the first and second kind respectively.

Formulae for derivatives of Bessel functions

$$\left(\frac{1}{x} \frac{d}{dx}\right)^k [x^n C_n(x)] = x^{n-k} C_{n-k}(x)$$

$$\left(\frac{1}{x} \frac{d}{dx}\right)^k [x^{-n} C_n(x)] = (-1)^k x^{-n-k} C_{n+k}(x), \quad k = 0, 1, 2$$

where C denotes J, Y or any linear combination of these functions, the coefficients of which are independent of x and n. As an example

$$J_0'(x) = -J_1(x) \quad \text{and} \quad Y_0'(x) = -Y_1(x)$$

The values of Bessel function derivatives can be deduced through the data given in Figs. IV.1 and IV.2 in conjunction with the above formulae.

Some useful constants and temperature interrelation

Constants

Boltzmann's constant (k_b) = $1 \cdot 3806 \times 10^{-23}$ J/K

Dielectric constant of free space (ϵ_0) = $8 \cdot 854 \times 10^{-12}$ F/m

Permeability of free space (μ_0) = $4\pi \times 10^{-7}$ H/m

Velocity of light in free space (c) = $2 \cdot 9979 \times 10^8$ m/s

Intrinsic impedance of free space $\sqrt{(\mu_0/\epsilon_0)}$ = $377 \, \Omega$

Latent heat of vaporisation of water (L_h) = $2260 \cdot 98$ kJ/kg

Latent heat of fusion of water = 334 kJ/kg

Relation between temperatures

$$T_K = T_C + 273 \cdot 15 = \tfrac{5}{9}(T_F + 459 \cdot 67) = \tfrac{5}{9} T_R$$

and

$$T_C = (T_K - 273 \cdot 15) = \tfrac{5}{9}(T_F - 32) = \tfrac{5}{9}(T_R - 491 \cdot 67)$$

where T_K, T_C, T_F and T_R are the temperatures in Kelvin, degrees Celcius, degrees Fahrenheit and degrees Rankine respectively.

Relation between dB and power and voltage ratios following the function
$$dB = 20 \log_{10} \frac{V_A}{V_B} = 10 \log_{10} \frac{P_A}{P_B}$$

dB	V_A/V_B	P_A/P_B	dB	V_A/V_B	P_A/P_B
0·1	1·012	1·023	14·0	5·01	25·1
0·2	1·023	1·047	15·0	5·62	31·6
0·3	1·035	1·072	16·0	6·31	39·8
0·4	1·047	1·096	17·0	7·08	50·1
0·5	1·059	1·122	18·0	7·94	63·1
0·6	1·072	1·148	19·0	8·91	79·4
0·7	1·084	1·175	20·0	10·00	100
0·8	1·096	1·202	21·0	11·22	126
0·9	1·109	1·230	22·0	12·59	158
1·0	1·122	1·259	23·0	14·13	200
1·1	1·135	1·288	24·0	15·85	251
1·2	1·148	1·318	25·0	27·78	316
1·3	1·162	1·349	26·0	19·95	398
1·4	1·175	1·380	27·0	22·4	501
1·5	1·188	1·413	28·0	25·1	631
1·6	1·202	1·445	29·0	28·2	794
1·7	1·216	1·479	30·0	31·6	1,000
1·8	1·230	1·514	31·0	35·5	1,260
1·9	1·245	1·549	32·0	39·8	1,580
2·0	1·259	1·585	33·0	44·7	2,000
3·0	1·413	1·995	34·0	50·1	2,510
4·0	1·585	2·51	35·0	56·2	3,160
5·0	1·778	3·16	36·0	63·1	3,980
6·0	1·995	3·98	37·0	70·8	5,010
7·0	2·24	5·01	38·0	79·4	6,310
8·0	2·51	6·31	39·0	89·1	7,940
9·0	2·82	7·94	40·0	100·0	$1·00 \times 10^4$
10·0	3·16	10·0	50·0	316·0	$1·00 \times 10^5$
11·0	3·55	12·6	60·0	1,000·0	$1·00 \times 10^6$
12·0	3·98	15·8	70·0	3,160·0	$1·00 \times 10^7$
13·0	4·47	20·0	80·0	10,000·0	$1·00 \times 10^8$

Waveguide frequency ranges, dimensions and official designations

Frequency range	Internal dimensions			Official designations			Frequency bands
(GHz)	(inches)	(mm. approx)	IEC	U.K. (RCSC)	U.S. (EIA)	U.S. (JAN)	U.K.
0·32–0·49	23·0 x 11·0	584·0 x 292·0		WG00	WR2300		
0·35–0·53	21·0 x 10·5	533·0 x 267·0		WG0	WR2100		
0·41–0·625	18·0 x 9·0	457·0 x 229·0		WG1	WR1800	RG-201/U	
0·49–0·75	15·0 x 7·5	381·0 x 191·0		WG2	WR1500	RG-202/U	
0·64–0·96	11·5 x 5·75	292·0 x 146·0		WG3	WR1150	RG-203/U	
0·75–1·12	9·75 x 4·875	248·0 x 124·0		WG4	WR975	RG-204/U	
0·96–1·45	7·7 x 3·85	196·0 x 98·0		WG5	WR770	RG-205/U	
1·12–1·7	6·5 x 3·25	165·0 x 83·0	R14	WG6	WR650	RG-69/U	L
1·45–2·2	5·1 x 2·55	131·0 x 65·0	R18	WG7	WR510		
1·7–2·6	4·3 x 2·15	109·0 x 55·0	R22	WG8	WR430	RG-104/U	
2·2–3·3	3·4 x 1·7	86·0 x 43·0	R26	WG9A	WR340	RG-112/U	
2·6–3·95	2·84 x 1·34	72·0 x 34·0	R32	WG10	WR284	RG-48/U	S
3·3–4·9	2·29 x 1·145	59·0 x 29·0	R40	WG11A	WR229		
3·95–5·85	1·872 x 0·872	48·0 x 22·0	R48	WG12	WR187	RG-49/U	C
4·9–7·05	1·59 x 0·795	40·0 x 20·0	R58	WG13	WR159		
5·85–8·2	1·372 x 0·622	35·0 x 16·0	R70	WG14	WR137	RG-50/U	
7·05–10·0	1·122 x 0·497	29·0 x 13·0	R84	WG15	WR112	RG-51/U	
8·2–12·4	0·9 x 0·4	23·0 x 10·0	R100	WG16	WR90	RG-52/U	X
10·0–15·0	0·75 x 0·375	19·0 x 9·5	R120	WG17	WR75		
12·4–18·0	0·622 x 0·311	16·0 x 7·9	R140	WG18	WR62	RG-91/U	J
15·0–22·0	0·510 x 0·255	13·0 x 5·8	R180	WG19	WR51		

896MHz

2450MHz

Glossary of radio frequency and microwave heating terms used in this book

a.c.	Alternating current
Anode	The positive electrode of an electrical device. Since in m.w. and r.f. devices the anode collects the electrons after they have performed their useful work, it must absorb their kinetic energy which is dissipated, usually by forced cooling.
Antenna	A device to transform electromagnetic energy from a conducted mode (e.g., in a transmission line) to a radiated mode or vice-versa.
Aperture	A hole in the walls of a resonant applicator for in-feed of m.w. energy whose dimensions are critical in affording a good impedance match, often called a coupling hole, *or* a hole in an applicator through which products to be processed are conveyed.
Applicator	A device for applying m.w. or r.f. energy to a product.
Arc	See corona.
Attenuation	When electromagnetic energy passes through a medium, part of the energy is absorbed and usually converted into heat. This process is called attenuation.
Attenuation duct	Short tunnels, of restricted aperture, placed at the inlet and outlet of continuous r.f. and m.w. heating ovens. They prevent excessive leakage of r.f. and m.w. energy.
Breakdown field strength	(of a material). The electric field strength at which excessive ionisation in the material occurs. The conductive paths may carbonize or cause arcs and consequential damage.
c.w.	Continuous wave (continuous operation as opposed to repeated pulsed operation).

Cathode	The negative electrode of an electrical device. The cathode is the source of electrons and is often heated to enhance this emission.
Cavity oscillator	A type of r.f. generator construction where the resonant circuit elements are lumped together. The valve is often enclosed within the cavity.
Choke	A structure connected at the input or output of a continuous high frequency applicator to limit the radiated energy.
Circulator or iso-circulator	A three port ferrite device allowing transmission of energy in one direction but directing reflected energy into a water load connected at the third port.
Co-axial transmission line	A co-axial pair of conductors acting as a transmission line.
Conductance	The real part of admittance.
Corona	Gas breakdown (ionisation) due to intense high frequency electric fields.
Coupling	The transfer of energy from one portion of a circuit to another.
Critical coupling	That value of coupling between an energy source or transmission line and a resonant structure such that all the energy is transferred to the resonant structure and none is reflected. Under such conditions, the resonant structure is said to be matched to the source or transmission line.
Crystal	The electromechanically resonant quartz crystal used to control some r.f. generators or the diode detector used to monitor m.w. power levels.
Cutoff wavelength	The limiting wavelength above which no transmission through a waveguide is permitted.
d.c.	Direct current.
Dielectric	A non-conducting medium in which polarisation takes place in an electric field.
Dielectric constant (relative)*	The ratio of the capacity of a condenser with the material as dielectric to that of same condenser *in vacuo*, *or* the real part of the complex dielectric constant, ϵ^*.
Dielectric loss factor (relative)	See loss factor.

* The word relative is omitted throughout this book.

Dispersion	The variation of the relative dielectric constant or the variation of the square of the refractive index at infra-red or optical frequencies.
Effective loss factor	The total loss in a medium at radio and microwave frequencies made up of d.c. conductivity, Maxwell–Wagner and dipolar contributing mechanisms.
Energy density	The density of the energy distribution in an electromagnetic wave measured in J/m^3.
Fundamental mode	The lowest mode established in a waveguide or applicator.
Gigahertz (GHz)	10^9 Hz $= 10^9$ cycle/second.
Hertz (Hz)	1 Hz $=$ 1 cycle/second.
Impedance matching	Adjusting the impedance of a circuit, e.g., an applicator with the dielectric inserted in it to the characteristic impedance of the connecting waveguide.
Interlock	A device (usually mechanical or electrical) intended to prevent certain actions unless the necessary conditions for safety are met.
Klystron	An electron tube for amplifying or generating microwave energy.
Loss factor	The loss factor is a parameter determining the energy dissipated in the dielectric and equals the product of the dielectric constant and loss tangent.
Loss tangent	See $\tan \delta_{eff}$.
Lumped circuit	An r.f. circuit with inductance and capacitance in the same physical component.
mm-waves	Electromagnetic radiation above 30 GHz.
Magnetron	A thermionic device for generating m.w. power, widely used for industrial applications.
Matching	See impedance matching.
Meander applicator	A type of travelling wave applicator used particularly for applying microwave energy to flat materials.
Megahertz (MHz)	10^6 Hz $= 10^6$ cycles/second.
Microwave radiation (abbreviation m.w.)	Electromagnetic radiation in the band 300 MHz to 30 GHz.
Mismatch	Unequal impedance due to two microwave components joined together causing a reflection of the incident energy.

Mode	A particular pattern of electromagnetic energy distribution within a confining structure caused by the interaction of two or more travelling waves.
Mode spectral density	The density of modes in a given cavity within a frequency band.
Mode stirrer	A device which alters the modes in a cavity to give more uniform heating effect.
Moding	Malfunction of a magnetron giving incorrect frequency due to operation in an undesired internal wave pattern. It results in inefficient operation and possible damage.
Multimode cavity	A cavity, large in relation to the free wavelength, which allows a number of different standing waves to be established.
Network analyser	Equipment for measuring transmitted and reflected signals and impedance of an applicator.
Parasitic oscillation	Any oscillation apart from the fundamental.
Penetration depth	The distance into a dielectric at which the power has decayed to $0 \cdot 368$ $(1/e)$ of its original value.
Polarisation	The ability of a dielectric to respond to an alternating electric field due to permanent or induced dipoles in its structure.
Q (quality factor)	The ratio of the energy stored in a device to the energy dissipated in that device.
Radio frequency radiation (abbreviated r.f.)	Electromagnetic radiation in the band 3 MHz to 300 MHz.
Relaxation	The variation of the dielectric constant with frequency of a medium exhibiting strong dipolar loss mechanism.
Resonant cavity	A space containing standing waves often used as an applicator of microwave energy.
Serpentine applicator	See meander applicator.
Skin depth	R.f. currents tend only to flow in the surface of conductors and penetrate into the metal to a distance known as the skin depth, at which the magnetic field strength is $(1/e)$ of its surface value.
Slotted line	A waveguide with a longitudinal slot for detecting standing waves due to reflections from a terminating applicator.

Standing waves	See mode.
Standing wave applicator	See resonant cavity.
Susceptance	The imaginary part of the admittance.
$\tan \delta_{eff}$	The ratio of the (relative)* effective loss factor to the (relative) dielectric constant.
Thermal runaway	The uncontrolled temperature rise in a material heated by high frequency energy due to a positive rate of change of the effective loss factor with temperature $(d\,\epsilon''_{eff}/dt)$.
Transmission line	A device which allows the propagation of electromagnetic energy along it.
Travelling wave applicator	A m.w. applicator where all the input power is either absorbed by the work or by a terminating water load having negligible reflected power and hence no standing waves.
Triode	A thermionic valve used as an amplifier or oscillator at r.f.
Tuning	A process of adjusting the frequency of the applicator to that of the r.f. or m.w. generator.
VSWR	The Voltage Standing Wave Ratio due to the superposition of the incident and reflected waves.
Wall loss	Power losses on the applicator internal walls due to the flow of skin currents.
Water load	A load for absorbing r.f. and m.w. power in which water is the dissipative as well as the cooling element.
Waveguide	A device for channelling r.f. or m.w. energy from one place to another with little loss and no leakage hazard.
Waveguide wavelength	The wavelength of electromagnetic radiation in a waveguide.
X-band	Electromagnetic radiation between 8 and 10 GHz.

---

* The word relative is omitted throughout this book.

Subject index

Author index